Fourth Edition

Introduction to Statistics and Data Analysis for Physicists

Fourth Edition

Introduction to Statistics and Data Analysis for Physicists

Gerhard Bohm
Deutsches Elektronen-Synchrotron (DESY), Germany

Günter Zech
University of Siegen, Germany

World Scientific

NEW JERSEY • LONDON • SINGAPORE • BEIJING • SHANGHAI • TAIPEI • CHENNAI

Published by

World Scientific Publishing Co. Pte. Ltd.
5 Toh Tuck Link, Singapore 596224
USA office: 27 Warren Street, Suite 401-402, Hackensack, NJ 07601
UK office: 57 Shelton Street, Covent Garden, London WC2H 9HE

Library of Congress Cataloging-in-Publication Data
Names: Bohm, Gerhard (Data scientist) author | Zech, Günter author.
Title: Introduction to statistics and data analysis for physicists /
Gerhard Bohm, Deutsches Elektronen-Synchrotron (DESY), Germany,
Günter Zech, University of Siegen, Germany.
Description: Fourth edition. | Hackensack, New Jersey : World Scientific, [2025] |
Includes bibliographical references and index.
Identifiers: LCCN 2025012314 | ISBN 9789819814312 hardcover |
ISBN 9789819814329 ebook for institutions | ISBN 9789819814336 ebook for individuals
Subjects: LCSH: Statistical physics | Mathematical statistics
Classification: LCC QC174.8 .B65 2025 | DDC 519.5--dc23/eng/20250415
LC record available at https://lccn.loc.gov/2025012314

British Library Cataloguing-in-Publication Data
A catalogue record for this book is available from the British Library.

For any available supplementary material, please visit
https://www.worldscientific.com/worldscibooks/10.1142/14343#t=suppl

Desk Editors: Eshak Nabi Akbar Ali/Joseph Ang

Typeset by Stallion Press
Email: enquiries@stallionpress.com

Preface

There exist a large number of excellent statistic books. Nevertheless, we think that it is justified to complement them by another textbook with the focus on modern applications in nuclear and particle physics. To this end we have included new topics and presented a large number of related examples and figures. We emphasize less the mathematical foundations but appeal to the intuition of the reader.

Data analysis in modern experiments is unthinkable without simulation techniques. We discuss in some detail how to apply Monte Carlo simulation to parameter estimation, deconvolution and goodness-of-fit tests. New developments have been included to the following topics: Parameter inference with constrained parameters, treatment of distorted data in parameter inference, unfolding, subtraction of background contamination, statistics of weighted events, goodness of fit in high dimensions. Resampling methods like bootstrap are introduced and their relevance for the estimation of uncertainties of fitted parameters is demonstrated. The elimination of nuisance parameters is discussed. We sketch machine learning with artificial neural nets, boosted decision trees and support vector machines.

Likelihood is a central concept of statistical analysis and its foundation is the likelihood principle. We discuss this concept in more detail than usually done in textbooks and base the treatment of inference problems as far as possible on the likelihood function only, as is common in the majority of the nuclear and particle physics community. In this way, point and interval estimation, error propagation, combining results, inference of discrete and continuous parameters are consistently treated. We apply Bayesian methods where the likelihood function is not sufficient to proceed to sensible results, for instance in handling systematic errors, deconvolution problems and in some cases when nuisance parameters have to be eliminated, but we avoid improper prior densities. Goodness-of-fit and significance tests, where no likelihood function exists, are based on standard frequentist methods.

Our book is based on lecture notes from a course given to master physics students at the University of Siegen, Germany, some years ago. The content has been considerably extended since. A preliminary German version is available as an electronic book at the DESY library and an updated compact text book has been published with Springer. The present book is addressed mainly to master and Ph.D. students but also to physicists who are interested to get an introduction into recent developments in statistical methods of data analysis. In a first reading many parts can be skipped.

We welcome comments, suggestions and indications of mistakes and typing errors. We are prepared to discuss or answer questions to specific statistical problems.

We acknowledge the technical support provided by DESY and the University of Siegen.

January 2025, Günter Zech

Contents

Chapter 1

Introduction: Probability and Statistics

Though it is exaggerated to pretend that in our life only the taxes and the death are certain, it is true that the majority of all predictions suffer from uncertainties. Thus the occupation with probabilities and statistics is useful for everybody, for scientists of experimental and empirical sciences it is indispensable.

1.1 The Purpose of Statistics

Whenever we perform an experiment and want to interpret the collected data, we need statistical tools. The accuracy of measurements is limited by the precision of the equipment which we use, and thus the results emerge from a random process. In many cases also the processes under investigation are of stochastic nature, i.e. not predictable with arbitrary precision, such that we are forced to present the results in form of estimates with error intervals. Estimates accompanied by an uncertainty interval allow us to test scientific hypotheses and by averaging the results of different experiments to improve continuously the accuracy. It is by this procedure that a constant progress in experimental sciences and their applications was made possible.

Inferential statistics provides mathematical methods to infer the properties of a population from a randomly selected sample taken from it. A population is an arbitrary collection of elements, a sample just a subset of it.

A trivial, qualitative case of an application of statistics in every day life is the following: To test whether a soup is too salted, we taste it with a spoon. To obtain a reliable result, we have to stir the soup thoroughly and the sample contained in the spoon should not be too small: Samples have to be representative and large enough to achieve a sufficiently precise estimate of the properties of the population.

Scientific measurements are subject to the same scheme. Let us look at a few statistical problems:

1. From the results of an exit poll the allocation of seats among the different parties in the parliament is predicted. The population is the total of the votes of all electors, the sample a representative selection from it. It is relatively simple to compute the distribution of the seats from the results of the poll, but one wants to know in addition the accuracy of the prognosis, respectively how many electors have to be asked in order to issue a reasonably precise statement.
2. In an experiment we record the lifetimes of 100 decays of an unstable nucleus. To determine the mean life of the nucleus, we take the average of the observed times. Here the uncertainty has its origin in the quantum mechanical random process. The laws of physics tell us, that the lifetimes follow a random exponential distribution. The sample is assumed to be representative of the total of the infinitely many decay times that could have occurred.
3. From 10 observations the period of a pendulum is to be determined. We will take as estimate the mean value of the replicates. Its uncertainty has to be evaluated from the dispersion of the individual observations. The actual observations form a sample from the infinite number of all possible observations.

These examples are related to *parameter inference*. Further statistical topics are *testing, deconvolution*, and *classification*.

4. A bump is observed in a mass distribution. Is it a resonance or just a background fluctuation?
5. An angular distribution is predicted to be linear in the cosine of the polar angle. Are the observed data compatible with this hypothesis?
6. It is to be tested whether two experimental setups perform identically. To this end, measurement samples from both are compared to each other. It is tested whether the samples belong to the same population, while the populations themselves are not identified explicitly.
7. A frequency spectrum is distorted by the finite resolution of the detector. We want to reconstruct the true distribution.

8. In a test beam the development of shower cascades produced by electrons and pions is investigated. The test samples are characterized by several variables like penetration depth and shower width. The test samples are used to develop procedures which predict the identity of unknown particles from their shower parameters.

A further, very important part of statistics is *decision theory.* It is not covered in this book, because it is of minor relevance in physics research.

1.2 Some Statistical Terms

Each discipline has its own terminology which sometimes has a meaning that is different from that in every day languish.

A *population* is a collection of elements. The pupils in a classroom form a finite population with *categorical* attributes. The decay times of an unstable particle form a population that consists of an infinite number of elements. We are mainly concerned with ordered elements that can be characterized by numerical variables. The population can often be described by an analytic statistical distribution. A *sample* is a random collection of elements of a population. The elements are drawn from the population with replacement. The same element can occur several times in the sample.

A *random event*, or in short an event or observation, is the outcome of a random process, like a decay time. It can also be a combination of observations.[1] (In the statistical literature the attributes of the population are denoted by capital letters and those of the observations by lower case letters. We will not follow this convention.)

When we infer properties, i.e. parameters characterizing the population, from a sample, we talk about an *estimate* or a *measurement.* The decay times of 10 pion decays correspond to a sample of observations from the population of all possible events, the decay times. The estimation of the mean life of pions from the sample is a measurement. An observation as such – the reading of a meter, a decay time, the number of detected cosmic muons – has no error associated with it. Its value is fixed by a random process. On the contrary, the measurement which corresponds to parameter inference is afflicted with an uncertainty. In many simple situations, observation and measurement coincide numerically, in other cases the measurement is the result of an extensive analysis based on a large amount of observations.

[1]At some occasions we will use the term *event* in the way physicists do, e.g. to describe certain kinematic reactions. This will become clear from the context.

1.3 How to Define Probability

Statistics is at least partially based on experience which is manifest in fields like deconvolution and pattern recognition. It applies probability theory but should not be confounded with it. Probability theory, contrary to statistics, is a purely mathematical discipline and based on simple axioms. On the other hand, all statistical methods use probability theory. Therefore, we will deal in the first part of this book with simple concepts and computational rules of this field.

In statistics, there exist several different notions on what probability means. In the *Dictionary of Statistical Terms* of Kendall and Buckland [1] we find the following definition:

> "**probability,** a basic concept which may be taken as undefinable, expressing in some way a *degree of belief*, or as the limiting frequency in an infinite random series. Both approaches have their difficulties and the preference for one or the other depends to some extent on the field of application. Fortunately both lead to much the same calculus of probability."

We will try to extend this short explanation:

In the *frequentist statistics*, sometimes also called *classical statistics*, the probability of an event, the possible outcome of an experiment, is defined as the frequency with which it occurs in the limit of an infinite number of repetitions of the experiment. If in throwing dice the result *five* occurs with frequency 1/6 in an infinite number of trials, the probability to obtain *five* is defined to be 1/6.

In the more modern, so-called *Bayesian statistics*[2] this narrow notion of probability is extended. Probability is also ascribed to fixed but incompletely known facts and to processes that cannot be repeated. It may be assigned to deterministic physical phenomena when we lack sufficient information. We may roll a dice and before looking at the result, state that the result is "5" with probability 1/6. Similarly, a probability can be attributed to the fact that the electron mass is located within a certain mass interval. That in the context of a constant like the electron mass probability statements are applied, is due to our limited knowledge of the true facts. It would be more correct, but rather clumsy, to formulate: "The probability that we are right with the proposition that the electron mass is located

[2]Thomas Bayes was a mathematician and theologian who lived in the 18th century.

in that error interval is such and such." The assignment of probabilities sometimes relies on assumptions which cannot be proved but usually they are well founded on symmetry arguments, physical laws or on experience.[3] The results obviously depend on these assumptions and can be interpreted only together with those.

The frequentist concept as compared to the Bayesian one has the advantage that additional not provable assumptions are obsolete but the disadvantages that its field of application is rather restricted. Important parts of statistics, like deconvolution, pattern recognition and decision theory are outside its reach. The Bayesian statistics exists in different variants. Its extreme version permits very subjective assignments of probabilities and thus its results are sometimes vulnerable and useless for scientific applications. Anyway, these very speculative probabilities do not play a significant role in the scientific practice.

Both schools, the classical frequentist oriented and the modern Bayesian have developed important statistical concepts. In most applications the results are quite similar. A very instructive and at the same time amusing article comparing the Bayesian and the frequentist statistical philosophies is presented in Ref. [2].

For completeness we mention a third *classical* interpretation of probability which is appreciated by mathematicians[4]: If an experiment has N equally likely and mutually exclusive outcomes, and if the event A can occur in P of them, then the probability of event A is equal to P/N. It has the difficulty that it can hardly be translated into real situations and a slight logical problem in that the term *equally likely* already presumes some idea of what probability means.

Independent of the statistical approach, in order to be able to apply the results of probability theory, it is necessary that the statistical probability follows the axioms of the mathematical probability theory, i.e. it has to obey Kolmogorov's axioms. For example, probabilities have to be positive and smaller or equal to one. We will discuss these axioms below.

In this book we will adopt a moderately Bayesian point of view. This means that in some cases we will introduce sensible assumptions without

[3]Remark, also probability assignments based on experience have a frequency background.

[4]For two reasons: The proof that the Kolmogorov's axioms are fulfilled is rather easy, and the calculation of the probability for complex events is possible by straight forward combinatorics.

being able to prove their validity. However, we will establish fixed, simple rules that have to be applied in data analysis. In this way we achieve an objective parametrization of the data. This does not exclude that in some occasions as in goodness-of-fit tests we favor methods of frequentist statistics.

1.4 Assignment of Probabilities to Events

The mathematician assumes that the assignment of probabilities to events exists. To achieve practical, useful results in the natural sciences, in sociology, economics or medicine, statistical methods are required and a sensible assignment has to be made.

There are various possibilities to do so:

- *Symmetry properties* are frequently used to assign equal probabilities to events. This is done in gambling, examples are rolling dice, roulette and card games. The isotropy of space predicts equal probabilities for radiation from a point source into different directions.
- *Laws of nature* like the Boltzmann's law of thermodynamics, the exponential decay law of quantum mechanics or Mendel's laws allow us to calculate the probabilities for certain events.
- From the observed frequencies of certain events in empirical studies we can estimate their probabilities, like those of female and male births, of muons in cosmic rays, or of measurement errors in certain repeatable experiments. Here we derive frequencies from a large sample of observations and approximate the probability of future events.
- In some situations we are left with educated guesses or we rely on the opinion of experts, when for example the weather is to be predicted or the risk of an accident of a new oil-ship has to be evaluated.
- In case of absolute ignorance often a *uniform probability distribution* is assumed. This is known as *Bayes' postulate.* When we watch a tennis game and do not know the players, we might assign equal probabilities of winning to player A and B.

To illustrate the last point in a more scientific situation, let us look at a common example in particle physics:

Example 1.1. Uniform prior for a particle mass

Before a precise measurement of a particle mass is performed, we only know that a particle mass m lies between the values m_1 and m_2. We may assume initially that all values of the mass inside the interval are equally likely. Then the a priori probability $P\{m_0 \leq m < m_2\}$ (or prior probability) that m it is larger than m_0, with m_0 located between m_1 and m_2, is equal to:

$$P\{m_0 \leq m < m_2\} = \frac{m_2 - m_0}{m_2 - m_1} .$$

This assertion relies on the assumption of a uniform distribution of the mass within the limits and is obviously assailable, because, had we assumed – with equal right – a uniform distribution for the mass squared, we had obtained a different result:

$$P\{m_0^2 \leq m^2 < m_2^2\} = \frac{m_2^2 - m_0^2}{m_2^2 - m_1^2} \neq P\{m_0 \leq m < m_2\} .$$

Of course, the difference is small, if the interval is small, $m_2 - m_1 \ll m$, for then we have:

$$\begin{aligned}
\frac{m_2^2 - m_0^2}{m_2^2 - m_1^2} &= \frac{m_2 - m_0}{m_2 - m_1} \times \frac{m_2 + m_0}{m_2 + m_1} \\
&= \frac{m_2 - m_0}{m_2 - m_1} \left(1 + \frac{m_0 - m_1}{m_2 + m_1}\right) \\
&\approx \frac{m_2 - m_0}{m_2 - m_1} .
\end{aligned}$$

When the Z_0 mass and its error were determined, a uniform prior probability in the mass was assumed. If instead a uniform probability in the mass squared had been used, the result had changed only by about 10^{-3} times the uncertainty of the mass determination. This means that applying Bayes' assumption to either the mass or the mass squared makes no difference within the precision of the measurement in this specific case.

In other situations prior probabilities which we will discuss in detail in Chapter 6 can have a considerable influence on a result.

1.5 Outline of This Book

After a short chapter on probability axioms and theorems we present properties of probability distributions in Chapter 3, and its application to simple error calculus and Monte Carlo simulation in Chapters 4 and 5.

The statistics part starts in Chapters 6 and 7 with point estimation followed by interval estimation, Chapter 8.

Chapter 9 deals with deconvolution problems.

In Chapter 10 significance and goodness-of-fit tests are discussed.

Chapter 11 with the title *Statistical Learning* summarizes some approximation and classification techniques.

In Chapter 12 a short introduction into probability density estimation and bootstrap techniques is given.

Finally, the appendix deals with some useful mathematical or technical subjects.

Recommendations for Ancillary Literature

(1) The standard book of Kendall and Stuart *The Advanced Theory of Statistics* [3], consisting of several volumes provides a excellent and rather complete presentation of classical statistics with all necessary proofs and many references. It is a sort of bible of conservative statistics, well suited to look up specific topics. Modern techniques, like Monte Carlo methods are not included.

(2) The books of Brandt *Data Analysis* [4] and Frodesen *et al. Probability and Statistics in Particle Physics* [5] give a pretty complete overview of the standard statistical methods as used by physicists.

(3) For a introduction into statistics for physicists, we highly recommend the book by Barlow *A Guide to the Use of Statistical Methods in the Physical Sciences* [6].

(4) Very intuitive and also well suited for beginners is the book by Lyons [7], *Statistics for Nuclear and Particle Physicists.* It reflects the long practical experience of the author.

(5) Larger, very professional and more ambitious is the book of Eadie *et al. Statistical Methods in Experimental Physics* [8], also intended mainly for particle and nuclear physicists and written by particle physicists and statisticians. A new edition has appeared recently [9]. Modern techniques of data analysis are not discussed.

(6) Recently a practical guide to data analysis in high energy physics [10] has been published. The chapters are written by different experienced physicists and presents the actual state of the art. As is common in statistics publications, some parts are slightly biased by the personal preferences of the authors. More specialized is [11] which emphasizes particularly probability density estimation and machine learning.
(7) Very useful especially for the solution of numerical problems is a book by Blobel and Lohrman *Statistische und numerische Methoden der Datenanalyse* [12] written in German.
(8) Other useful books written by particle physicists are found in Refs. [13–15]. The book by Roe is more conventional while Cowan and D'Agostini favor a Bayesian view.
(9) Modern techniques of statistical data analysis are presented in a book written by professional statisticians for non-professional users, Hastie *et al. The Elements of Statistical Learning* [16]
(10) A modern professional treatment of Bayesian statistics is the textbook by Box and Tiao *Bayesian Inference in Statistical Analysis* [17].

The interested reader will find work on the foundations of statistics, on basic principles and on the standard theory in the following books:

(1) Fisher's book [18] *Statistical Method, Experimental Design and Scientific Inference* provides an interesting overview of his complete work.
(2) Edward's book *Likelihood* [19] stresses the importance of the likelihood function, contains many useful references and the history of the likelihood concept.
(3) Many basic considerations and a collection of personal contributions from a moderate Bayesian view are contained in the book *Good Thinking* by Good [20]. A collection of work by Savage [21], presents a more extreme Baysian point of view.
(4) Somewhat old fashioned textbooks of Bayesian statistic which are of historical interest are the books of Jeffreys [22] and Savage [23].

Recent statistical work by particle physicists and astrophysicists can be found in the proceedings of the PHYSTAT Conferences [24] held during the past few years. Many interesting and well written articles can be found also in the internet.

This personal selection of literature is obviously in no way exhaustive.

Chapter 2

Basic Probability Relations

2.1 Random Events and Variables

Events are processes or facts that are characterized by a specific property, like obtaining a "3" with a dice. A goal in a soccer play or the existence of fish in a lake are events. Events can also be complex facts like rolling two times a dice with the results *three* and *five* or the occurrence of a certain mean value in a series of measurements or the estimate of the parameter of a theory. There are elementary events, which mutually exclude each other but also events that correspond to a class of elementary events, like the result *greater than three* when throwing a dice. We are concerned with random events which emerge from a stochastic process as already introduced above.

When we consider several events, then there are events which exclude each other and events which are compatible. We stick to our standard example *dice.* The elementary events *three* and *five* exclude each other, the events *greater than two* and *five* are of course compatible. An other common example: We select an object from a bag containing blue and red cubes and spheres. Here the events *sphere* and *cube* exclude each other, the events *sphere* and *red* may be compatible.

The event $\overline{A}$ is called the complement of event A if either event A *or* event $\overline{A}$ applies, but *not* both at the same time (exclusive *or*). Complementary to the event *three* in the dice example is the event *less than three or larger than three* (inclusive *or*). Complementary to the event *red sphere* is the event *cube or blue sphere.*

The event consisting of the fact that an arbitrary event out of all possible events applies, is called the certain event. We denote it with Ω. The complementary event is the impossible event, that none of all considered events applies: It is denoted with $\emptyset$, thus $\emptyset = \overline{\Omega}$.

Some further definitions are useful:

Definition 1: $A \cup B$ means A *or* B.

The event $A \cup B$ has the attributes of event A or event B or those of both A and B (inclusive *or*). (The attribute *cube* $\cup$ *red* corresponds to the complementary event *blue sphere.*)

Definition 2: $A \cap B$ means A *and* B.

The event $A \cap B$ has the attributes of event A as well as those of event B. (The attribute *cube* $\cap$ *red* corresponds to *red cube.*) If $A \cap B = \emptyset$, then A, B mutually exclude each other.

Definition 3: $A \subset B$ means that A implies B.

It is equivalent to both $A \cup B = B$ and $A \cap B = A$.

From these definitions follow the trivial relations

$$A \cup \overline{A} = \Omega, \quad A \cap \overline{A} = \emptyset,$$

and

$$\emptyset \subset A \subset \Omega\,. \tag{2.1}$$

For any A, B we have

$$\overline{A \cup B} = \overline{A} \cap \overline{B}, \quad \overline{A \cap B} = \overline{A} \cup \overline{B}.$$

To the random event A we associate the probability $P\{A\}$ as discussed above. In all practical cases random events can be identified with a variable, the *random variable* or in short *variate* following [19]. When the random events involve values that cannot be ordered, like shapes or colors, then they are associated to classes and categorical variates.

2.2 Probability Axioms and Theorems

2.2.1 *Axioms*

The assignment of probabilities $P\{A\}$ to members A, B, C, ... of a set of events has to satisfy the following axioms.[1] Only then the rules of probability theory are applicable.

- **Axiom 1** $\quad 0 \leq P\{A\}$
  The probability of an event is a positive real number.
- **Axiom 2** $\quad P\{\Omega\} = 1$
  The probability of the certain event is *one*.

---

[1] They are called *Kolmogorov* axioms, after the Russian mathematician A. N. Kolmogorov (1903–1987).

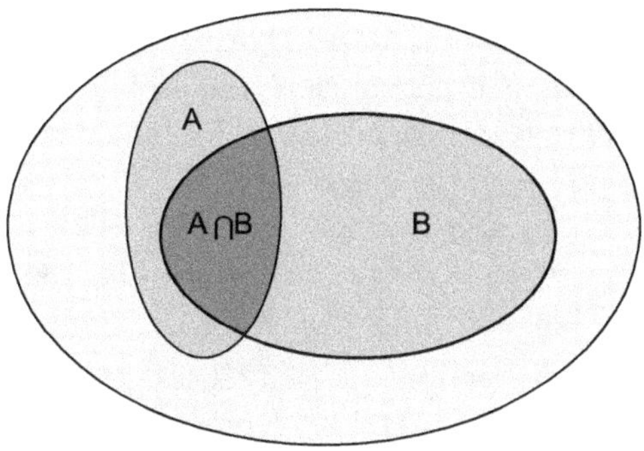

Fig. 2.1 Venn-diagram.

- **Axiom 3** $P\{A \cup B\} = P\{A\} + P\{B\}$ if $A \cap B = \emptyset$
  The probability that A *or* B applies is equal to the sum of the probabilities that A or that B applies, if the events A and B are mutually exclusive.

These axioms and definitions imply the following theorems whose validity is rather obvious. They can be illustrated with so-called *Venn diagrams*, Fig. 2.1.

$$\begin{aligned} P\{\overline{A}\} &= 1 - P\{A\},\ P\{\emptyset\} = 0, \\ P\{A \cup B\} &= P\{A\} + P\{B\} - P\{A \cap B\}, \\ \text{if } A \subset B &\Rightarrow P\{A\} \leq P\{B\}. \end{aligned} \tag{2.2}$$

Relation (2.1) together with (2.2) and axioms 1, 2 imply $0 \leq P\{A\} \leq 1$. For arbitrary events we have

$$P\{A \cup B\} \geq P\{A\}, P\{B\} \ ; \ P\{A \cap B\} \leq P\{A\}, P\{B\}.$$

If all events with the attribute A possess also the attribute B, $A \subset B$, then we have $P\{A \cap B\} = P\{A\}$, and $P\{A \cup B\} = P\{B\}$.

2.2.2 *Conditional Probability, Independence and Bayes' Theorem*

In the following we need two further definitions:

Definition: $P\{A \mid B\}$ is the *conditional probability* of event A under the condition that B applies. It is given, as is obvious from Fig. 2.1, by:

$$P\{A \mid B\} = \frac{P\{A \cap B\}}{P\{B\}}, \quad P\{B\} \neq 0. \tag{2.3}$$

A conditional probability is, for example, the probability to find a *sphere* among the *red* objects. The notation $A \mid B$ expresses that B is considered as fixed, while A is the random event to which the probability refers. Contrary to $P\{A\}$, which refers to arbitrary events A, we require that also B is valid and therefore $P\{A \cap B\}$ is normalized to $P\{B\}$.

Among the events $A \mid B$ the event $A = B$ is the certain event, thus $P\{B \mid B\} = 1$. More generally, from definition 3 of the last section and (2.3) follows $P\{A|B\} = 1$ if B implies A:

$$B \subset A \Rightarrow A \cap B = B \Rightarrow P\{A|B\} = 1.$$

Definition: If $P\{A \cap B\} = P\{A\} \times P\{B\}$, the events A and B (more precisely: the probabilities for their occurrence) are *independent.*

From (2.3) then follows $P\{A \mid B\} = P\{A\}$, i.e. the conditioning on B is irrelevant for the probability of A. Likewise $P\{B \mid A\} = P\{B\}$.

In Relation (2.3) we can exchange A and B and thus $P\{A \mid B\}P\{B\} = P\{A \cap B\} = P\{B \mid A\}P\{A\}$ and we obtain the famous Bayes' theorem:

$$P\{A \mid B\}P\{B\} = P\{B \mid A\}P\{A\}. \tag{2.4}$$

Bayes' theorem is frequently used to relate the conditional probabilities $P\{A \mid B\}$ and $P\{B \mid A\}$, and, as we will see, is of some relevance in parameter inference.

The following simple example illustrates some of our definitions. It assumes that each of the considered events is composed of a certain number of elementary events which mutually exclude each other and have the same probability.

Example 2.1. Card game, independent events

The following table summarizes some probabilities for randomly selected cards from a card set consisting of 32 cards and 4 colors.

$P\{\text{king}\}$:	$4/32 = 1/8$	(prob. for king)
$P\{\text{heart}\}$:	$1/4$	(prob. for heart)
$P\{\text{heart} \cap \text{king}\}$:	$1/8 \cdot 1/4 = 1/32$	(prob. for heart king)
$P\{\text{heart} \cup \text{king}\}$:	$1/8 + 1/4 - 1/32 = 11/32$	(prob. for heart or king)
$P\{\text{heart} \mid \text{king}\}$:	$1/4$	(prob. for heart if king)

The probabilities $P\{\text{heart}\}$ and $P\{\text{heart} \mid \text{king}\}$ are equal as required from the independence of the events A and B.

The following example illustrates how we make use of independence.

Example 2.2. Random coincidences, measuring the efficiency of a counter

When we want to measure the efficiency ϵ_1 of a particle counter (1), we combine it with a second counter (2) in such a way that a particle beam crosses both detectors. We record the number of events n_1, n_2 in the two counters and in addition the number of coincidences $n_{1\cap 2}$. The corresponding efficiencies relate these numbers, ignoring the statistical fluctuations of the observations, to the unknown number of particles n crossing the detectors.

$$n_1 = \varepsilon_1 n, \; n_2 = \varepsilon_2 n, \; n_{1\cap 2} = \varepsilon_{1\cap 2} n .$$

For *independent* counting efficiencies we have $\varepsilon_{1\cap 2} = \varepsilon_1 \varepsilon_2$ and we get

$$\varepsilon_1 = \frac{n_{1\cap 2}}{n_2}, \; \varepsilon_2 = \frac{n_{1\cap 2}}{n_1}, \; n = \frac{n_1 n_2}{n_{1\cap 2}} .$$

This scheme is used in many analog situations.

Bayes' theorem is applied in the next two examples, where the attributes are not independent.

Example 2.3. Bayes' theorem, fraction of women among students

From the proportion of students and women in the population and the fraction of students among women we compute the fraction of women among students:

$P\{A\}$	$= 0.02$	(fraction of students in the population)
$P\{B\}$	$= 0.5$	(fraction of women in the population)
$P\{A \mid B\}$	$= 0.018$	(fraction of students among women)
$P\{B \mid A\}$	$=?$	(fraction of women among students)

The dependence of the events A and B manifests itself in the difference of $P\{A\}$ and $P\{A \mid B\}$. Applying Bayes' theorem we obtain

$$\begin{aligned} P\{B \mid A\} &= \frac{P\{A \mid B\} P\{B\}}{P\{A\}} \\ &= \frac{0.018 \cdot 0.5}{0.02} = 0.45 . \end{aligned}$$

About 45% of the students are women.

Example 2.4. Bayes' theorem, beauty filter

The probability $P\{A\}$ that beauty quark production occurs in a colliding beam reaction be 0.0001. A filter program selects beauty reactions A with efficiency $P\{b \mid A\} = 0.98$ and the probability that it falsely assumes that beauty is present if it is not, be $P\{b \mid \overline{A}\} = 0.01$. What is the probability $P\{A \mid b\}$ to have genuine beauty production in a selected event? To solve the problem, first the probability $P\{b\}$ that a random event is selected has to be evaluated,

$$\begin{aligned} P\{b\} &= P\{b\} \left[P\{A \mid b\} + P\{\overline{A} \mid b\}\right] \\ &= P\{b \mid A\}P\{A\} + P\{b \mid \overline{A}\}P\{\overline{A}\}, \end{aligned}$$

where the bracket in the first line is equal to 1. In the second line Bayes' theorem is applied. Applying it once more, we get

$$\begin{aligned} P\{A \mid b\} &= \frac{P\{b \mid A\}P\{A\}}{P\{b\}} \\ &= \frac{P\{b \mid A\}P\{A\}}{P\{b \mid A\}P\{A\} + P\{b \mid \overline{A}\}P\{\overline{A}\}} \\ &= \frac{0.98 \cdot 0.0001}{0.98 \cdot 0.0001 + 0.01 \cdot 0.9999} = 0.0097. \end{aligned}$$

About 1% of the selected events corresponds to b quark production.

Bayes' theorem is rather trivial, thus the results of the last two examples could have easily been written down without referring to it.

Chapter 3

Probability Distributions and Their Properties

A probability distribution assigns probabilities to random variables. As an example we show in Fig. 3.1 the distribution of the sum s of the points obtained by throwing three ideal dice. Altogether there are $6^3 = 216$ different combinations. The random variable s takes values between 3 and 18 with different probabilities. The sum $s = 6$, for instance, can be realized in 10 different ways, all of which are equally probable. Therefore the probability for $s = 6$ is $P\{s = 6\} = 10/216 \approx 4.6\,\%$. The distribution is symmetric with respect to its mean value 10.5. It is restricted to discrete values of s, namely natural numbers.

In our example the random variable is discrete. In other cases the random variables are continuous. Then the probability for any fixed value is zero, we have to describe the distribution by a probability density and we obtain a finite probability when we integrate the density over a certain interval.

3.1 Definition of Probability Distributions

We define a distribution function, also called cumulative or integral distribution function, $F(t)$, which specifies the probability P to find a value of x smaller than t:

$$F(t) = P\{x < t\}, \text{ with } \; -\infty < t < \infty.$$

The probability axioms require the following properties of the distribution function:

(a) $F(t)$ is a non-decreasing function of t,
(b) $F(-\infty) = 0$,
(c) $F(\infty) = 1$.

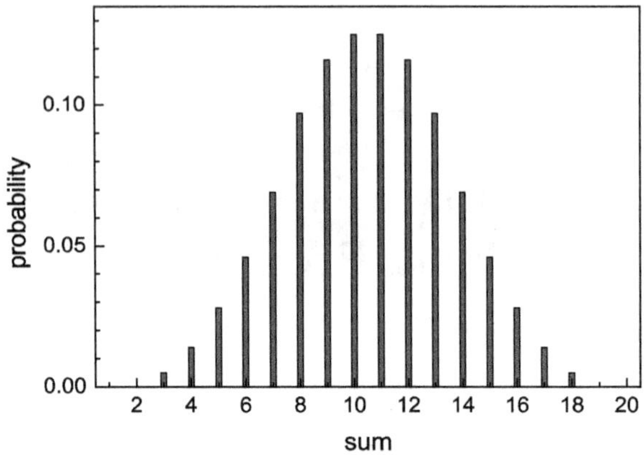

Fig. 3.1 Probability distribution of the sum of the points of three dice.

We distinguish between

- Discrete distributions (Fig. 3.2)
- Continuous distributions (Fig. 3.3)

3.1.1 *Discrete Distributions*

If not specified differently, we assume in the following that discrete distributions assign probabilities $p(x_i)$ to an enumerable set of different events, which are characterized by an ordered, real number x_i, with $i = 1, \ldots, N$, where N may be finite or infinite. The probabilities $p(x_i)$ to observe the values x_i satisfy the normalization condition:

$$\sum_{i=1}^{N} p(x_i) = 1 .$$

It is defined by

$$p(x_i) = P\{x = x_i\} = F(x_i + \epsilon) - F(x_i - \epsilon) ,$$

with ϵ positive and smaller than the distance to neighboring variate values.

Example 3.1. Discrete probability distribution (dice)

For a fair die, the probability to throw a certain number k is just one-sixth: $p(k) = 1/6$ for $k = 1, 2, 3, 4, 5, 6$.

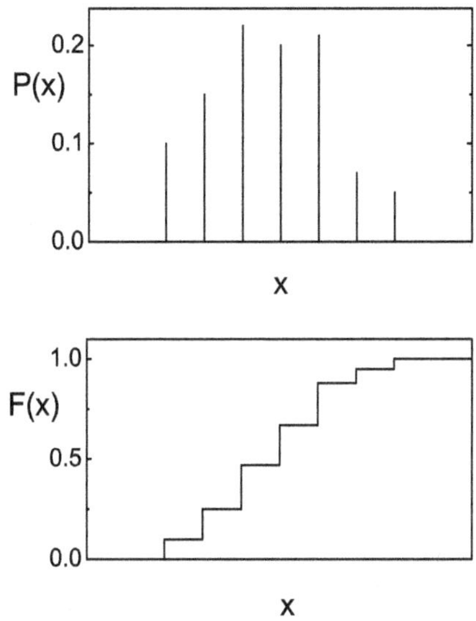

Fig. 3.2 Discrete probability distribution and distribution function.

It is possible to treat discrete distributions with the help of Dirac's δ-function like continuous ones. Therefore we will often consider only the case of continuous variates.

3.1.2 *Continuous Distributions*

We replace the discrete probability distribution by a *probability density*[1] $f(x)$, abbreviated as *probability density function* (*p.d.f.*). It is defined as follows:

$$f(x) = \frac{\mathrm{d}F(x)}{\mathrm{d}x}. \tag{3.1}$$

Remark that the p.d.f. is defined in the full range $-\infty < x < \infty$. It may be zero in certain regions. Hereafter, if not stated differently, we assume that the variable space is simple connected and that the p.d.f. is differentiable.

It has the following properties:

(a) $f(-\infty) = f(+\infty) = 0$,
(b) $\int_{-\infty}^{\infty} f(x)\mathrm{d}x = 1$.

---

[1]We will, however, use the notations *probability distribution* and *distribution* for discrete as well as for continuous distributions.

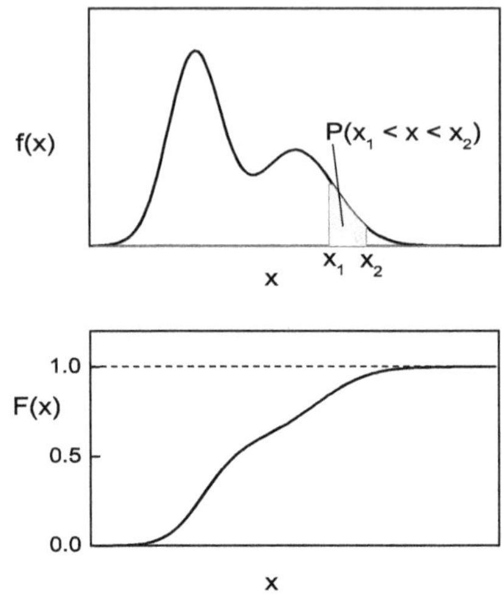

Fig. 3.3 Probability density and distribution function of a continuous distribution.

The probability $P\{x_1 \leq x \leq x_2\}$ to find the random variable x in the interval $[x_1, x_2]$ is given by

$$P\{x_1 \leq x \leq x_2\} = F(x_2) - F(x_1) = \int_{x_1}^{x_2} f(x)\mathrm{d}x.$$

We will discuss specific distributions in Section 3.6 but we introduce two common distributions already here. They will serve us as examples in the following sections.

Example 3.2. Probability density of an exponential distribution

The decay time t of an instable particles follows an exponential distribution with the p.d.f.

$$f(t) \equiv f(t|\lambda) = \lambda \mathrm{e}^{-\lambda t} \quad \text{for } t \geq 0, \tag{3.2}$$

where the parameter $\lambda > 0$, the decay constant, is the inverse of the mean lifetime $\tau = 1/\lambda$. The distribution function is

$$F(t) = \int_{-\infty}^{t} f(t')\mathrm{d}t' = 1 - \mathrm{e}^{-\lambda t}.$$

The probability of observing a lifetime longer than τ is

$$P\{t > \tau\} = F(\infty) - F(\tau) = \mathrm{e}^{-1}.$$

The exponential function is shown in Fig. 3.4.

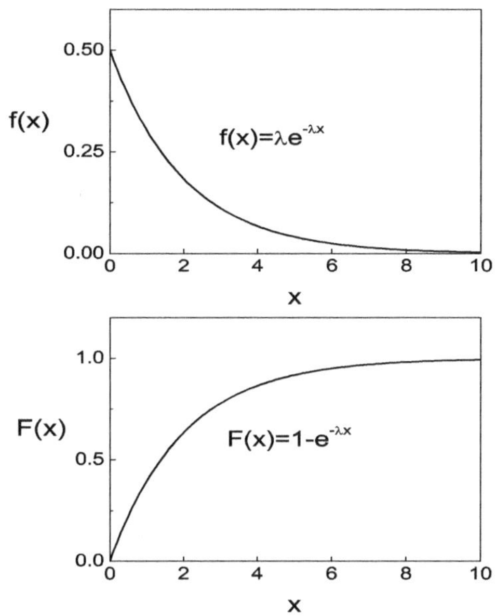

Fig. 3.4 Probability density and distribution function of an exponential distribution.

Example 3.3. Probability density of the normal distribution

An oxygen atom is drifting in argon gas, driven by thermal scattering. It starts at the origin. After a certain time its position is (x, y, z). Each projection, for instance x, has approximately a normal distribution (see Fig. 3.5), also called Gauss distribution.

$$f(x) = \mathcal{N}(x|0, s),$$
$$\mathcal{N}(x|x_0, s) = \frac{1}{\sqrt{2\pi}s}\mathrm{e}^{-(x-x_0)^2/(2s^2)}. \tag{3.3}$$

The width constant s is, as will be shown later, proportional to the square root of the number of scattering processes or the square root of time. When we descent by the factor $1/\sqrt{\mathrm{e}}$ from the maximum, the full width is just $2s$.

A statistical drift motion, or more generally a random walk, is met frequently in science and also in every day life. The normal distribution also

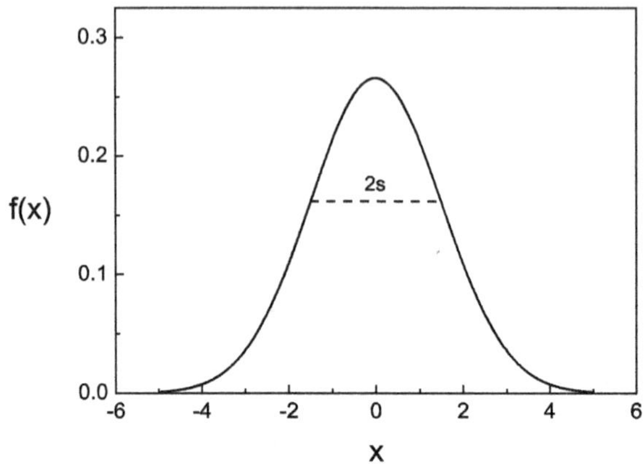

Fig. 3.5 Normal distribution.

describes approximately the motion of snow flakes or the erratic movements of a drunkard in the streets.

3.1.3 *Empirical Distributions*

Many processes are too complex or not well enough understood to be described by a distribution in form of a simple algebraic formula. In these cases it may be useful to approximate the underlying distribution using an experimental data sample. The simplest way to do this, is to histogram the observations and to normalize the frequency histogram. More sophisticated methods of probability density estimation will be sketched in Chapter 12. The quality of the approximation depends of course on the available number of observations.

3.2 Expected Values

In this section we will consider some general characteristic quantities of distributions, like mean value, width, and asymmetry or skewness which are defined through expected values.

The expected value $\mathrm{E}(u)$ of a quantity $u(x)$, which depends on the random variable x, is the average of an infinite number of values $u_i = u(x_i, i = 1, ..., \infty$ where x_i is drawn randomly from the distribution $f(x)$. Obviously, we have to assume the existence of such a limiting value.

In quantum mechanics, expected values of physical quantities are the main results of theoretical calculations and experimental investigations, and provide the connection to classical mechanics. Also in statistical mechanics and thermodynamics the calculation of expected values is frequently needed. We can, for instance, calculate from the velocity distribution of gas molecules the expected value of their kinetic energy, that means essentially their temperature. In probability theory and statistics expected values play a central role.

3.2.1 *Definition and Properties of the Expected Value*

Definition:

$$\mathrm{E}(u(x)) = \sum_{i=1}^{\infty} u(x_i)p(x_i) \quad \text{(discrete distribution)}, \tag{3.4}$$

$$\mathrm{E}(u(x)) = \int_{-\infty}^{\infty} u(x)f(x)\,\mathrm{d}x \quad \text{(continuous distribution)}. \tag{3.5}$$

Here and in what follows, we assume the existence of integrals and sums. This condition restricts the choice of the allowed functions u, p, f. If the number N of discrete variables x_i is finite, we set $p(x_i) = 0$ for $i > N$ to simplify formulas. From the definition of the expected value follow the relations (c is a constant, u, v are functions of x):

$$\mathrm{E}(c) = c, \tag{3.6}$$

$$\mathrm{E}(\mathrm{E}(u)) = \mathrm{E}(u), \tag{3.7}$$

$$\mathrm{E}(u + v) = \mathrm{E}(u) + \mathrm{E}(v), \tag{3.8}$$

$$\mathrm{E}(cu) = c\mathrm{E}(u). \tag{3.9}$$

They characterize E as a linear functional.

For *independent* (see also Chapter 2 and Section 3.5) variates x, y the following important relation holds:

$$\mathrm{E}\left(u(x)v(y)\right) = \mathrm{E}(u)\mathrm{E}(v). \tag{3.10}$$

Often expected values are denoted by angular brackets,

$$\mathrm{E}(u) \equiv \langle u \rangle,$$

to simplify the appearance of the formulas. We will use both notations.

3.2.2 *Mean Value*

The expected value of the variate x is also called the *mean value*. It can be visualized as the center of gravity of the distribution. Usually it is denoted by the Greek letter μ. Both names, mean value, and expected value[2] of the corresponding distribution are used synonymously.

Definition:

$$\mathrm{E}(x) \equiv \langle x \rangle = \mu = \sum_{i=1}^{\infty} x_i p(x_i) \text{ (discrete distribution)},$$

$$\mathrm{E}(x) \equiv \langle x \rangle = \mu = \int_{-\infty}^{\infty} x\, f(x)\, \mathrm{d}x \text{ (continuous distribution)}.$$

The mean value of the exponential distribution (3.2) is

$$\langle t \rangle = \int_0^{\infty} \lambda t e^{-\lambda t}\, \mathrm{d}t = 1/\lambda = \tau .$$

We will distinguish $\langle x \rangle$ from the average value of a sample, consisting of a *finite* number N of variate values, $x_1, \ldots, x_N$, which will be denoted by $\overline{x}$:

$$\overline{x} = \frac{1}{N} \sum_i x_i .$$

It is called sample mean. It is a random variable and has the expected value

$$\langle \overline{x} \rangle = \frac{1}{N} \sum_i \langle x_i \rangle = \langle x \rangle ,$$

as follows from (3.8), (3.9).

3.2.3 *Variance*

The variance var(x), usually denoted by σ^2, measures the spread of a distribution, defined as the mean quadratic deviation of the variate from its mean value. Usually, we want to know not only the mean value of a stochastic quantity, but require also information on the dispersion of the individual random values relative to it. When we buy a laser, we are of course interested in its mean energy per pulse, but also in the variation of the single energies around that mean value. The mean value alone does not provide information about the shape of a distribution. The mean height with respect to sea level of Switzerland is about 700 m, but this alone does not say much about the beauty of that country, which, to a large degree, depends on the spread of the height distribution.

[2]The notation *expected value* is somewhat misleading, as the probability to obtain it can be zero (see the example "dice" in Section 3.2.7).

The square root σ of the variance is called *standard deviation*, and is the standard measure of stochastic uncertainties.

A mechanical analogy to the variance is the moment of inertia for a mass distribution along the x-axis for a total mass equal to unity.

Definition:

$$\mathrm{var}(x) = \sigma^2 = \mathrm{E}\left[(x-\mu)^2\right].$$

From this definition follows immediately

$$\mathrm{var}(cx) = c^2\mathrm{var}(x),$$

and σ/μ is independent of the scale of x.

Very useful is the following expression for the variance which is easily derived from its definition and (3.8), (3.9):

$$\begin{aligned}\sigma^2 &= \mathrm{E}(x^2 - 2x\mu + \mu^2)\\ &= \mathrm{E}(x^2) - 2\mu^2 + \mu^2\\ &= \mathrm{E}(x^2) - \mu^2.\end{aligned}$$

Sometimes this is written more conveniently as

$$\sigma^2 = \langle x^2\rangle - \langle x\rangle^2 = \langle x^2\rangle - \mu^2. \tag{3.11}$$

In analogy to Steiner's theorem for moments of inertia, we have

$$\begin{aligned}\langle (x-a)^2\rangle &= \langle (x-\mu)^2\rangle + \langle (\mu-a)^2\rangle\\ &= \sigma^2 + (\mu-a)^2,\end{aligned}$$

implying (3.11) for $a = 0$.

The variance is invariant against a translation of the distribution by a:

$$x \to x+a,\ \mu \to \mu + a \Rightarrow \sigma^2 \to \sigma^2.$$

3.2.3.1 *Variance of a Sum of Random Numbers*

Let us calculate the variance σ^2 for the distribution of the sum $x = x_1 + x_2$ of two *independent* random numbers x_1 and x_2, which follow different distributions with mean values μ_1, μ_2 and variances σ_1^2, σ_2^2:

$$\begin{aligned}\sigma^2 &= \langle (x - \langle x\rangle)^2\rangle\\ &= \langle ((x_1-\mu_1) + (x_2-\mu_2))^2\rangle\\ &= \langle (x_1-\mu_1)^2 + (x_2-\mu_2)^2 + 2(x_1-\mu_1)(x_2-\mu_2)\rangle\\ &= \langle (x_1-\mu_1)^2\rangle + \langle (x_2-\mu_2)^2\rangle + 2\langle x_1-\mu_1\rangle\langle x_2-\mu_2\rangle\\ &= \sigma_1^2 + \sigma_2^2.\end{aligned}$$

In the fourth step the independence of the variates (3.10) was used.

This result is important for all kinds of error estimation. For a sum of two independent measurements, their standard deviations add quadratically. We can generalize the last relation to a sum $x = \sum x_i$ of N variates or measurements:

$$\sigma^2 = \sum_{i=1}^{N} \sigma_i^2 . \tag{3.12}$$

Example 3.4. Variance of the convolution of two distributions

We consider a quantity x with the p.d.f. $g(x)$ with variance σ_g^2 which is measured with a device which produces a smearing with a p.d.f. $h(y)$ with variance σ_h^2. We want to know the variance of the "smeared" value $x' = x + y$. According to 3.12, this is the sum of the variances of the two p.d.f.s:

$$\sigma^2 = \sigma_g^2 + \sigma_h^2 .$$

3.2.3.2 *Variance of the Sample Mean of Independent Identically Distributed Variates*

From the last relation we obtain the variance $\sigma_{\overline{x}}^2$ of the sample mean $\overline{x}$ from N independent random numbers x_i, which all follow the same distribution[3] $f(x)$, with expected value μ and variance σ^2:

$$\begin{aligned} \overline{x} &= \sum_{i=1}^{N} x_i/N , \\ \mathrm{var}(N\overline{x}) &= N^2\, \mathrm{var}(\overline{x}) = N\sigma^2 , \\ \sigma_{\overline{x}} &= \frac{\sigma}{\sqrt{N}} . \end{aligned} \tag{3.13}$$

The last two relations (3.12), (3.13) have many applications, for instance in random walk, diffusion, and error propagation. The root mean square distance reached by a diffusing molecule after N scatters is proportional to $\sqrt{N}$ and therefore also to $\sqrt{t}$, t being the diffusion time. The total length of 100 aligned objects, all having the same standard deviation σ of their nominal length, will have a standard deviation of only 10 σ. To a certain degree, random fluctuations compensate each other.

[3]The usual abbreviation is i.i.d. variates for independent identically distributed.

3.2.3.3 *Width v of a Sample and Variance of the Distribution*

Often, as we will see in Chapter 6, a sample is used to estimate the variance σ^2 of the underlying distribution. In case the mean value μ is known, we calculate the quantity

$$v_\mu^2 = \frac{1}{N}\sum_{i=1}^{N}(x_i - \mu)^2$$

which has the correct expected value $\langle v_\mu^2 \rangle = \sigma^2$. Usually, however, the true mean value μ is unknown – except perhaps in calibration measurements – and must be estimated from the same sample as is used to derive v_μ^2. We then are obliged to use the sample mean $\overline{x}$ instead of μ and calculate the mean quadratic deviation v^2 of the sample values relative to $\overline{x}$. In this case the expected value of v^2 will depend not only on σ, but also on N. In a first step we find

$$\begin{aligned} v^2 &= \frac{1}{N}\sum_i (x_i - \overline{x})^2 \\ &= \frac{1}{N}\sum_i \left(x_i^2 - 2x_i\overline{x} + \overline{x}^2\right) \\ &= \frac{1}{N}\sum_i x_i^2 - \overline{x}^2. \end{aligned} \tag{3.14}$$

To calculate the expected value, we use (3.11) and (3.13),

$$\begin{aligned} \langle x^2 \rangle &= \sigma^2 + \mu^2 \\ \langle \overline{x}^2 \rangle &= \mathrm{var}(\overline{x}) + \langle \overline{x} \rangle^2 \\ &= \frac{\sigma^2}{N} + \mu^2 \end{aligned}$$

and get with (3.14)

$$\langle v^2 \rangle = \langle x^2 \rangle - \langle \overline{x}^2 \rangle = \sigma^2\left(1 - \frac{1}{N}\right),$$

$$\sigma^2 = \frac{N}{N-1}\langle v^2 \rangle = \frac{\langle \sum_i (x_i - \overline{x})^2 \rangle}{N-1}. \tag{3.15}$$

The expected value of the mean squared deviation $\langle v^2 \rangle$ is smaller than the variance of the distribution by a factor of $(N-1)/N$.

The relation (3.15) is widely used for the estimation of measurement errors, when several independent measurements are available. The variance $\sigma_{\overline{x}}^2$ of the sample mean $\overline{x}$ itself is approximated, according to (3.13), by

$$\frac{v^2}{N-1} = \frac{\sum_i (x_i - \overline{x})^2}{N(N-1)}.$$

3.2.3.4 *Mean Value and Variance of a Superposition of Two Distributions*

Frequently a distribution consists of a superposition of elementary distributions. Let us compute the mean μ and variance σ^2 of a linear superposition of two distributions

$$f(x) = \alpha f_1(x) + \beta f_2(x), \ \alpha + \beta = 1,$$

where f_1, f_2 may have different mean values μ_1, μ_2 and variances σ_1^2, σ_2^2:

$$\begin{aligned}
\mu &= \alpha\mu_1 + \beta\mu_2, \\
\sigma^2 &= \mathrm{E}\left((x - \mathrm{E}(x))^2\right) \\
&= \mathrm{E}(x^2) - \mu^2 \\
&= \alpha\mathrm{E}_1(x^2) + \beta\mathrm{E}_2(x^2) - \mu^2 \\
&= \alpha(\mu_1^2 + \sigma_1^2) + \beta(\mu_2^2 + \sigma_2^2) - \mu^2 \\
&= \alpha\sigma_1^2 + \beta\sigma_2^2 + \alpha\beta(\mu_1 - \mu_2)^2 .
\end{aligned}$$

Here E, E_1, E_2 denote expected values related to the p.d.f.s f, f_1, f_2. In the last step the relation $\alpha + \beta = 1$ has been used. Of course, the width increases with the distance of the mean values. The result for σ^2 could have been guessed by considering the limiting cases ($\mu_1 = \mu_2$, $\sigma_1 = \sigma_2 = 0$).

3.2.4 *Skewness*

The skewness coefficient γ_1 measures the asymmetry of a distribution with respect to its mean. It is zero for the normal distribution, but quite sizable for the exponential distribution. There it has the value $\gamma_1 = 2$, see Section 3.3.4 below.

Definition:

$$\gamma_1 = \mathrm{E}\left[(x - \mu)^3\right] / \sigma^3 .$$

Similarly to the variance, γ_1 can be expressed by expected values of powers of the variate x:

$$\begin{aligned}
\gamma_1 &= \mathrm{E}\left[(x - \mu)^3\right] / \sigma^3 \\
&= \mathrm{E}\left[x^3 - 3\mu x^2 + 3\mu^2 x - \mu^3\right] / \sigma^3 \\
&= \left\{\mathrm{E}(x^3) - 3\mu\left[\mathrm{E}(x^2) - \mu\mathrm{E}(x)\right] - \mu^3\right\} / \sigma^3 \\
&= \frac{\mathrm{E}(x^3) - 3\mu\sigma^2 - \mu^3}{\sigma^3} .
\end{aligned}$$

The skewness coefficient is defined in such a way that it satisfies the requirement of invariance under translation and dilatation of the distribution. Its square value is usually denoted by $\beta_1 = \gamma_1^2$.

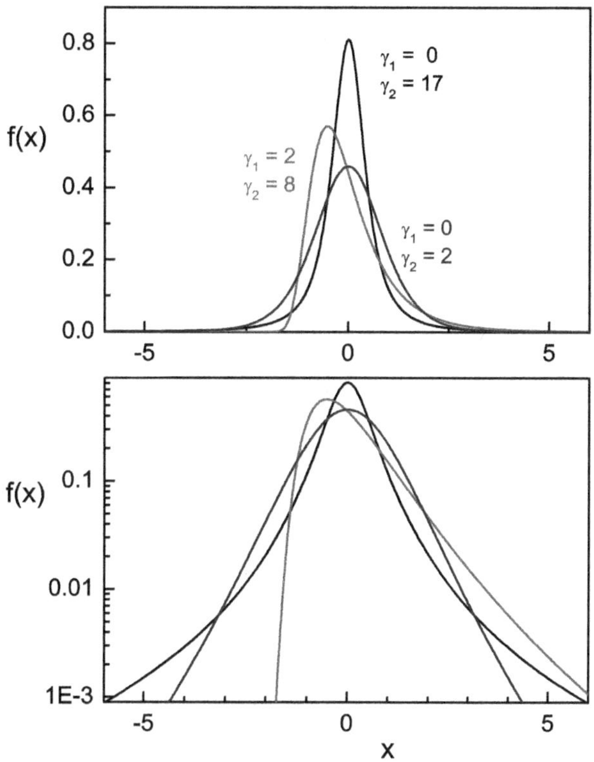

Fig. 3.6 Three distribution with equal mean and variance but different skewness and kurtosis.

3.2.5 *Kurtosis (Excess)*

A fourth parameter β_2, the kurtosis, measures the tails of a distribution.

Definition:

$$\beta_2 = \mathrm{E}[(x-\mu)^4]/\sigma^4 .$$

A kurtosis coefficient or excess $\gamma_2 = \beta_2 - 3$ is defined such that it is equal to zero for the normal distribution which is used as a reference. (see Section 3.6.5).

3.2.6 *Discussion*

The mean value of a distribution is a *position* or *location* parameter, the standard deviation is a *scale parameter*. A translation of the variate

$x \to y = x + a$ changes the mean value correspondingly, $\langle y \rangle = \langle x \rangle + a$. This parameter is therefore sensitive to the location of the distribution (like the center of gravity for a mass distribution). The variance (corresponding to the moment of inertia for a mass distribution), respectively the standard deviation remain unchanged. A change of the scale (dilatation) $x \to y = cx$ entails, besides $\langle y \rangle = c\langle x \rangle$ also $\sigma(y) = c\sigma(x)$. Skewness and kurtosis remain unchanged under both transformations. They are *shape parameters.*

The four parameters mean, variance, skewness, and kurtosis, or equivalently the expected values of x, x^2, x^3 and x^4, fix a a unimodal[4] distribution quite well if in addition the range of the variates and the behavior of the distribution at the limits is given. Then the distribution can be reconstructed quite accurately.

Figure 3.6 shows three probability densities, all with the same mean $\mu = 0$ and standard deviation $\sigma = 1$, but different skewness and kurtosis. The apparently narrower curve has clearly longer tails, as seen in the lower graph with logarithmic scale.

Mainly in cases, where the type of the distribution is not well known, i.e. for empirical distributions, other location and scale parameters are common. These are the *mode* x_{mod}, the variate value, at which the distribution has its maximum, and the *median*, defined as the variate value $x_{0.5}$, at which $P\{x < x_{0.5}\} = F(x_{0.5}) = 0.5$, i.e. the median subdivides the domain of the variate into two regions with equal probability of 50%. More generally, we define a *quantile* x_a of order a by the requirement $F(x_a) = a$.

A well known example for a median is the half-life $t_{0.5}$ which is the time at which 50% of the nuclei of an unstable isotope have decayed. From the exponential distribution (3.2) follows the relation between the half-life and the mean lifetime τ

$$t_{0.5} = \tau \ln 2 \approx 0.693\,\tau .$$

The median is invariant under monotonic transformations $y = y(x)$ of the variate, $y_{0.5} = y(x_{0.5})$ while for the mean value μ and the mode x_{mod} this is usually not the case, $\mu_y \neq y(\mu_x)$, $y_{mod} \neq y(x_{mod})$. The reason for these properties is that probabilities but not probability densities are invariant

[4]A unimodal distribution has a single maximum

under variate transformations. Thus the mode should not be considered as the "most probable value". The probability to obtain exactly the mode value is zero. To obtain finite probabilities, we have to integrate the p.d.f. over some range of the variate as is the case for the median.

In statistical analyses of data contaminated by background the sample median is more "robust" than the sample mean as estimator of the distribution mean. (see Appendix, Section A.16). Instead of the sample width v, often the *full width at half maximum* (f.w.h.m.) is used to characterize the spread of a distribution. It ignores the tails of the distribution. This makes sense for empirical distributions, e.g. in the investigation of spectral lines above a sizable background. For a normal distribution the f.w.h.m. is related to the standard deviation by

$$f.w.h.m._{gauss} \approx 2.36\,\sigma_{gauss}.$$

This relation is often used to estimate quickly the standard deviation σ for an empirical distribution given in form of a histogram. As seen from the examples in Fig.3.6, which, for the same variance, differ widely in their f.w.h.m., this procedure may lead to wrong results for non-Gaussian distributions.

3.2.7 *Examples*

In this section we compute expected values of some quantities for different distributions.

Example 3.5. Expected values, dice

We have $p(k) = 1/6$, $k = 1, \dots, 6$.

$$\begin{aligned}
\langle x \rangle &= (1+2+3+4+5+6)\,1/6 = 7/2, \\
\langle x^2 \rangle &= (1+4+9+16+25+36)\,1/6 = 91/6, \\
\sigma^2 &= 91/6 - (7/2)^2 = 35/12, \\
\sigma &\approx 1.71, \\
\gamma_1 &= 0.
\end{aligned}$$

The expected value has probability zero.

Example 3.6. Expected values, lifetime distribution $f(t) = \frac{1}{\tau}\,\mathrm{e}^{-t/\tau}$, $t \geq 0$,

$$\langle t^n \rangle = \int_0^\infty \frac{t^n}{\tau}\,\mathrm{e}^{-t/\tau}\,\mathrm{d}t = n!\;\tau^n,$$

$$\begin{aligned} \langle t \rangle &= \tau, \\ \langle t^2 \rangle &= 2\tau^2, \\ \langle t^3 \rangle &= 6\tau^3, \\ \sigma &= \tau, \\ \gamma_1 &= 2. \end{aligned}$$

Example 3.7. Mean value of the volume of a sphere with a normally distributed radius

The normal distribution is given by

$$f(x) = \frac{1}{\sqrt{2\pi}s}\,\mathrm{e}^{-(x-x_0)^2/(2s^2)}.$$

It is symmetric with respect to x_0. Thus the mean value is $\mu = x_0$, and the skewness is zero. For the variance we obtain

$$\begin{aligned} \sigma^2 &= \frac{1}{\sqrt{2\pi}s}\int_{-\infty}^{\infty} \mathrm{d}x (x-x_0)^2\,\mathrm{e}^{-(x-x_0)^2/(2s^2)} \\ &= s^2. \end{aligned}$$

The parameters x_0, s of the normal distribution are simply the mean value and the standard deviation μ, σ, and the p.d.f. with these parameters is abbreviated as $\mathcal{N}(x|\mu,\sigma)$. We now assume that the radius r_0 of a sphere is smeared according to a truncated normal distribution $r > 0$. We calculate the expected value of the volume $V(r) = 4/3\,\pi r^3$:

$$\begin{aligned} \langle V \rangle &= \int_{-\infty}^{\infty} \mathrm{d}r\,V(r) f(r) \\ &= \frac{4}{3}\frac{\pi}{\sqrt{2\pi}s}\int_{-\infty}^{\infty} \mathrm{d}r\,r^3 \mathrm{e}^{-\frac{(r-r_0)^2}{2s^2}} \\ &= \frac{4}{3}\pi(r_0^3 + 3s^2 r_0). \end{aligned}$$

The mean volume is larger than the volume calculated using the mean radius.

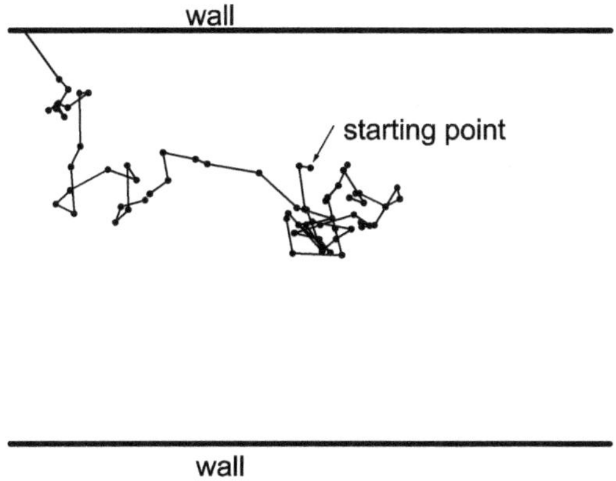

Fig. 3.7 Brownian motion.

Example 3.8. Playing poker until the bitter end

Two players are equally clever, but dispose of different capital K_1, respectively K_2. They play, until one of the players is left without money. We denote the probabilities for player 1, (2) to win finally with w_1 (w_2). The probability, that one of the two players wins, is unity,

$$w_1 + w_2 = 1.$$

Player 1 gains the capital K_2 with probability w_1 and looses K_1 with probability w_2. Thus his mean gain is $w_1K_2 - w_2K_1$. The same is valid for player two, only with reversed sign. As both players play equally well, the expected gain should be zero for both

$$w_1K_2 - w_2K_1 = 0.$$

From the two relation follows:

$$w_1 = \frac{K_1}{K_1 + K_2}; \quad w_2 = \frac{K_2}{K_1 + K_2}.$$

The probability to win is proportional to the capital disposed of. However, the greater risk of the player with the smaller capital comes along with the possibility of a higher gain.

Example 3.9. Diffusion (random walk)

A particle is moving randomly according to the Brownian motion, where every step is independent of the previous ones (Fig. 3.7). The starting point has a distance d_1 from the wall 1 and d_2 from the opposite wall 2. We want to know the probabilities w_1, w_2 to hit wall 1 or 2. The direct calculation of w_1 and w_2 is a quite involved problem. However, using the properties of expected values, it can be solved quite simply, without even knowing the probability density. The problem here is completely analogous to the previous one:

$$w_1 = \frac{d_2}{d_1 + d_2}, \qquad w_2 = \frac{d_1}{d_1 + d_2}.$$

Example 3.10. Mean kinetic energy of a gas molecule

The velocity of a particle in x-direction v_x is given by a normal distribution

$$f(v_x) = \frac{1}{s\sqrt{2\pi}} \mathrm{e}^{-v_x^2/(2s^2)},$$

with

$$s^2 = \frac{kT}{m},$$

where k, T, m are the Boltzmann constant, the temperature, and the mass of the molecule. The kinetic energy is

$$\epsilon_{kin} = \frac{m}{2}(v_x^2 + v_y^2 + v_z^2)$$

with the expected value

$$\mathrm{E}(\epsilon_{kin}) = \frac{m}{2}\left(\mathrm{E}(v_x^2) + \mathrm{E}(v_y^2) + \mathrm{E}(v_z^2)\right) = \frac{3m}{2}\mathrm{E}(v_x^2),$$

where in the last step the velocity distribution was assumed to be isotropic. It follows:

$$\mathrm{E}(v_x^2) = \frac{1}{s\sqrt{2\pi}} \int_{-\infty}^{\infty} dv_x v_x^2 \mathrm{e}^{-v_x^2/(2s^2)} = s^2 = kT/m,$$

$$\mathrm{E}(\epsilon_{kin}) = \frac{3}{2}kT.$$

Example 3.11. Reading accuracy of a digital clock

For an accurate digital clock which displays the time in seconds, the deviation of the reading from the true time is maximally $\pm$ 0.5 seconds. After the reading, we may associate to the true time a uniform distribution with the actual reading as its central value. To simplify the calculation of the variance, we set the reading equal to zero. We thus have

$$f(t) = \begin{cases} 1 & \text{if} - 0.5 < t < 0.5 \\ 0 & \text{else} \end{cases}$$

and

$$\sigma^2 = \int_{-0.5}^{0.5} t^2 \, \mathrm{d}t = \frac{1}{12} . \tag{3.16}$$

The root mean square measurement uncertainty (standard deviation) is $\sigma = 1\,\mathrm{s}/\sqrt{12} \approx 0.29\,\mathrm{s}$. The variance of a uniform distribution, which covers a range of a, is accordingly $\sigma^2 = a^2/12$. This result is widely used for the error estimation of digital measurements. A typical example from particle physics is the position measurement of ionizing particles with wire chambers or micro strip detectors.

Example 3.12. Efficiency fluctuations of a detector

A counter registers on average the fraction $\varepsilon = 0.9$ of all traversing electrons. How large is the relative fluctuation σ of the the registered number N_1 for N particles passing the detector? The exact solution of this problem requires the knowledge of the probability distribution, in this case the binomial distribution. But also without this knowledge we can derive the dependence on N with the help of relation (3.13):

$$\sigma\left(\frac{N_1}{N}\right) \sim \frac{1}{\sqrt{N}} .$$

The whole process can be split into single processes, each being associated with the passing of a single particle. Averaging over all processes leads to the above result. (The binomial distribution gives $\sigma(N_1/N) = \sqrt{\varepsilon(1-\varepsilon)/N}$, see Section 3.6.1).

All stochastic processes, which can be split into N identical, independent elementary processes, show the typical $1/\sqrt{N}$ behavior of their relative fluctuations.

3.3 Moments and Characteristic Function

The characteristic quantities of distributions considered up to now, mean value, variance, skewness, and kurtosis, have been calculated from expected values of the lower four powers of the variate. Now we will investigate the expected value of arbitrary powers of the random variable x for discrete and continuous probability distributions $p(x)$, $f(x)$, respectively. They are called *moments* of the distribution. Their calculation is particularly simple, if the characteristic function of the distribution is known. The latter is the Fourier transform of the distribution.

3.3.1 *Moments*

Definition: The n-th moments of $f(x)$, respectively $p(x)$ are

$$\mu_n = \mathrm{E}(x^n) = \int_{-\infty}^{\infty} x^n f(x)\,\mathrm{d}x,$$

and

$$\mu_n = \mathrm{E}(x^n) = \sum_{k=1}^{\infty} x_k^n p(x_k),$$

where n is a natural number.[5]

Apart from these moments, called *moments about the origin*, we consider also the moments about an arbitrary point a where x^n is replaced by $(x-a)^n$. Of special importance are the moments about the expected value of the distribution. They are called *central moments.*

Definition: The n-th central moment about $\mu = \mu_1$ of $f(x)$ or $p(x)$, respectively, is:

$$\mu_n' = \mathrm{E}\left((x-\mu)^n\right) = \int_{-\infty}^{\infty} (x-\mu)^n f(x)\,\mathrm{d}x,$$

respectively

$$\mu_n' = \mathrm{E}\left((x-\mu)^n\right) = \sum_{k=1}^{\infty} (x_k-\mu)^n p(x_k).$$

[5]In one dimension the zeroth moment is irrelevant. Formally, it is equal to *one*.

Accordingly, the first central moment is zero: $\mu'_1 = 0$. Generally, the moments are related to the expected values introduced before as follows:

First central moment: $\mu'_1 = 0$
Second central moment: $\mu'_2 = \sigma^2$
Third central moment: $\mu'_3 = \gamma_1 \sigma^3$
Fourth central moment: $\mu'_4 = \beta_2 \sigma^4$

Under conditions usually met in practice, a distribution is uniquely fixed by its moments. This means, if two distributions have the same moments in all orders, they are identical. Below plausibility arguments for the validity of this important assertion will be presented.

3.3.2 *Characteristic Function*

We define the characteristic function $\phi(t)$ of a distribution as follows:

Definition: The characteristic function $\phi(t)$ of a probability density $f(x)$ is

$$\phi(t) = \mathrm{E}(\mathrm{e}^{\mathrm{i}tx}) = \int_{-\infty}^{\infty} \mathrm{e}^{\mathrm{i}tx} f(x)\,\mathrm{d}x, \tag{3.17}$$

and, respectively for a discrete distribution $p(x_k)$

$$\phi(t) = \mathrm{E}(\mathrm{e}^{\mathrm{i}tx}) = \sum_{k=1}^{\infty} \mathrm{e}^{\mathrm{i}tx_k} p(x_k). \tag{3.18}$$

For continuous distributions, $\phi(t)$ is the Fourier transform of the p.d.f. $f(x)$.

From the definition of the characteristic function follow several useful properties.

$\phi(t)$ is a continuous, in general complex function of t, $-\infty < t < \infty$ with $|\phi(t)| \leq 1$, $\phi(0) = 1$ and $\phi(-t) = \phi^*(t)$. $\phi(t)$ is a real function, if and only if the distribution is symmetric, $f(x) = f(-x)$. Especially for continuous distributions $\lim_{t\to\infty} \phi(t) = 0$. A linear transformation of the variate $x \to y = ax + b$ induces a transformation of the characteristic function of the form

$$\phi_x(t) \to \phi_y(t) = \mathrm{e}^{\mathrm{i}bt} \phi_x(at). \tag{3.19}$$

Further properties are found in handbooks on the Fourier transform.

The transformation is invertible: With (3.17) it is

$$\begin{aligned}\int_{-\infty}^{\infty} \phi(t)\mathrm{e}^{-\mathrm{i}tx}\,\mathrm{d}t &= \int_{-\infty}^{\infty} \mathrm{e}^{-\mathrm{i}tx} \int_{-\infty}^{\infty} \mathrm{e}^{\mathrm{i}tx'} f(x')\,\mathrm{d}x'\,\mathrm{d}t \\ &= \int_{-\infty}^{\infty} f(x') \left(\int_{-\infty}^{\infty} \mathrm{e}^{\mathrm{i}t(x'-x)}\,\mathrm{d}t\right) \mathrm{d}x' \\ &= 2\pi \int_{-\infty}^{\infty} f(x')\,\delta(x'-x)\,\mathrm{d}x' \\ &= 2\pi f(x),\end{aligned}$$

$$f(x) = \frac{1}{2\pi}\int_{-\infty}^{\infty} \phi(t)\mathrm{e}^{-\mathrm{i}tx}\,\mathrm{d}t.$$

The same is true for discrete distributions, as may be verified by substituting (3.18):

$$p(x_k) = \lim_{T\to\infty} \frac{1}{2T}\int_{-T}^{T} \phi(t)\mathrm{e}^{-\mathrm{i}tx_k}\,\mathrm{d}t.$$

In all cases of practical relevance, the probability distribution is uniquely determined by its characteristic function.

Knowing the characteristic functions simplifies considerably the calculation of moments and of the distributions of sums or linear combinations of variates. For continuous distributions moments are found by n-fold derivation of $\phi(t)$:

$$\frac{\mathrm{d}^n\phi(t)}{\mathrm{d}t^n} = \int_{-\infty}^{\infty} (\mathrm{i}x)^n \mathrm{e}^{\mathrm{i}tx} f(x)\,\mathrm{d}x.$$

With $t = 0$ follows

$$\frac{\mathrm{d}^n\phi(0)}{\mathrm{d}t^n} = \int_{-\infty}^{\infty} (\mathrm{i}x)^n f(x)\,\mathrm{d}x = \mathrm{i}^n\mu_n. \tag{3.20}$$

The Taylor expansion of $\phi(t)$,

$$\phi(t) = \sum_{n=0}^{\infty} \frac{1}{n!}t^n\frac{\mathrm{d}^n\phi(0)}{\mathrm{d}t^n} = \sum_{n=0}^{\infty} \frac{1}{n!}(\mathrm{i}t)^n\mu_n, \tag{3.21}$$

generates the moments of the distribution.

The characteristic function $\phi(t)$ is closely related to the *moment generating function* which is defined through $M(t) = \mathrm{E}(\mathrm{e}^{tx})$. In some textbooks M is used instead of ϕ for the evaluation of the moments.

We realize that the moments determine ϕ uniquely, and, since the Fourier transform is uniquely invertible, the moments also determine the probability density, as stated above.

In the same way we obtain the central moments:

$$\phi'(t) = \mathrm{E}(\mathrm{e}^{\mathrm{i}t(x-\mu)}) = \int_{-\infty}^{\infty} \mathrm{e}^{\mathrm{i}t(x-\mu)} f(x)\,\mathrm{d}x = \mathrm{e}^{-\mathrm{i}t\mu}\phi(t), \tag{3.22}$$

$$\frac{\mathrm{d}^n\phi'(0)}{\mathrm{d}t^n} = \mathrm{i}^n \mu_n'. \tag{3.23}$$

The Taylor expansion is

$$\phi'(t) = \sum_{n=0}^{\infty} \frac{1}{n!}(\mathrm{i}t)^n \mu_n'. \tag{3.24}$$

The results (3.20), (3.21), (3.23), (3.24) remain valid also for discrete distributions. The Taylor expansion of the right hand side of relation (3.22) allows us to calculate the central moments from the moments about the origin and vice versa:

$$\mu_n' = \sum_{k=0}^{n} (-1)^k \binom{n}{k} \mu_{n-k}\mu^k, \quad \mu_n = \sum_{k=0}^{n} \binom{n}{k} \mu_{n-k}'\mu^k.$$

Note, that for $n = 0$, $\mu_0 = \mu_0' = 1$.

In some applications we have to compute the distribution $f(z)$ where z is the sum $z = x + y$ of two *independent* random variables x and y with the probability densities $g(x)$ and $h(y)$. The result is given by the convolution integral, see Section 3.5.4,

$$f(z) = \int g(x)h(z-x)\,\mathrm{d}x = \int h(y)g(z-y)\,\mathrm{d}y$$

which often is difficult to evaluate analytically. In most cases, it is simpler to proceed indirectly via the characteristic functions $\phi_g(t)$, $\phi_h(t)$ and $\phi_f(t)$ of the three p.d.f.s which obey the simple relation $\phi_f(t) = \phi_g(t)\phi_h(t)$

Proof:

Because of (3.10) we find for expected values

$$\begin{aligned}\phi_f(t) &= \mathrm{E}(\mathrm{e}^{\mathrm{i}t(x+y)})\\ &= \mathrm{E}(\mathrm{e}^{\mathrm{i}tx})\mathrm{E}(\mathrm{e}^{\mathrm{i}ty})\\ &= \phi_g(t)\phi_h(t).\end{aligned}$$

The third step requires the independence of the two variates. We apply the inverse Fourier transform to $\phi_f(t)$ and obtain

$$f(z) = \frac{1}{2\pi}\int \mathrm{e}^{-\mathrm{i}tz}\phi_f(t)\,\mathrm{d}t.$$

The solution of this integral is not always simple. For some functions it can be found in tables of Fourier transforms.

In the general case where x is a linear combination of independent random variables, $x = \sum c_j x_j$, we find in an analogous way:

$$\phi(t) = \prod \phi_j(c_j t).$$

3.3.3 *Cumulants*

As we have seen, the characteristic function simplifies in many cases the calculation of moments and the convolution of two distributions. Interesting relations between the moments of the three distributions $g(x)$, $h(y)$ and $f(z)$ with $z = x + y$ are obtained from the expansion of the logarithm $K(t)$ of the characteristic functions into powers of it:

$$K(t) = \ln \phi(t) = \ln \langle \exp(\mathrm{i}tx) \rangle = \kappa_1(\mathrm{i}t) + \kappa_2 \frac{(\mathrm{i}t)^2}{2!} + \kappa_3 \frac{(\mathrm{i}t)^3}{3!} + \cdots .$$

Since $\phi(0) = 1$ there is no constant term. The coefficients κ_i, defined in this way, are called *cumulants* or *semiinvariants*. The denotation *semiinvariant* indicates that the cumulants κ_i, with the exception of κ_1, remain invariant under the translations $x \to x + b$ of the variate x. Of course, the cumulant of order i can be expressed by moments about the origin or by central moments μ_k, μ'_k up to the order i. We do not present the general analytic expressions for the cumulants which can be derived from the power expansion of $\exp K(t)$ and present only the remarkably simple relations for the first six cummulants as a function of the central moments:

$$\begin{aligned}
\kappa_1 &= \mu_1 \equiv \mu = \langle x \rangle , \\
\kappa_2 &= \mu'_2 \equiv \sigma^2 = \mathrm{var}(x) , \\
\kappa_3 &= \mu'_3 , \\
\kappa_4 &= \mu'_4 - 3\mu'^2_2 , \\
\kappa_5 &= \mu'_5 - 10\mu'_2\mu'_3 , \\
\kappa_6 &= \mu'_6 - 15\mu'_2\mu'_4 - 10\mu'^2_3 + 30\mu'^3_2 .
\end{aligned} \tag{3.25}$$

Besides expected value and variance, also skewness and excess are easily expressed by cumulants:

$$\gamma_1 = \frac{\kappa_3}{\kappa_2^{3/2}} , \qquad \gamma_2 = \frac{\kappa_4}{\kappa_2^2} . \tag{3.26}$$

Since the product of the characteristic functions $\phi(t) = \phi^{(1)}(t)\phi^{(2)}(t)$ turns into the sum $K(t) = K^{(1)}(t) + K^{(2)}(t)$, the cumulants are additive, $\kappa_i = \kappa_i^{(1)} + \kappa_i^{(2)}$. In the general case, where x is a linear combination of independent variates, $x = \sum c_j x^{(j)}$, the cumulants of the resulting x-distribution, κ_i, are derived from those of the various $x^{(j)}$-distributions according to

$$\kappa_i = \sum_j c_j^i \kappa_i^{(j)} . \tag{3.27}$$

We have met examples for this useful relation already in Section 3.2.3 where we have computed the variance of the distribution of a sum of variates. We will use it again in the discussion of the Poisson distribution in the following example and in Section 3.6.3.

3.3.4 *Examples*

Example 3.13. Characteristic function of the Poisson distribution

The Poisson distribution

$$\mathcal{P}_\lambda(k) = \frac{\lambda^k}{k!}\mathrm{e}^{-\lambda}$$

has the characteristic function

$$\phi(t) = \sum_{k=0}^{\infty} \mathrm{e}^{\mathrm{i}tk}\,\frac{\lambda^k}{k!}\mathrm{e}^{-\lambda}$$

which can be simplified to

$$\begin{aligned}\phi(t) &= \sum_{k=0}^{\infty}\frac{1}{k!}(\mathrm{e}^{\mathrm{i}t}\lambda)^k\mathrm{e}^{-\lambda}\\ &= \exp(\mathrm{e}^{\mathrm{i}t}\lambda)\mathrm{e}^{-\lambda}\\ &= \exp\left(\lambda(\mathrm{e}^{\mathrm{i}t}-1)\right),\end{aligned}$$

from which we derive the moments:

$$\begin{aligned}\frac{\mathrm{d}\phi}{\mathrm{d}t} &= \exp\left(\lambda(\mathrm{e}^{\mathrm{i}t}-1)\right)\lambda\mathrm{i}\mathrm{e}^{\mathrm{i}t},\\ \frac{\mathrm{d}\phi(0)}{\mathrm{d}t} &= \mathrm{i}\lambda,\\ \frac{\mathrm{d}^2\phi}{\mathrm{d}t^2} &= \exp\left(\lambda(\mathrm{e}^{\mathrm{i}t}-1)\right)\left((\lambda\mathrm{i}\mathrm{e}^{\mathrm{i}t})^2-\lambda\mathrm{e}^{\mathrm{i}t}\right),\\ \frac{\mathrm{d}^2\phi(0)}{\mathrm{d}t^2} &= -(\lambda^2+\lambda).\end{aligned}$$

Thus, the two lowest moments are

$$\begin{aligned}\mu &= \langle k\rangle = \lambda,\\ \mu_2 &= \langle k^2\rangle = \lambda^2+\lambda\end{aligned}$$

and the mean value and the standard deviation are given by

$$\begin{aligned}\langle k\rangle &= \lambda,\\ \sigma &= \sqrt{\lambda}.\end{aligned}$$

Expanding

$$K(t) = \ln \phi(t) = \lambda(\mathrm{e}^{\mathrm{i}t} - 1) = \lambda[(\mathrm{i}t) + \frac{1}{2!}(\mathrm{i}t)^2 + \frac{1}{3!}(\mathrm{i}t)^3 + \cdots],$$

for the cumulants we get the simple result

$$\kappa_1 = \kappa_2 = \kappa_3 = \cdots = \lambda.$$

The calculation of the lower central moments is then trivial. For example, skewness and excess are simply given by

$$\gamma_1 = \kappa_3/\kappa_2^{3/2} = 1/\sqrt{\lambda}, \ \gamma_2 = \kappa_4/\kappa_2^2 = 1/\lambda.$$

Example 3.14. Distribution of a sum of independent, Poisson distributed variates

We start from the distributions

$$\mathcal{P}_1(k_1) = \mathcal{P}_{\lambda_1}(k_1),$$
$$\mathcal{P}_2(k_2) = \mathcal{P}_{\lambda_2}(k_2)$$

and calculate the probability distribution $\mathcal{P}(k)$ for $k = k_1 + k_2$. When we write down the characteristic function for $\mathcal{P}(k)$,

$$\begin{aligned}\phi(t) &= \phi_1(t)\phi_2(t)\\ &= \exp\left(\lambda_1(\mathrm{e}^{\mathrm{i}t} - 1)\right)\exp\left(\lambda_2(\mathrm{e}^{\mathrm{i}t} - 1)\right)\\ &= \exp\left((\lambda_1 + \lambda_2)(\mathrm{e}^{\mathrm{i}t} - 1)\right),\end{aligned}$$

we observe that $\phi(t)$ is just the characteristic function of the Poisson distribution $\mathcal{P}_{\lambda_1+\lambda_2}(k)$. The sum of two Poisson distributed variates is again Poisson distributed, because the mean value is the sum of the mean values of the two original distributions. This property is sometimes called *stability.*

Example 3.15. Characteristic function and moments of the exponential distribution

For the p.d.f.

$$f(x) = \lambda \mathrm{e}^{-\lambda x}$$

we obtain the characteristic function

$$\begin{aligned}\phi(t) &= \int_0^\infty \mathrm{e}^{\mathrm{i}tx}\,\lambda \mathrm{e}^{-\lambda x}\,\mathrm{d}x \\ &= \frac{\lambda}{-\lambda+\mathrm{i}t}\mathrm{e}^{(-\lambda+\mathrm{i}t)x}\Big|_0^\infty \\ &= \frac{\lambda}{\lambda-\mathrm{i}t}\end{aligned}$$

and deriving it with respect to t, we get from

$$\begin{aligned}\frac{\mathrm{d}\phi(t)}{\mathrm{d}t} &= \frac{\mathrm{i}\lambda}{(\lambda-\mathrm{i}t)^2}, \\ \frac{\mathrm{d}^n\phi(t)}{\mathrm{d}t^n} &= \frac{n!\,\mathrm{i}^n\lambda}{(\lambda-\mathrm{i}t)^{n+1}}, \\ \frac{\mathrm{d}^n\phi(0)}{\mathrm{d}t^n} &= \frac{n!\,\mathrm{i}^n}{\lambda^n}\end{aligned}$$

the moments of the distribution:

$$\mu_n = n!\,\lambda^{-n}.$$

From these we obtain the mean value

$$\mu = 1/\lambda,$$

the standard deviation

$$\sigma = \sqrt{\mu_2 - \mu^2} = 1/\lambda,$$

and the skewness

$$\gamma_1 = (\mu_3 - 3\sigma^2\mu - \mu^3)/\sigma^3 = 2.$$

Contrary to the Poisson example, here we do not gain in using the characteristic function, since the moments can be calculated directly:

$$\int_0^\infty x^n \lambda \mathrm{e}^{-\lambda x}\mathrm{d}x = n!\,\lambda^{-n}.$$

3.4 Transformation of Variables

In one of the examples of Section 3.2.7 we had calculated the expected value of the energy from the distribution of velocity. Of course, for certain applications it may be necessary to know not only the mean value of the energy but its complete distribution. To derive it, we have to perform a variable transformation.

For discrete distributions, this is a trivial exercise: The probability that the event "u has the value $u(x_k)$" occurs, where u is an uniquely invertible function of x, is of course the same as for "x has the value x_k":

$$P\{u = u(x_k)\} = P\{x = x_k\}\,.$$

For continuous distributions, the probability densities are transformed according to the usual rules as applied for example for mass or charge densities.

3.4.1 *Calculation of the Transformed Density*

We consider a probability density $f(x)$ and a monotone, i.e. uniquely invertible function $u(x)$. We we are interested in the p.d.f. of u, $g(u)$ (Figure 3.8).

The relation $P\{x_1 < x < x_2\} = P\{u_1 < u < u_2\}$ with $u_1 = u(x_1)$, $u_2 = u(x_2)$ has to hold, and therefore

$$\begin{aligned} P\{x_1 < x < x_2\} &= \int_{x_1}^{x_2} f(x')\,\mathrm{d}x' \\ &= \int_{u_1}^{u_2} g(u')\,\mathrm{d}u'. \end{aligned}$$

This may be written in differential form as

$$\begin{aligned} |g(u)\mathrm{d}u| &= |f(x)\mathrm{d}x|, \qquad (3.28) \\ g(u) &= f(x)\left|\frac{\mathrm{d}x}{\mathrm{d}u}\right|. \end{aligned}$$

Taking the absolute value guarantees the positivity of the probability density. Integrating (3.28), we find numerical equality of the distribution functions, $F(x) = G(u(x))$.

If $u(x)$ is not a monotone function, then, contrary to the above assumption, $x(u)$ is not a unique function (Fig. 3.9) and we have to sum over the contributions of the various branches of the inverse function:

$$g(u) = \left\{f(x)\left|\frac{\mathrm{d}x}{\mathrm{d}u}\right|\right\}_{branch1} + \left\{f(x)\left|\frac{\mathrm{d}x}{\mathrm{d}u}\right|\right\}_{branch2} + \cdots\,. \qquad (3.29)$$

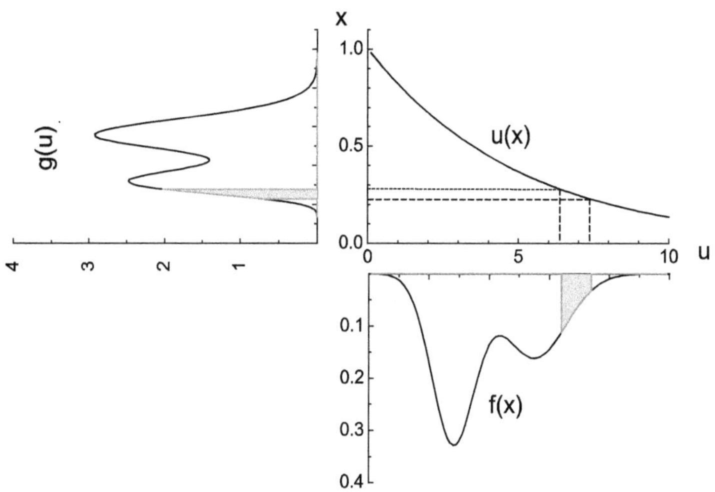

Fig. 3.8 Transformation of a probability density $f(x)$ into $g(u)$ given $u(x)$. The shaded areas are equal.

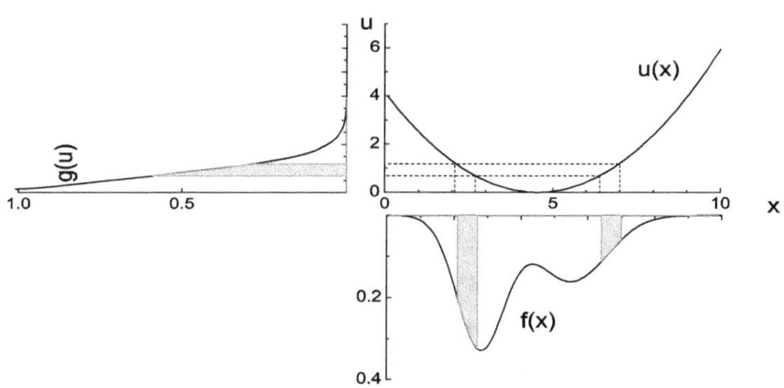

Fig. 3.9 Transformation of a p.d.f. $f(x)$ into $g(u)$ with $u = x^2$. The sum of the shaded areas below $f(x)$ is equal to the shaded area below $g(u)$.

Example 3.16. Calculation of the p.d.f. for the volume of a sphere from the p.d.f. of the radius

Given a uniform distribution for the radius r

$$f(r) = \begin{cases} 1/(r_2 - r_1) & \text{if } r_1 < r < r_2, \\ 0 & \text{else.} \end{cases}$$

we are interested in the distribution $g(V)$ of the volume $V(r)$. With

$$g(V) = f(r)\left|\frac{\mathrm{d}r}{\mathrm{d}V}\right|, \quad \frac{\mathrm{d}V}{\mathrm{d}r} = 4\pi r^2$$

we get

$$g(V) = \frac{1}{r_2 - r_1}\frac{1}{4\pi\, r^2} = \frac{1}{V_2^{1/3} - V_1^{1/3}}\frac{1}{3}V^{-2/3}.$$

The result is shown in Fig. 3.10.

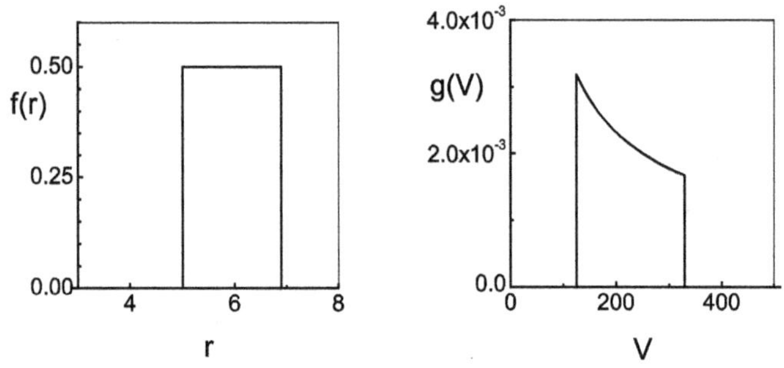

Fig. 3.10 Transformation of a uniform distribution of the radius into the distribution of the volume of a sphere.

Example 3.17. Distribution of the quadratic deviation

Given a normal distributed variate x with mean value x_0 and variance s^2 we determine the distribution $g(u)$, where

$$u = (x - x_0)^2/s^2$$

is the normalized quadratic deviation. The expected value of u is unity, since the expected value of $(x-\mu)^2$ per definition equals σ^2 for any distribution. The function $x(u)$ has two branches. With

$$f(x) = \frac{1}{s\sqrt{2\pi}}\mathrm{e}^{-(x-x_0)^2/(2s^2)}$$

and

$$\frac{\mathrm{d}x}{\mathrm{d}u} = \frac{s}{2\sqrt{u}}$$

we find

$$g(u)=\left\{\frac{1}{2\sqrt{2\pi u}}\mathrm{e}^{-u/2}\right\}_{branch1}+\left\{\cdots\right\}_{branch2}.$$

The contributions from both branches are the same, thus

$$g(u)=\frac{1}{\sqrt{2\pi u}}\mathrm{e}^{-u/2}. \tag{3.30}$$

The function $g(u)$ is the so-called χ^2 - *distribution* (chi-squared distribution) for one degree of freedom.

Example 3.18. Distribution of the kinetic energy in the one-dimensional ideal gas

Be v the velocity of a particle in x direction with probability density

$$f(v)=\sqrt{\frac{m}{2\pi kT}}\mathrm{e}^{-mv^2/(2kT)}.$$

Its kinetic energy $E=v^2/(2m)$ of which we want to know the distribution $g(E)$. The function $v(E)$ has again two branches. We get, in complete analogy to the example above,

$$\frac{\mathrm{d}v}{\mathrm{d}E}=\frac{1}{\sqrt{2mE}},$$

$$g(E)=\left\{\frac{1}{2\sqrt{\pi kTE}}\mathrm{e}^{-E/kT}\right\}_{branch1}+\left\{\cdots\right\}_{branch2}.$$

The contributions of both branches are the same, hence

$$g(E)=\frac{1}{\sqrt{\pi kTE}}\mathrm{e}^{-E/kT}.$$

3.4.2 *Determination of the Variable Transformation for Given Distributions*

In the computer simulation of stochastic processes we are frequently confronted with the problem that we have to transform the uniform distribution of a random number generator into a desired distribution, e.g. a normal or exponential distribution. More generally, we want to obtain for two given distributions $f(x)$ and $g(u)$ the transformation $u(x)$ that relates u and x. We assume that the relation is monotone.

We have

$$\int_{-\infty}^{x} f(x')\,\mathrm{d}x' = \int_{-\infty}^{u} g(u')\,\mathrm{d}u'.$$

Integrating, we get $F(x)$ and $G(u)$:

$$F(x) = G(u),$$

$$u(x) = G^{-1}\left(F(x)\right).$$

G^{-1} is the inverse function of G. The problem can be solved analytically, only if f and g can be integrated analytically and if the inverse of G can be derived.

Let us consider now the special case mentioned above, where the primary distribution $f(x)$ is uniform, $f(x) = 1$ for $0 \leq x \leq 1$. This implies $F(x) = x$ and

$$G(u) = x,$$

$$u = G^{-1}(x). \tag{3.31}$$

Example 3.19. Generation of an exponential distribution starting from a uniform distribution

Given are the p.d.f.s

$$f(x) = \begin{cases} 1 & \text{for } 0 < x < 1, \\ 0 & \text{else,} \end{cases}$$

$$g(u) = \begin{cases} \lambda \mathrm{e}^{-\lambda u} & \text{for } 0 < u, \\ 0 & \text{else.} \end{cases}$$

The desired transformation $u(x)$, as demonstrated above in the general case, is obtained by integration and inversion:

$$\int_0^u g(u')\,\mathrm{d}u' = \int_0^x f(x')\,\mathrm{d}x',$$

$$\int_0^u \lambda \mathrm{e}^{-\lambda u'}\,\mathrm{d}u' = \int_0^x f(x')\,\mathrm{d}x',$$

$$1 - \mathrm{e}^{-\lambda u} = x,$$

$$u = -\ln(1-x)/\lambda.$$

We could have used, of course, also the relation (3.31) directly. Obviously in the last relation we could substitute $1 - x$ by x, since both quantities are uniformly distributed. When we transform the uniformly distributed random numbers x delivered by our computer according to the last relation into the variable u, the latter will be exponentially distributed. This is the usual way to simulate the lifetime distribution of unstable particles and other decay processes (see Chapter 5).

3.5 Multivariate Probability Densities

The results of the last sections are easily extended to multivariate distributions. We restrict ourselves here to the case of continuous distributions.

3.5.1 *Probability Density of Two Variables*

3.5.1.1 *Definitions*

As in the one-dimensional case, we define an integral distribution function $F(x, y)$, now taken to be the probability to find values of the variates x', y' smaller than x, respectively y

$$F(x, y) = P\{(x' < x) \cap (y' < y)\} . \tag{3.32}$$

The following properties of this distribution function are satisfied:

$$F(\infty, \infty) = 1,$$
$$F(-\infty, y) = F(x, -\infty) = 0.$$

In addition, F has to be a monotone increasing function of both variables. We define a two-dimensional probability density, the so-called joined probability density, as the partial derivation of F with respect to the variables x, y:

$$f(x, y) = \frac{\partial^2 F}{\partial x \, \partial y} .$$

From these definitions follows the normalization condition

$$\int_{-\infty}^{\infty} \int_{-\infty}^{\infty} f(x, y) \, \mathrm{d}x \, \mathrm{d}y = 1.$$

The projections $f_x(x)$ respectively $f_y(y)$ of the joined probability density onto the coordinate axes are called marginal distributions :

$$f_x(x) = \int_{-\infty}^{\infty} f(x,y)\,\mathrm{d}y,$$
$$f_y(y) = \int_{-\infty}^{\infty} f(x,y)\,\mathrm{d}x.$$

The marginal distributions are one-dimensional (univariate) probability densities.

The conditional probability densities for fixed values of the second variate and normalized with respect to the first one are denoted by $f_x(x|y)$ and $f_y(y|x)$ for given values of y or x, respectively. We have the following relations:

$$\begin{aligned} f_x(x|y) &= \frac{f(x,y)}{\int_{-\infty}^{\infty} f(x,y)\,\mathrm{d}x} \\ &= \frac{f(x,y)}{f_y(y)}, \end{aligned} \tag{3.33}$$

$$\begin{aligned} f_y(y|x) &= \frac{f(x,y)}{\int_{-\infty}^{\infty} f(x,y)\,\mathrm{d}y} \\ &= \frac{f(x,y)}{f_x(x)}. \end{aligned} \tag{3.34}$$

Together, (3.33) and (3.34) express again Bayes' theorem:

$$f_x(x|y)f_y(y) = f_y(y|x)f_x(x) = f(x,y).$$

Example 3.20. Superposition of two two-dimensional normal distributions

(The two-dimensional normal distribution will be discussed in Section 3.6.5.) The marginal distributions $f_x(x)$, $f_y(y)$ and the conditional p.d.f.

$$f_y(y|x=1)$$

for the joined two-dimensional p.d.f.

$$f(x,y) = \frac{1}{2\pi}\left[0.6\exp\left(-\frac{x^2}{2}-\frac{y^2}{2}\right) + \frac{0.4}{\sqrt{3}}\exp\left(-\frac{(x-2)^2}{3}-\frac{(y-2.5)^2}{4}\right)\right]$$

are

$$
\begin{aligned}
f_x(x) &= \frac{1}{\sqrt{2\pi}}\left[0.6\exp\left(-\frac{x^2}{2}\right) + \frac{0.4}{\sqrt{1.5}}\exp\left(-\frac{(x-2)^2}{3}\right)\right], \\
f_y(y) &= \frac{1}{\sqrt{2\pi}}\left[0.6\exp\left(-\frac{y^2}{2}\right) + \frac{0.4}{\sqrt{2}}\exp\left(-\frac{(y-2.5)^2}{4}\right)\right], \\
f(y, x=1) &= \frac{1}{2\pi}\left[0.6\exp\left(-\frac{1}{2}-\frac{y^2}{2}\right) + \frac{0.4}{\sqrt{3}}\exp\left(-\frac{1}{3}-\frac{(y-2.5)^2}{4}\right)\right], \\
f_y(y|x=1) &= 2/3\left[0.6\exp\left(-\frac{1}{2}-\frac{y^2}{2}\right) + \frac{0.4}{\sqrt{3}}\exp\left(-\frac{1}{3}-\frac{(y-2.5)^2}{4}\right)\right].
\end{aligned}
$$

$f_y(y|1)$ and $f(y,1)$ differ in the normalization factor, which results from the requirement $\int f_y(y|1)\,\mathrm{d}y = 1$.

3.5.1.2 *Graphical Presentation*

Figure 3.11 shows a similar superposition of two Gaussians together with its marginal distributions and one conditional distribution. The chosen form of the graphical representation as a contour plot for two-dimensional distributions is usually to be favored over three-dimensional surface plots.

3.5.2 *Moments*

Analogously to the one-dimensional case we define moments of two-dimensional distributions:

$$
\begin{aligned}
\mu_x &= \mathrm{E}(x), \\
\mu_y &= \mathrm{E}(y), \\
\sigma_x^2 &= \mathrm{E}\left[(x-\mu_x)^2\right], \\
\sigma_y^2 &= \mathrm{E}\left[(y-\mu_y)^2\right], \\
\sigma_{xy} &= \mathrm{E}\left[(x-\mu_x)(y-\mu_y)\right], \\
\mu_{lm} &= \mathrm{E}(x^l y^m), \\
\mu'_{lm} &= \mathrm{E}\left[(x-\mu_x)^l (y-\mu_y)^m\right].
\end{aligned}
$$

Explicitly,

$$
\begin{aligned}
\mu_x &= \int_{-\infty}^{\infty}\int_{-\infty}^{\infty} x f(x,y)\,\mathrm{d}x\,\mathrm{d}y = \int_{-\infty}^{\infty} x f_x(x)\,\mathrm{d}x, \\
\mu_y &= \int_{-\infty}^{\infty}\int_{-\infty}^{\infty} y f(x,y)\,\mathrm{d}x\,\mathrm{d}y = \int_{-\infty}^{\infty} y f_y(y)\,\mathrm{d}y,
\end{aligned}
$$

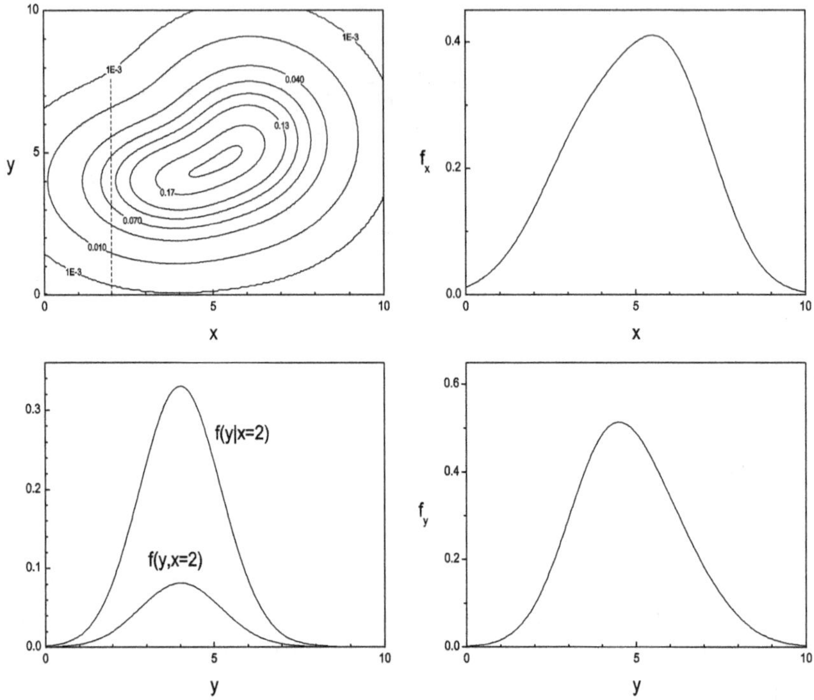

Fig. 3.11 Two-dimensional probability density. The lower left-hand plot shows the conditional p.d.f. of y for $x = 2$. The lower curve is the p.d.f. $f(y, 2)$. It corresponds to the dashed line in the upper plot. The right-hand side displays the marginal distributions.

$$\mu_{lm} = \int_{-\infty}^{\infty}\int_{-\infty}^{\infty} x^l y^m f(x,y)\,\mathrm{d}x\,\mathrm{d}y,$$

$$\mu'_{lm} = \int_{-\infty}^{\infty}\int_{-\infty}^{\infty} (x-\mu_x)^l (y-\mu_y)^m f(x,y)\,\mathrm{d}x\,\mathrm{d}y.$$

Obviously, μ'_x, μ'_y $(= \mu'_{10}, \mu'_{01})$ are zero.

3.5.2.1 *Correlations, Covariance, Independence*

The mixed moment σ_{xy} is called *covariance* of x and y, and sometimes also denoted as $\mathrm{cov}(x, y)$. If σ_{xy} is different from zero, the variables x and y are said to be *correlated*. The mean value of y depends on the value chosen for x and vice versa. Thus, for instance, the weight of a man is positively correlated with its height.

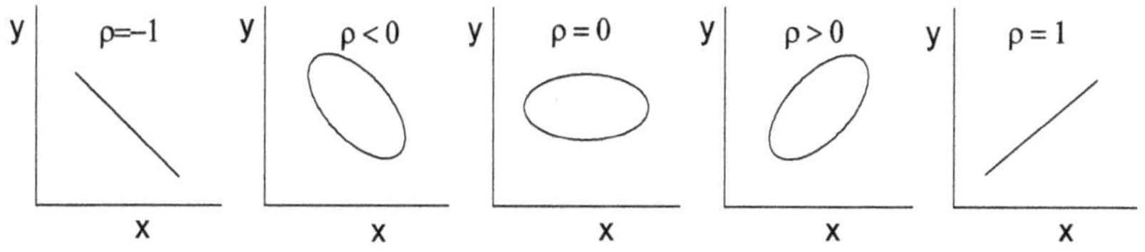

Fig. 3.12 Curves $f(x, y) = const$ with different correlation. coefficients.

The degree of correlation is quantified by

$$\rho_{xy} = \frac{\sigma_{xy}}{\sigma_x \sigma_y},$$

the *correlation coefficient.* Schwarz' inequality insures $|\rho_{xy}| \leq 1$.

Figure 3.12 shows lines of constant probability for various kinds of correlated distributions. In the extreme case $|\rho| = 1$ the variates are linearly related.

If the correlation coefficient is zero, this does not necessarily mean statistical *independence* of the variates. The dependence may be more subtle, as we will see shortly. As defined in Chapter 2, two random variables x, y are called *independent* or *orthogonal*, if the probability to observe one of the two variates x, y is independent from the value of the other one, i.e. the conditional distributions are equal to the marginal distributions, $f_x(x|y) = f_x(x)$, $f_y(y|x) = f_y(y)$. Independence is realized only if the joined distribution $f(x, y)$ factorizes into its marginal distributions (see Chapter 2):

$$f(x, y) = f_x(x) f_y(y).$$

Clearly, correlated variates cannot be independent.

Example 3.21. Correlated variates

A measurement uncertainty of a point in the xy-plane follows independent normal distributions in the polar coordinates r, φ (the errors are assumed small enough to neglect the regions $r < 0$ and $|\varphi| > \pi$). A line of constant probability in the xy-plane would look similar to the second graph of Fig. 3.12. The Cartesian coordinates are negatively correlated, although the original polar coordinates have been chosen as uncorrelated, in fact they are even independent.

Example 3.22. Dependent variates with correlation coefficient zero

For the probability density

$$f(x,y) = \frac{1}{2\pi\sqrt{x^2+y^2}} \mathrm{e}^{-\sqrt{x^2+y^2}}$$

we find $\sigma_{xy} = 0$. The curves $f = const.$ are circles, but x and y are *not* independent, the conditional distribution $f_y(y|x)$ of y depends on x.

3.5.3 *Transformation of Variables*

The probability densities $f(x,y)$ and $g(u,v)$ are transformed given the transformation functions $u(x,y)$, $v(x,y)$, analogously to the univariate case

$$g(u,v)\,\mathrm{d}u\,\mathrm{d}v = f(x,y)\,\mathrm{d}x\,\mathrm{d}y,$$

$$g(u,v) = f(x,y)\left|\frac{\partial(x,y)}{\partial(u,v)}\right|.$$

The Jacobian determinant replaces the differential quotient.

Example 3.23. Transformation of a normal distribution from Cartesian into polar coordinates

A two-dimensional normal distribution

$$f(x,y) = \frac{1}{2\pi}\mathrm{e}^{-(x^2+y^2)/2}$$

is to be transformed into polar coordinates

$$x = r\cos\varphi,$$
$$y = r\sin\varphi.$$

The Jacobian is

$$\frac{\partial(x,y)}{\partial(r,\varphi)} = r.$$

We get

$$g(r,\varphi) = \frac{1}{2\pi} r\mathrm{e}^{-r^2/2}$$

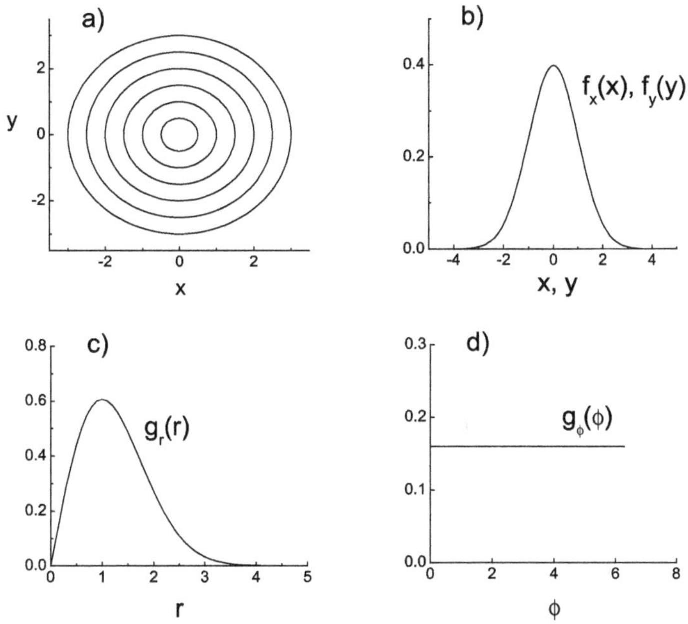

Fig. 3.13 Transformation of a two-dimensional normal distribution of Cartesian coordinates into the distribution of polar coordinates: (a) lines of constant probability; (b) Cartesian marginal distributions; (c) and (d) marginal distributions of the polar coordinates.

with the marginal distributions

$$g_r = \int_0^{2\pi} g(r, \varphi)\, d\varphi = re^{-r^2/2},$$

$$g_\varphi = \int_0^{\infty} g(r, \varphi)\, dr = \frac{1}{2\pi}.$$

The joined distribution factorizes into its marginal distributions (Fig. 3.13). Not only x, y, but also r, φ are independent.

3.5.4 *Reduction of the Number of Variables*

Frequently, we are faced with the problem to find from a given joined distribution $f(x, y)$ the distribution $g(u)$ of a dependent random variable $u(x, y)$. We can reduce it to that of a usual transformation: We invent a second variable $v = v(x, y)$, perform the transformation $f(x, y) \longrightarrow h(u, v)$ and,

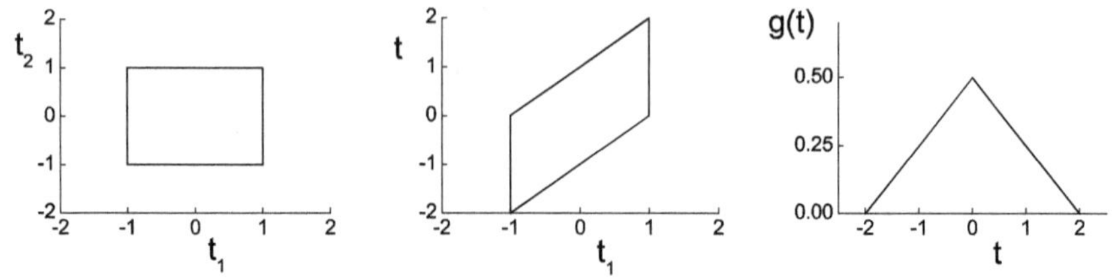

Fig. 3.14 Distribution of the difference t between two times t_1 and t_2 which have both clock readings equal to zero.

finally, we calculate the marginal distribution of u,

$$g(u) = \int_{-\infty}^{\infty} h(u, v)\, \mathrm{d}v.$$

In most cases, the choice $v = x$ is suitable. More formally, we might use the equivalent reduction formula

$$g(u) = \int_{-\infty}^{\infty} f(x, y)\delta\left(u - u(x, y)\right)\, \mathrm{d}x\, \mathrm{d}y. \tag{3.35}$$

For the distribution of a sum $u = x + y$ of two independent variates x, y, i.e. $f(x, y) = f_x(x)f_y(y)$, after integration over y follows

$$g(u) = \int f(x, u - x)\, \mathrm{d}x = \int f_x(x)f_y(u - x)\, \mathrm{d}x.$$

This is called the convolution integral or convolution product of f_x and f_y.

Example 3.24. Distribution of the difference of two digitally measured times

The true times t_1, t_2 are taken to follow a uniform distribution

$$f(t_1, t_2) = \begin{cases} 1/\Delta^2 & \text{for } |t_1 - T_1|,\ |t_2 - T_2| < \Delta/2 \\ 0 & \text{else} \end{cases}$$

around the readings T_1, T_2. We are interested in the probability density of the difference $t = t_1 - t_2$. To simplify the notation, we choose the case $T_1 = T_2 = 0$ and $\Delta = 2$ (Fig. 3.14). We keep $t_1 = t_1$ and set $t = t_1 - t_2$, with the Jacobian

$$\frac{\partial(t_1, t_2)}{\partial(t_1, t)} = 1.$$

The new distribution is also uniform:

$$h(t_1, t) = f(t_1, t_2) = 1/\Delta^2,$$

and has the boundaries shown in Fig. 3.14. The form of the marginal distribution is found by integration over t_1, or directly by reading it off from the figure:

$$g(t) = \begin{cases} (\ t - T + \Delta)/\Delta^2 & \text{for } t - T < 0, \\ (-t + T + \Delta)/\Delta^2 & \text{for } 0 < t - T, \\ 0 & \text{else.} \end{cases}$$

where $T = T_1 - T_2$ now for arbitrary values of T_1 and T_2.

Example 3.25. Distribution of the transverse momentum squared of particle tracks

The projections of the momenta transverse to the dominant beam direction are assumed to be independent and normally distributed,

$$f(p_x, p_y) = \frac{1}{2\pi s^2} \mathrm{e}^{-(p_x^2+p_y^2)/(2s^2)},$$

with equal variances $\langle p_x^2 \rangle = \langle p_y^2 \rangle = s^2$. For the transverse momentum squared we set $q = p^2$ and calculate its distribution. We transform the distributions into polar coordinates

$$p_x = \sqrt{q}\cos\varphi,$$
$$p_y = \sqrt{q}\sin\varphi$$

with

$$\frac{\partial(p_x, p_y)}{\partial(q, \varphi)} = \frac{1}{2}$$

and obtain

$$h(q, \varphi) = \frac{1}{4\pi s^2} \mathrm{e}^{-q/(2s^2)}$$

with the marginal distribution

$$h_q(q) = \int_0^{2\pi} \frac{1}{4\pi s^2} \mathrm{e}^{-q/(2s^2)}\, \mathrm{d}\varphi$$
$$= \frac{1}{2s^2} \mathrm{e}^{-q/(2s^2)},$$
$$g(p^2) = \frac{1}{\langle p^2 \rangle} \mathrm{e}^{-p^2/\langle p^2 \rangle}.$$

The result is an exponential distribution in p^2 with mean $\langle p^2 \rangle = \langle p_x^2 \rangle + \langle p_y^2 \rangle$.

Example 3.26. Quotient of two normally distributed variates

For variates x, y, independently and identically normally distributed, i.e.

$$f(x,y) = f(x)f(y) = \frac{1}{2\pi}\exp\left(-\frac{x^2+y^2}{2}\right),$$

we want to find the distribution $g(u)$ of the quotient $u = y/x$. Again, we transform first into new variates $u = y/x$, $v = x$, or, inverted, $x = v$, $y = uv$ and get

$$h(u,v) = f(x(u,v), y(u,v))\frac{\partial(x,y)}{\partial(u,v)},$$

with the Jacobian

$$\frac{\partial(x,y)}{\partial(u,v)} = -v.$$

Hence

$$\begin{aligned} g(u) &= \int h(u,v)\,\mathrm{d}v \\ &= \frac{1}{2\pi}\int_{-\infty}^{\infty}\exp\left(-\frac{v^2+u^2v^2}{2}\right)|v|\,\mathrm{d}v \\ &= \frac{1}{\pi}\int_0^{\infty}\mathrm{e}^{-(1+u^2)z}\,\mathrm{d}z \\ &= \frac{1}{\pi}\frac{1}{1+u^2}, \end{aligned}$$

where the substitution $z = v^2/2$ has been used. The result $g(u)$ is the Cauchy distribution (see Section 3.6.9). Its long tails are caused here by the finite probability of arbitrary small values in the denominator. This effect is quite important in experimental situations when we estimate the uncertainty of quantities which are the quotients of normally distributed variates in cases, where the p.d.f. in the denominator is not negligible at the value zero.

The few examples given above should not lead to the impression that transformations of variates always yield more or less simple analytical expressions for the resulting distributions. This is rather the exception. However, as we will learn in Chapter 5, a simple, straight forward numerical solution is provided by Monte Carlo methods.

3.5.5 *Determination of the Variable Transformation for Given Distributions*

As in the one-dimensional case, need to generate the required distribution from the uniformly distributed random numbers delivered by the computer. The method integrating and inverting the cumulative distribution can be used only if we deal with independent variates. Then the problem can sometimes be solved by a transformation of the correlated variates to independent variates. Here we consider a special example, which we need later in Chapter 5.

Example 3.27. Generation of a two-dimensional normal distribution starting from uniform distributions

We use the result from example 3.23 and start with the representation of the two-dimensional Gaussian in polar coordinates

$$g(\rho, \varphi)\,\mathrm{d}\rho\,\mathrm{d}\varphi = \frac{1}{2\pi}\,\mathrm{d}\varphi\,\rho\,\mathrm{e}^{-\rho^2/2}\,\mathrm{d}\rho,$$

which factorizes in φ and ρ. With two in the interval $[0,1]$ uniformly distributed variates r_1, r_2, we obtain the function $\rho(r_1)$:

$$\begin{aligned}
\int_0^\rho \rho' \mathrm{e}^{-\rho'^2/2}\,\mathrm{d}\rho' &= r_1, \\
-\mathrm{e}^{-\rho'^2/2}\Big|_0^\rho &= r_1, \\
1 - \mathrm{e}^{-\rho^2/2} &= r_1, \\
\rho &= \sqrt{-2\ln(1 - r_1)}.
\end{aligned}$$

In the same way we get $\varphi(r_2)$:

$$\varphi = 2\pi r_2.$$

Finally we find x and y:

$$x = \rho\cos\varphi = \sqrt{-2\ln(1 - r_1)}\cos(2\pi r_2), \tag{3.36}$$

$$y = \rho\sin\varphi = \sqrt{-2\ln(1 - r_1)}\sin(2\pi r_2). \tag{3.37}$$

These variables are independent and distributed normally about the origin with variance unity:

$$f(x,y) = \frac{1}{2\pi} e^{-(x^2+y^2)/2}.$$

(We could replace $1-r_1$ by r_1, since $1-r_1$ is uniformly distributed as well.)

3.5.6 *Distributions of More Than Two Variables*

It is not difficult to generalize the relations just derived for two variables to multivariate distributions, of N variables. We define the distribution function $F(x_1, \ldots, x_N)$ as the probability to find values of the variates smaller than $x_1, \ldots, x_N$,

$$F(x_1, \ldots, x_N) = P\{(x'_1 < x) \cap \cdots \cap (x'_N < x_N)\},$$

and the p.d.f.

$$f(x_1, \ldots, x_N) = \frac{\partial^N F}{\partial x_1, \ldots, \partial x_N}.$$

Often it is convenient to use the vector notation, $F(\boldsymbol{x})$, $f(\boldsymbol{x})$ with

$$\boldsymbol{x} = \{x_1, x_2, \ldots, x_N\}.$$

The variate vector can be represented by a point in an N-dimensional space.

3.5.6.1 *Expected Values and Correlation Matrix*

The expected value of a function $u(\boldsymbol{x})$ is

$$\mathrm{E}(u) = \int_{-\infty}^{\infty} \cdots \int_{-\infty}^{\infty} u(\boldsymbol{x}) f(\boldsymbol{x}) \prod_{i=1}^{N} \mathrm{d}x_i.$$

Because of the additivity of expected values, this relation also holds for vector functions $\boldsymbol{u}(\boldsymbol{x})$.

The dispersion of multivariate distributions is now described by the so-called covariance matrix V:

$$V_{ij} = \langle (x_i - \langle x_i \rangle)(x_j - \langle x_j \rangle) \rangle = \langle x_i x_j \rangle - \langle x_i \rangle \langle x_j \rangle.$$

The correlation matrix is given by

$$\rho_{ij} = \frac{V_{ij}}{\sqrt{V_{ii} V_{jj}}}.$$

3.5.6.2 *Transformation of Variables*

We multiply with the absolute value of the N-dimensional Jacobian

$$g(\boldsymbol{y}) = f(\boldsymbol{x}) \left| \frac{\partial(x_1, \dots, x_N)}{\partial(y_1, \dots, y_N)} \right| .$$

3.5.6.3 *Correlation and Independence*

As in the two-dimensional case two variables x_i, x_j are called uncorrelated if their correlation coefficient ρ_{ij} is equal to zero. The two variates x_i, x_j are independent if the conditional p.d.f. of x_i conditioned on all other variates does not depend on x_j. The combined density f then has to factorize into two factors where one of them is independent of x_i and the other one is independent of x_j.[6] All variates are independent of each other, if f factorizes into the marginal distributions:

$$f(x_1, x_2, \dots, x_N) = \prod_{i=1}^{N} f_{x_i}(x_i).$$

3.5.7 *Independent, Identically Distributed Variables*

One of the main topics of statistics is the estimation of free parameters of a distribution from a random sample of observations all drawn from the same population. For example, we might want to estimate the mean lifetime τ of a particle from N independent measurements t_i where t follows an exponential distribution depending on τ. The probability density $\tilde{f}$ for N *independent* and *identically distributed* variates (abbreviated as i.i.d. variates) x_i, each distributed according to $f(x)$, is, according to the definition of independence,

$$\tilde{f}(x_1, \dots, x_N) = \prod_{i=1}^{N} f(x_i).$$

The covariance matrix of i.i.d. variables is diagonal, with $V_{ii} = \mathrm{var}(x_i) = \mathrm{var}(x_1)$.

3.5.8 *Angular Distributions*

In physics applications we are often interested in spatial distributions. Fortunately our problems often exhibit certain symmetries which facilitate the

[6] The formulas are omitted because they are very clumsy.

description of the phenomena. Depending on the kind of symmetry of the physical process or the detector, we choose appropriate coordinates, spherical, cylindrical or polar. These coordinates are especially well suited to describe processes where radiation is emitted by a local source or where the detector has a spherical or cylindrical symmetry. Then the distance, i.e. the radius vector, is not the most interesting parameter and we often describe the process solely by angular distributions. In other situations, only directions enter, for example in particle scattering, when we investigate the polarization of light crossing an optically active medium, or of a particle decaying in flight into a pair of secondaries where the orientation of the normal of the decay plane contains relevant information. Similarly, distributions of physical parameters at the surface of the earth are expressed as functions of the angular coordinates.

3.5.8.1 *Distribution of the Polar Angle*

The expressions $x = r\cos\varphi$, $y = r\sin\varphi$ relate the polar coordinates r, φ to the Cartesian coordinates x, y. Since we have periodic functions, we restrict the angle φ to the interval $[-\pi, \pi]$. This choice is arbitrary to a certain extent.

For an isotropic distribution all angles are equally likely and we obtain the uniform distribution of φ

$$g(\varphi) = \frac{1}{2\pi} .$$

Since we have to deal with periodic functions, we have to be careful when we compute moments and in general expected values. For example the mean of the two angles $\varphi_1 = \pi/2$, $\varphi_2 = -\pi$ is not $(\varphi_1 + \varphi_2)/2 = -\pi/4$, but $3\pi/4$. To avoid this kind of mistake it is advisable to go back to the unit vectors $\{x_i, y_i\} = \{\cos\varphi_i, \sin\varphi_i\}$, to average those and to extract the resulting angle.

Example 3.28. The v. Mises distribution

We consider the Brownian motion of a particle at the surface of a liquid. Starting from a point $\boldsymbol{r}_0$ its position $\boldsymbol{r}$ after some time will be given by the expression

$$f(\boldsymbol{r}) = \frac{1}{2\pi\sigma^2} \exp\left(-\frac{|\boldsymbol{r} - \boldsymbol{r}_0|^2}{2\sigma^2}\right) .$$

Taking into account the Jacobian $\partial(x,y)/\partial(r,\varphi) = r$, the distribution in polar coordinates is:

$$g(r,\varphi) = \frac{r}{2\pi\sigma^2}\exp\left(-\frac{r^2 + r_0^2 - 2rr_0\cos\varphi}{2\sigma^2}\right).$$

For convenience we have chosen the origin of φ such that $\varphi_0 = 0$. For fixed r we obtain the conditional distribution

$$\tilde{g}(\varphi) = g(\varphi|r) = c_N(\kappa)\exp(\kappa\cos\varphi)$$

with $\kappa = rr_0/\sigma^2$ and $c_N(\kappa)$ the normalization constant. This is the v. Mises distribution. It is symmetric in φ, unimodal with its maximum at $\varphi = 0$. The normalization

$$c_N(\kappa) = \frac{1}{2\pi I_0(\kappa)}$$

contains I_0, the modified Bessel function of order zero [25]. For large values of κ the distribution approaches a Gaussian with variance $1/\kappa$. To demonstrate this feature, we rewrite the distribution in a slightly modified way,

$$\tilde{g}(\varphi) = c_N(\kappa)\mathrm{e}^{\kappa}\mathrm{e}^{[-\kappa(1-\cos\varphi)]},$$

and make use of the asymptotic form $\lim_{x\to\infty} I_0(x) \sim \mathrm{e}^x/\sqrt{2\pi x}$ (see [25]). The exponential function is suppressed for large values of $(1 - \cos\varphi)$, and small values can be approximated by $\varphi^2/2$. Thus the asymptotic form of the distribution is

$$\tilde{g} = \sqrt{\frac{\kappa}{2\pi}}\mathrm{e}^{-\kappa\varphi^2/2}. \tag{3.38}$$

In the limit $\kappa = 0$, which is the case for $r_0 = 0$ or $\sigma \to \infty$, the distribution becomes uniform, as it should.

3.5.8.2 *Distribution of Spherical Angles*

Spatial directions are described by the polar angle θ and the azimuthal angle φ which we define through the transformation relations from the Cartesian coordinates:

$$\begin{aligned} x &= r\sin\theta\cos\varphi, \quad -\pi \le \varphi \le \pi \\ y &= r\sin\theta\sin\varphi, \quad 0 \le \theta \le \pi \\ z &= r\cos\theta. \end{aligned}$$

The Jacobian is $\partial(x,y,z)/\partial(r,\theta,\varphi) = r^2\sin\theta$. A uniform distribution inside a sphere of radius R in Cartesian coordinates

$$f_u(x,y,z) = \begin{cases} 3/(4\pi R^3) & \text{if } x^2+y^2+z^2 \le R^2, \\ 0 & \text{else} \end{cases}$$

thus transforms into

$$h_u(r,\theta,\varphi) = \frac{3r^2}{4\pi R^3}\sin\theta \;\; \text{if } r \le R.$$

We obtain the isotropic angular distribution by marginalizing or conditioning on r:

$$h_u(\theta,\varphi) = \frac{1}{4\pi}\sin\theta. \tag{3.39}$$

Spatial distributions are usually expressed in the coordinates $\tilde{z} = \cos\theta$ and φ, because then the uniform distribution simplifies further to

$$g_u(\tilde{z},\varphi) = \frac{1}{4\pi}$$

with $|\tilde{z}| \le 1$.

The p.d.f. $g(\tilde{z},\varphi)$ of an arbitrary distribution of $\tilde{z},\varphi$ is defined in the standard way through the probability $\mathrm{d}^2P = g(\tilde{z},\varphi)\mathrm{d}\tilde{z}\mathrm{d}\varphi$. The product $\mathrm{d}\tilde{z}\mathrm{d}\varphi = \sin\theta\mathrm{d}\theta\mathrm{d}\varphi = \mathrm{d}^2\Omega$ is called solid angle element and corresponds to an infinitesimal area at the surface of the unit sphere. A solid angle Ω defines a certain area at this surface and contains all directions pointing into this area.

Example 3.29. Fisher's spherical distribution

Instead of the uniform distribution considered in the previous example we now investigate the angular distribution generated by a three-dimensional rotationally symmetric Gaussian distribution with variances $\sigma^2 = \sigma_x^2 = \sigma_y^2 = \sigma_z^2$. We put the center of the Gaussian at the z-axis, $\boldsymbol{r}_0 = \{0,0,1\}$. In spherical coordinates we then obtain the p.d.f.

$$f(r,\theta,\varphi) = \frac{1}{(2\pi)^{3/2}\sigma^3} r^2 \sin\theta \exp\left(-\frac{r^2+r_0^2-2rr_0\cos\theta}{2\sigma^2}\right).$$

For fixed distance r we obtain a function of θ and φ only which for our choice of $\boldsymbol{r}_0$ is also independent of φ:

$$g(\theta, \varphi) = c_N(\kappa) \sin\theta \exp(\kappa \cos\theta).$$

The parameter κ is again given by $\kappa = rr_0/\sigma^2$. Applying the normalization condition $\int g \mathrm{d}\theta \mathrm{d}\varphi = 1$ we find $c_N(\kappa) = \kappa/(4\pi \sinh\kappa)$ and

$$g(\theta, \varphi) = \frac{\kappa}{4\pi \sinh\kappa} \mathrm{e}^{\kappa \cos\theta} \sin\theta \tag{3.40}$$

a two-dimensional, unimodal distribution, known as Fisher's distribution. As in the previous example we get in the limit $\kappa \to 0$ the uniform distribution (3.39) and for large κ the asymptotic distribution

$$\tilde{g}(\theta, \varphi) \approx \frac{1}{4\pi} \kappa\theta \, \mathrm{e}^{-\kappa\theta^2/2},$$

which is an exponential distribution of θ^2. As a function of $\tilde{z} = \cos\theta$ the distribution (3.40) simplifies to

$$h(\tilde{z}, \varphi) = \frac{\kappa}{4\pi \sinh\kappa} \mathrm{e}^{\kappa\tilde{z}}.$$

which illustrates the spatial shape of the distribution much better than (3.40).

3.6 Some Important Distributions

Detailed descriptions of all common distributions can be found in the internet.

3.6.1 *The Binomial Distribution*

What is the probability to get with 10 dice just two times *six*? The answer is given by the binomial distribution:

$$\mathcal{B}^{10}_{1/6}(2) = \binom{10}{2} \left(\frac{1}{6}\right)^2 \left(1 - \frac{1}{6}\right)^8.$$

The probability to get with 2 particular dice *six* is $(1/6)^2$ and to get with the remaining 8 dice *not* the number *six*, is given by $(1 - 1/6)^8$. The binomial coefficient

$$\binom{10}{2} = \frac{10!}{2!\,8!}$$

counts the number of possibilities to distribute the 2 *six* over the 10 dice. There are 10! ways to distribute 10 different numbers, but to exchange the two *sixes* does not produce a new configuration. This explains the factor 2! in the denominator. Likewise we have the factor 1/8! to account for the number of ways to distribute the 8 numbers that are different from *six*. There remain just 45 configurations.. With the above formula we obtain a probability of about 0.29.

In the general case with n randomly chosen objects (or a sequence of n independent trials), which have with probability p the property A which we will call *success*, the probability to find k successes out of these n objects is $\mathcal{B}_p^n(k)$ is

$$\mathcal{B}_p^n(k) = \binom{n}{k} p^k (1-p)^{n-k}, \qquad k = 0, \dots, n.$$

We have the normalization condition

$$\sum_{k=0}^{n} \mathcal{B}_p^n(k) = 1. \tag{3.41}$$

Since the mean number of successes in one trial is given by p, we obtain, following the rules for expected values, for n independent trials

$$\mathrm{E}(k) = np.$$

With a similar argument we can find the variance: For a single trial we can directly compute the expected quadratic difference, i.e. the variance σ_1^2. Using $\langle k \rangle = p$ and that $k = 1$ is found with probability[7] $P\{1\} = p$ and $k = 0$ with $P\{0\} = 1 - p$, we find:

$$\begin{aligned}\sigma_1^2 &= \langle (k - \langle k \rangle)^2 \rangle \\ &= p(1-p)^2 + (1-p)(0-p)^2 \\ &= p(1-p).\end{aligned}$$

According to (3.12) the variance of the sum of n i.i.d. random numbers is

$$\sigma^2 = n\sigma_1^2 = np(1-p).$$

The characteristic function has the form:

$$\phi(t) = \left[1 + p\left(\mathrm{e}^{\mathrm{i}t} - 1\right)\right]^n. \tag{3.42}$$

It is easily derived by substituting in the expansion of (3.41) in the k_{th} term p^k with $\left(p\,\mathrm{e}^{\mathrm{i}t}\right)^k$. From (3.3.2) follows the property of stability, which is also convincing intuitively:

[7] For $n = 1$ the binomial distribution is also called two-point or Bernoulli distribution.

The distribution of a sum of numbers $k = k_1 + \ldots + k_N$ obeying binomial distributions $B_p^{n_i}(k_i)$, is again a binomial distribution $B_p^n(k)$ with $n = n_1 + \cdots + n_N$.

There is no particularly simple expression for higher moments; they can of course be calculated from the Taylor expansion of $\phi(t)$, as explained in Section 3.3.2. We give only the results for the coefficients of skewness and excess:

$$\gamma_1 = \frac{1-2p}{\sqrt{np(1-p)}}, \quad \gamma_2 = \frac{1-6p(1-p)}{np(1-p)}.$$

Example 3.30. Efficiency fluctuations of a Geiger counter

A Geiger counter with a registration probability of 90% ($p = 0.9$) detects n' out of $n = 1000$ particles crossing it. On average this will be $\langle n' \rangle = np = 900$. The mean fluctuation (standard deviation) of this number is $\sigma = \sqrt{np(1-p)} = \sqrt{90} \approx 9.5$. The observed efficiency $\varepsilon = n'/n$ will fluctuate by $\sigma_\varepsilon = \sigma/n = \sqrt{p(1-p)/n} \approx 0.0095$. For efficiencies close to one, we find $\sigma \approx \sqrt{n(1-p)}$ which is the Poisson fluctuation of the missing particles. For small efficiencies, we find $\sigma \approx \sqrt{np}$ which is the Poisson fluctuation of the detected particles.

Example 3.31. Accuracy of a Monte Carlo integration

We want to estimate the value of π by a Monte Carlo integration. We distribute randomly n points in a square of area $4\,\text{cm}^2$. The probability for a point to fall into a circle with radius 1 enclosed by the box is given by $p = \pi/4$, the ratio of the two areas. In average there will be $k = np$ points inside the circle with the standard deviation $\sigma = \sqrt{np(1-p)}$. The expected relative error $\sigma/(np)$ is equal to the relative error that is obtained for the determination of π. To reach an accuracy of 1% requires

$$\frac{\sigma}{np} = 0.01,$$

$$n = \frac{(1-p)}{0.01^2 p} = \frac{(4-\pi)}{0.01^2 \pi} \approx 2732.$$

We have to generate pairs of uniformly distributed random numbers with the computer.

Example 3.32. Acceptance fluctuations for weighted events

The acceptance of a complex detector is determined by Monte Carlo simulation which depends on a probability density $f_0(\boldsymbol{x})$ where $\boldsymbol{x}$ denotes all relevant kinematic variables. In order to avoid the repetition of the simulation for a different physical situation (e.g. a different cross section) described by a p.d.f. $f(\boldsymbol{x})$, it is customary to weight the individual events with $w_i = f(\boldsymbol{x})/f_0(\boldsymbol{x})$, $i = 1, \dots, N$ for N generated events. The acceptance ε_i for event i is either 1 or 0. Hence the overall acceptance is

$$\varepsilon_T = \frac{\sum w_i \varepsilon_i}{\sum w_i}.$$

The variance for each single term in the numerator is $w_i^2 \varepsilon_i (1 - \varepsilon_i)$. Then the variance σ_T^2 of ε_T becomes

$$\sigma_T^2 = \frac{\sum w_i^2 \varepsilon_i (1 - \varepsilon_i)}{\left(\sum w_i\right)^2}.$$

3.6.2 *The Multinomial Distribution*

When a single experiment or trial has not only two, but N possible outcomes with probabilities $p_1, p_2, \dots, p_N$, the probability to observe in n experiments $k_1, k_2, \dots, k_N$ trials belonging to the outcomes $1, \dots, N$ is equal to

$$\mathcal{M}^n_{p_1, p_2, \dots, p_N}(k_1, k_2, \dots, k_N) = \frac{n!}{\prod_{i=1}^N k_i!} \prod_{i=1}^N p_i^{k_i},$$

where $\sum_{i=1}^N p_i = 1$ and $\sum_{i=1}^N k_i = n$ are satisfied. Hence we have $N-1$ independent variates. The value $N = 2$ reproduces the binomial distribution.

The expected values are

$$\mathrm{E}(k_i) = np_i$$

and the covariance matrix is given by

$$V_{ij} = np_i(\delta_{ij} - p_j).$$

They can be derived from the characteristic function

$$\phi(t_1, \dots, t_{N-1}) = \left(1 + \sum_1^{N-1} p_i \left(\mathrm{e}^{\mathrm{i}t_i} - 1\right)\right)^n,$$

which is a straight forward generalization of the 1-dimensional case (3.42). The correlations are negative: If, for instance, more events k_i as expected fall into class i, the mean number of k_j for any other class will tend to be smaller than its expected value $\mathrm{E}(k_j)$.

The multinomial distribution describes the distribution of events into histogram bins. For total a number n of events with the probability p_i to collect an event in bin i, the expected number of events in that bin will be $n_i = np_i$ and the variance $V_{ii} = np_i(1-p_i)$. Normally, a histogram has many bins and $p_i \ll 1$ for all i. Then we approximate $V_{ij} \approx n_i\delta_{ij}$. The correlation between the bin entries can be neglected and the fluctuation of the entries in a bin is described by the Poisson distribution which we discuss in the following section.

3.6.3 *The Poisson Distribution*

When a certain reaction happens randomly in time with an average frequency λ in a given time interval, then the number k of reactions in that time interval will follow a Poisson distribution (Fig. 3.15)

$$\mathcal{P}_\lambda(k) = \mathrm{e}^{-\lambda}\frac{\lambda^k}{k!}.$$

Occasionally we will use also the notation $\mathcal{P}(k|\lambda)$. The expected value and variance have already been calculated above (see Section 3.3.4):

$$\mathrm{E}(k) = \lambda, \ \mathrm{var}(k) = \lambda.$$

The characteristic function and cumulants have also been derived in Section 3.3.2:

$$\phi(t) = \exp\left(\lambda(\mathrm{e}^{\mathrm{i}t} - 1)\right), \tag{3.43}$$

$$\kappa_i = \lambda, i = 1, 2, \ldots.$$

Skewness and excess,

$$\gamma_1 = \frac{1}{\sqrt{\lambda}}, \ \gamma_2 = \frac{1}{\lambda}$$

decrease with λ and indicate that the distribution approaches the normal distribution ($\gamma_1 = 0$, $\gamma_2 = 0$) with increasing λ (see Figs. 3.15, 3.16).

The Poisson distribution itself can be considered as the limiting case of a binomial distribution with $np = \lambda$, where n approaches infinity ($n \to \infty$) and, at the same time, p approaches zero, $p \to 0$. The corresponding limit

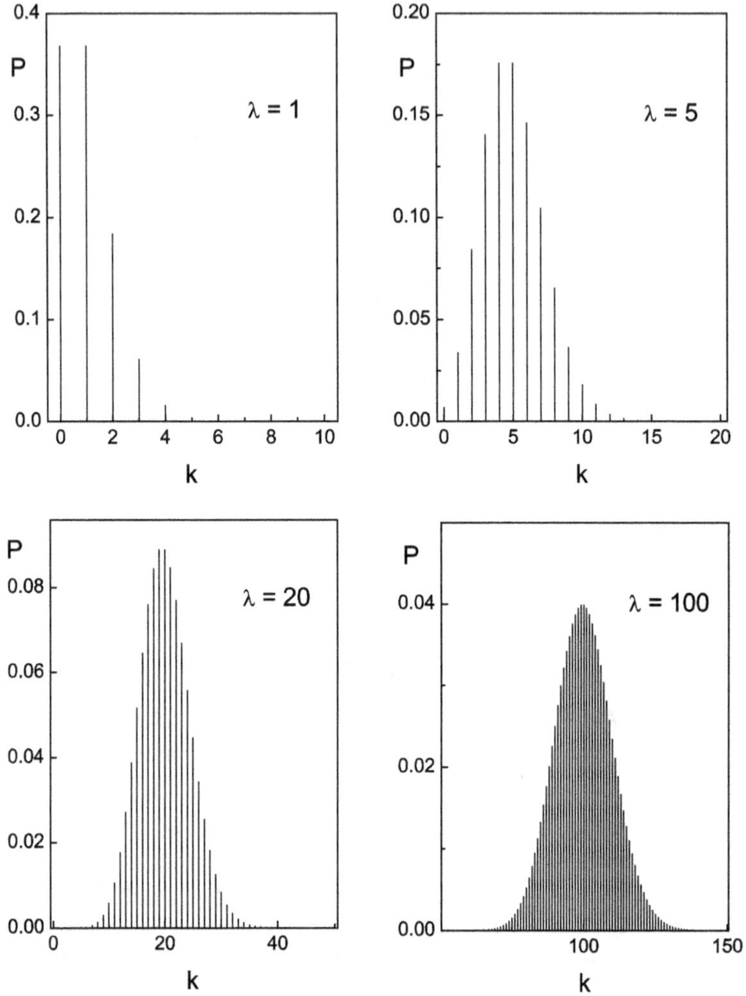

Fig. 3.15 Poisson distriutions with different expected values.

of the characteristic function of the binomial distribution (3.42) produces the characteristic function of the Poisson distribution (3.43): With $p = \lambda/n$ we then obtain

$$\lim_{n\to\infty}\left[1+\frac{\lambda}{n}(\mathrm{e}^{\mathrm{i}t}-1)\right]^n = \exp\left(\lambda(\mathrm{e}^{\mathrm{i}t}-1)\right).$$

For the Poisson distribution, the supply of potential events or number of trials is supposed to be infinite while the chance of a success, p, tends to zero. It is often used in cases where in principle the binomial distribution applies, but where the number of trials is very large.

Example 3.33. Poisson limit of the binomial distribution

A volume of $1\,\text{l}$ contains 10^{16} hydrogen ions. The mean number of ions in a sub-volume of $1\,\mu\text{m}^3$ is then $\lambda = 10$ and its standard deviation for a Poisson distribution is $\sigma = \sqrt{10} \approx 3$. The exact calculation of the standard deviation with the binomial distribution would change σ only by a factor $\sqrt{1-10^{-15}}$.

Also the number of radioactive decays in a given time interval follows a Poisson distribution, if the number of nuclei is big and the decay probability for a single nucleus is small.

The Poisson distribution is of exceptional importance in nuclear and particle physics, but also in the fields of microelectronics (noise), optics, and gas discharges it describes the statistical fluctuations.

3.6.3.1 *Specific Properties of the Poisson Distribution*

The sum $k = k_1 + k_2$ of Poisson distributed numbers k_1, k_2 with expected values λ_1, λ_2 is again a Poisson distributed number with expected value $\lambda = \lambda_1 + \lambda_2$. This property, which we called stability in connection with the binomial distribution follows formally from the structure of the characteristic function, or from the additivity of the cumulants given above. It is also intuitively obvious.

Example 3.34. Fluctuation of a counting rate with subtracted background

Expected are S signal events with an expected number of background events B. The mean fluctuation (standard deviation) of the observed number k is $\sqrt{S+B}$. This is also the fluctuation of $k - B$, because B is a constant. For a mean signal $S = 100$ and an expected background $B = 50$ we will observe on average 150 events with a fluctuation of $\sqrt{150}$. After subtracting the background, this fluctuation will remain. Hence, the background corrected signal is expected to be 100 with the standard deviation $\sigma = \sqrt{150}$. The uncertainty becomes even be larger, if also the mean value B is not known exactly.

If from a Poisson-distributed number n with expected value λ_0 on average only a fraction ε is registered, for instance when the size of a detector is reduced by a factor of ε, then the expected rate is $\lambda = \lambda_0\varepsilon$ and the number of observed events k follows the Poisson distribution $P_\lambda(k)$. This intuitive result is also obtained analytically: The number k follows a binomial distribution $\mathcal{B}_\varepsilon^n(k)$ where n is a Poisson-distributed number. The probability $p(k)$ is:

$$\begin{aligned}
p(k) &= \sum_{n=k}^{\infty} \mathcal{B}_\varepsilon^n(k)\mathcal{P}_{\lambda_0}(n) \\
&= \sum_{n=k}^{\infty} \frac{n!}{k!(n-k)!}\varepsilon^k(1-\varepsilon)^{n-k}\mathrm{e}^{-\lambda_0}\frac{\lambda_0^n}{n!} \\
&= \mathrm{e}^{-\lambda_0}\frac{(\varepsilon\lambda_0)^k}{k!}\sum_{n=k}^{\infty}\frac{1}{(n-k)!}(\lambda_0-\lambda_0\varepsilon)^{n-k} \\
&= \mathrm{e}^{-\varepsilon\lambda_0}\frac{(\varepsilon\lambda_0)^k}{k!} \\
&= \mathcal{P}_\lambda(k).
\end{aligned}$$

Of interest is also the following mathematical identity

$$\begin{aligned}
\sum_{i=0}^{k}\mathcal{P}_\lambda(i) &= \int_\lambda^\infty \mathrm{d}\lambda' \mathcal{P}_{\lambda'}(k), \\
\sum_{i=0}^{k}\frac{\lambda^i}{i!}\mathrm{e}^{-\lambda} &= \int_\lambda^\infty \frac{(\lambda')^k}{k!}\mathrm{e}^{-\lambda'}\,\mathrm{d}\lambda'
\end{aligned}$$

which allows us to calculate the probability $P\{i \le k\}$ to find a number i less or equal k using a well known integral (described by the incomplete gamma function). It is applied to estimate upper and lower interval limits in Chapter 8.

3.6.4 *The Uniform Distribution*

The uniform distribution is the simplest continuous distribution. It describes, for instance, digital measurements where the random variable is tied to a given interval and where inside the interval all its values are equally probable.

Given an interval of length α centered at the mean value ξ the p.d.f. reads

$$f(x|\xi,\alpha) = \begin{cases} 1/\alpha & \text{if } |x-\xi| < \alpha/2, \\ 0 & \text{else.} \end{cases} \tag{3.44}$$

Mean value and variance are $\langle x \rangle = \xi$ and $\sigma^2 = \alpha^2/12$, respectively. The characteristic function is

$$\phi(t) = \frac{1}{\alpha} \int_{\xi-\alpha/2}^{\xi+\alpha/2} \mathrm{e}^{\mathrm{i}tx} \mathrm{d}x = \frac{2}{\alpha t} \sin \frac{\alpha t}{2} \, \mathrm{e}^{\mathrm{i}\xi t}. \tag{3.45}$$

Using the power expansion of the sinus function, we find with (3.45) for $\xi = 0$ the even moments (the odd moments vanish):

$$\mu'_{2k} = \frac{1}{2k+1} \left(\frac{\alpha}{2} \right)^{2k}, \quad \mu'_{2k-1} = 0.$$

The uniform distribution is the basis for the computer simulation of all other distributions because computers generate random numbers uniformly distributed between 0 and 1. We will discuss simulations in some detail in Chapter 5.

3.6.5 *The Normal Distribution*

The normal or Gauss[8] distribution which we had introduced already in Section 3.2.7,

$$\mathcal{N}(x|\mu, \sigma) = \frac{1}{\sqrt{2\pi}\sigma} \mathrm{e}^{-(x-\mu)^2/(2\sigma^2)},$$

enjoys great popularity among statisticians. This has several reasons which, however, are not independent from each other.

1. The sum of normally distributed quantities is again normally distributed (stability), with $\mu = \sum \mu_i$, $\sigma^2 = \sum \sigma_i^2$, in obvious notation.
2. The discrete binomial- and Poisson distributions and also the χ^2-distribution, in the limit of a large number, a large mean value and many degrees of freedom, respectively, approach the normal distribution.
3. Many distributions met in natural sciences are well approximated by normal distributions. We have already mentioned some examples: velocity components of gas molecules, diffusion, Brownian motion and many measurement errors obey normal distributions to good accuracy.
4. Certain analytically simple statistical procedures for parameter estimation and propagation of errors are valid exactly only for normally distributed errors.

[8]Karl Friedrich Gauss, German mathematician and statistician (1777-1855).

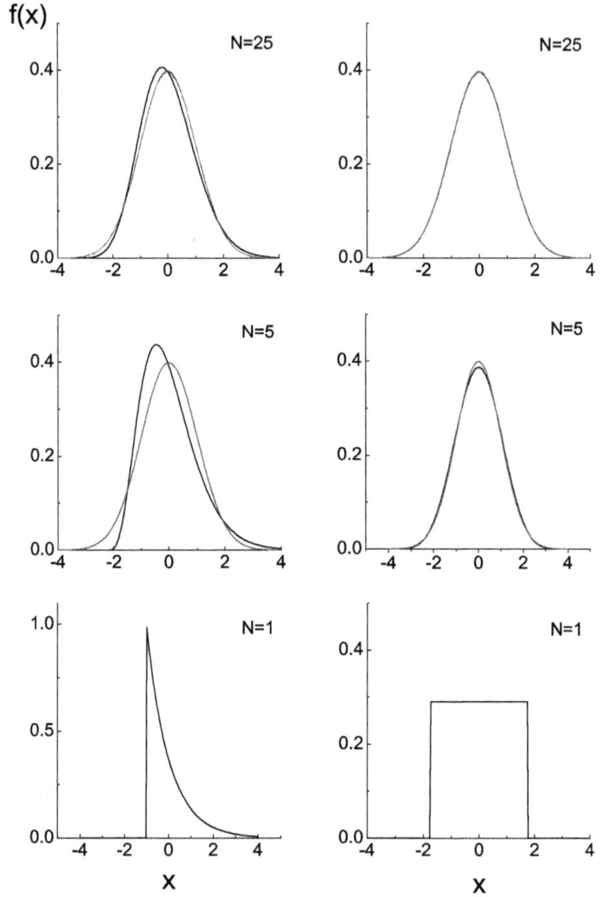

Fig. 3.16 Illustration of the central limit theorem. The mean values of N exponentially or uniformly distributed variates times $\sqrt{N}$ approach with increasing N the normal distribution.

The deeper reason for point 2 and 3 is explained by the *central limit theorem*: The mean value of a large number N of independent random variables, obeying the same distribution with variance σ_0^2, approaches a normal distribution with variance $\sigma^2 = \sigma_0^2/N$. The important point is that this theorem is valid for quite arbitrary distributions, provided they have a finite variance, a condition which practically always can be fulfilled, if necessary by cutting off large absolute values of the variates. Instead of a formal proof,[9] we show in Fig. 3.16, how with increasing number of variates the distribution of their mean value approaches the normal distribution better and better.

As example we have chosen the mean values for uniformly respectively exponentially distributed numbers. For the very asymmetric exponential distribution on the left-hand side of the figure the convergence to a normal distribution is not as fast as for the uniform distribution, where already the distribution of the mean of five uniform random numbers is in good agreement with the normal distribution. The central limit theorem applies also when the individual variates follow different distributions provided that the variances are of the same order of magnitude.

The characteristic function of the normal distribution is

$$\phi(t) = \exp(-\frac{1}{2}\sigma^2 t^2 + \mathrm{i}\mu t).$$

It is real and also of Gaussian shape for $\mu = 0$. The stability (see point 1 above) is easily proven, using the convolution theorem (3.3.2) and the exponential form of $\phi(t)$.

We obtain the central moments of the normal distribution by differentiation of the characteristic function, setting $\mu = 0$.

$$\mu'_{2j} = \frac{(2j)!}{2^j j!}\sigma^{2j}.$$

Cumulants, with the exception of $\kappa_1 = \mu$ and $\kappa_2 = \sigma^2$, vanish. Also the odd central moments are zero.

3.6.5.1 *The Normal Distribution in Higher Dimensions*

The normal distribution in two dimensions with its maximum at the origin has the general form

$$\mathcal{N}_0(x,y) = \frac{1}{\sqrt{1-\rho^2}2\pi s_x s_y}\exp\left[-\frac{1}{2(1-\rho^2)}\left(\frac{x^2}{s_x^2} - 2\rho\frac{xy}{s_x s_y} + \frac{y^2}{s_y^2}\right)\right]. \tag{3.46}$$

The moments are:

$$\begin{aligned}\langle x^2\rangle &= s_x^2,\\ \langle y^2\rangle &= s_y^2,\\ \langle xy\rangle &= \rho s_x s_y.\end{aligned}$$

[9] A simplified proof is presented in Appendix A.1.

We skip the explicit calculation. Integrating (3.46) over $y, (x)$, we obtain the marginal distributions of $x, (y)$. They are again normal distributions with standard deviations s_x and s_y. A characteristic feature of the normal distribution is that for a vanishing correlation $\rho = 0$ the two variables are independent, because in this case the p.d.f. $\mathcal{N}_0(x, y)$ factorizes into normal distributions of x and y.

Curves of equal probability are obtained by equating the exponent to a constant. The equations

$$\frac{1}{1-\rho^2}\left(\frac{x^2}{s_x^2} - 2\rho\frac{xy}{s_x s_y} + \frac{y^2}{s_y^2}\right) = const$$

describe concentric ellipses. For the special choice $const = 1$ we get the error or covariance ellipse for Gaussian errors. It is shown in Fig. 3.17. The function value at the ellipse is reduced by a factor $1/\sqrt{e}$ with respect to the maximum of the Gaussian.

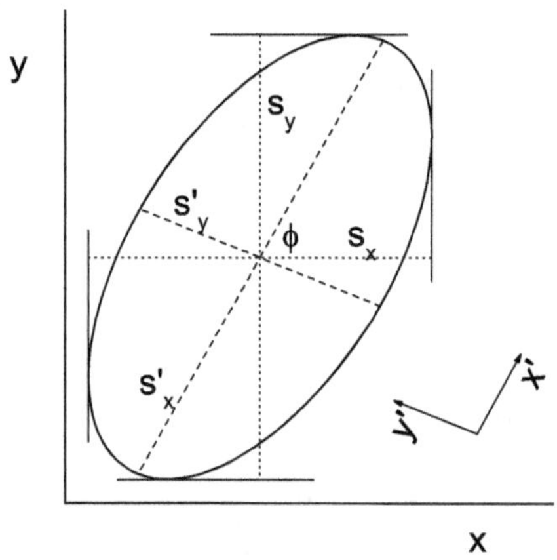

Fig. 3.17 Transformation of the error ellipse.

By a simple rotation of a normal distribution that is centered at $(0, 0)$ we achieve uncorrelated variables x' and y':

$$\begin{aligned} x' &= x\cos\phi + y\sin\phi, \\ y' &= -x\sin\phi + y\cos\phi, \\ \tan 2\phi &= \frac{2\rho s_x s_y}{s_x^2 - s_y^2}. \end{aligned}$$

The half-axes, i.e. the variances ${s'_x}^2$ and ${s'_y}^2$ of the uncorrelated variables x' and y' are

$$s'^2_x = \frac{s_x^2 + s_y^2}{2} + \frac{s_x^2 - s_y^2}{2\cos 2\phi},$$
$$s'^2_y = \frac{s_x^2 + s_y^2}{2} - \frac{s_x^2 - s_y^2}{2\cos 2\phi}.$$

In the new variables, the normal distribution has then the simple form

$$\mathcal{N}'_0(x', y') = \frac{1}{2\pi s'_x s'_y} \exp\left(-\frac{1}{2}\left(\frac{x'^2}{{s'_x}^2} + \frac{y'^2}{{s'_y}^2}\right)\right) = f(x')g(y').$$

The two-dimensional normal distribution with its maximum at (x_0, y_0) is obtained with the substitution $x \to x - x_0, \quad y \to y - y_0$ in (3.46).

We now generalize the normal distribution to n dimensions. We skip again the simple algebra and present directly the result. The variables are written in vector form $\boldsymbol{x}$ and with the symmetric and positive definite covariance matrix V, the p.d.f. is given by

$$\mathcal{N}(\boldsymbol{x}) = \frac{1}{\sqrt{(2\pi)^n \det(\mathsf{V})}} \exp\left(-\frac{1}{2}(\boldsymbol{x} - \boldsymbol{x}_0)^T \mathsf{C}(\boldsymbol{x} - \boldsymbol{x}_0)\right).$$

Here

$$\mathsf{C} = \mathsf{V}^{-1}$$

is the inverse of the covariance matrix and is called weight matrix. Small variances V_{ii} of components x_i lead to large weights C_{ii}.

In the two-dimensional case the matrices V and C are

$$\mathsf{V} = \begin{pmatrix} s_x^2 & \rho s_x s_y \\ \rho s_x s_y & s_y^2 \end{pmatrix},$$
$$\mathsf{C} = \frac{1}{1-\rho^2} \begin{pmatrix} \frac{1}{s_x^2} & -\frac{\rho}{s_x s_y} \\ -\frac{\rho}{s_x s_y} & \frac{1}{s_y^2} \end{pmatrix}$$

with the determinant $\det(\mathsf{V}) = s_x^2 s_y^2 (1 - \rho^2) = 1/\det(\mathsf{C})$.

3.6.6 *The Exponential Distribution*

Also the exponential distribution appears in many physical phenomena. Besides life time distributions (decay of instable particles, nuclei or excited states), it describes the distributions of intervals between Poisson distributed events like time intervals between decays or gap lengths in track chambers (bubble chambers, emulsion stacks) and of the penetration depth of radiation or particles in absorbing materials.

The main characteristics of processes described by the exponential distribution is lack of memory, i.e. processes which are not influenced by their history. For instance, the decay probability of an instable particle is independent of its age, and the scattering probability for a gas molecule is independent of the time that has passed since the last scattering event. The probability density for the decay of a particle at the time $t_1 + t_2$ must be equal to the probability density $f(t_2)$ multiplied with the probability $1 - F(t_1)$ to survive until t_1:

$$f(t_1 + t_2) = (1 - F(t_1))\, f(t_2) .$$

Since $f(t_1 + t_2)$ must be symmetric under exchanges of t_1 and t_2, the first factor has to be proportional to $f(t_1)$,

$$1 - F(t_1) = cf(t_1) , \tag{3.47}$$

$$f(t_1 + t_2) = cf(t_1)f(t_2) \tag{3.48}$$

with constant c. The property (3.48) is satisfied only for the exponential function: $f(t) = ae^{bt}$. If we require that the probability density is normalized, we get

$$f(t) = \lambda \mathrm{e}^{-\lambda t} .$$

(This result could also have been derived by differentiating (3.47) and solving the corresponding differential equation $f = -c\,\mathrm{d}f/\mathrm{d}t$.)

The characteristic function

$$\phi(t) = \frac{\lambda}{\lambda - \mathrm{i}t}$$

and the moments

$$\mu_n = n!\, \lambda^{-n}$$

have already been derived in Example 3.15 in Section 3.3.4.

3.6.7 *The χ^2-Distribution*

The Chi-square distribution (χ^2 distribution) plays an important role in the comparison of measurements with theoretical distributions (see Chapter 10). The corresponding tests allow us to discover systematic measurement errors and to check the validity of theoretical models. The variable χ^2 which we will define below, is certainly the quantity which is most frequently used to quantify the quality of the agreement of experimental data with the theory.

The variate χ^2 is defined by the sum

$$\chi^2 = \sum_{i=1}^{f} \frac{x_i^2}{\sigma_i^2},$$

where x_i are independent, normally distributed variates with zero mean and variance σ_i^2.

We have already come across the simplest case with $f = 1$ in Section 3.4.1: The transformation of a normally distributed variate x with expected value zero to $u = x^2/s^2$, where s^2 is the variance, yields

$$g_1(u) = \frac{1}{\sqrt{2\pi u}} \mathrm{e}^{-u/2} \qquad (f = 1).$$

(We have replaced the variable χ^2 by $u = \chi^2$ to simplify the writing.) Mean value and variance of this distribution are $\mathrm{E}(u) = 1$ and $\mathrm{var}(u) = 2$.

When we now add f independent summands, we obtain

$$g_f(u) = \frac{1}{\Gamma(f/2)2^{f/2}} u^{f/2-1} \mathrm{e}^{-u/2}. \qquad (3.49)$$

The only parameter of the χ^2-distribution is the number of degrees of freedom f, the meaning of which will become clear later. We will prove (3.49) when we discuss the gamma distribution, which includes the χ^2-distribution as a special case. Fig. 3.18 shows the χ^2-distribution for some values of f. The value $f = 2$ corresponds to an exponential distribution. As follows from the central limit theorem, for large values of f the χ^2-distribution approaches a normal distribution.

By differentiation of the p.d.f. we find for $f > 2$ the maximum at the mode $u_{mod} = f - 2$. The expected value of the variate u is equal to f and its variance is $2f$. These relations follow immediately from the definition of u.

$$\begin{aligned} u_{mod} &= f - 2 \text{ for } f > 2, \\ \mathrm{E}(u) &= f, \\ \mathrm{var}(u) &= 2f. \end{aligned}$$

3.6.7.1 *Distribution of the Sample Width*

We define the width v of a sample of N elements x_i as follows (see 3.2.3):

$$\begin{aligned} v^2 &= \frac{1}{N} \sum_{i=1}^{N} x_i^2 - \overline{x}^2 \\ &= \overline{x^2} - \overline{x}^2. \end{aligned}$$

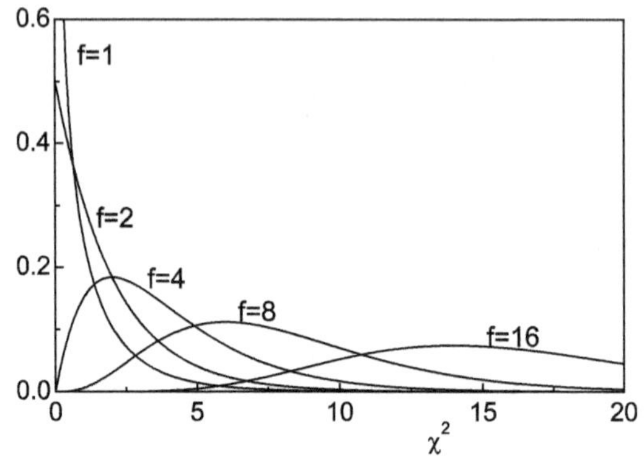

Fig. 3.18 χ^2-distribution for different degrees of freedom.

If the variates x_i of the sample are distributed normally with mean x_0 and variance σ^2, then Nv^2/σ^2 follows a χ^2-distribution with $f = N - 1$ degrees of freedom. We omit the formal proof; the result is plausible, however, from the expected value derived in Section 3.2.3:

$$\langle v^2 \rangle = \frac{N-1}{N}\sigma^2,$$

$$\left\langle \frac{Nv^2}{\sigma^2} \right\rangle = N - 1.$$

These relations can be used to estimate the variance of the underlying distribution from the sample width.

3.6.7.2 *Degrees of Freedom and Constraints*

In Section 6.8 we will discuss the method of *least squares* for parameter estimation. To adjust Z parameters λ_i of a function $t(x|\lambda_1 ..., \lambda_Z)$ with N measured values t_i at $x_i, ..., x_N$ with Gaussian errors δ_i, we minimize the quantity

$$\chi^2 = \sum_{i=1}^{N} \frac{(t_i - t(x_i|\lambda_1, \ldots, \lambda_Z))^2}{\delta_i^2}.$$

Large values of χ^2 signal a bad agreement between measured values and the fitted function. If the prediction $t(x|\lambda_1 ..., \lambda_Z)$ depends linearly on the parameters, the adjusted sum χ^2 obeys a χ^2 distribution with $f = N - Z$ degrees of freedom. The reduction of f accounts for the fact that the expected value of χ^2 is reduced when we allow for free parameters. Indeed, for $Z = N$ we could adjust the parameters such that χ^2 would vanish.

Generally, in statistics the term degrees of freedom[10] f denotes the number of independent predictions. For $N = Z$ we have no prediction for the observations x_i. For $Z = 0$ we predict all N observations, $f = N$. When we fit a straight line through 3 points with given abscissa and observed ordinate, we have $N = 3$ and $Z = 2$ because the line contains 2 parameters. The corresponding χ^2-distribution has 1 degree of freedom. The quantity Z is called the *number of constraints*, a somewhat misleading term. In the case of the sample width discussed above, one quantity, the mean, is adjusted. Consequently, we have $Z = 1$ and the sample width follows a χ^2-distribution of $f = N - 1$ degrees of freedom.

3.6.8 *The Gamma Distribution*

The distributions considered in the last two sections, the exponential- and the χ^2-distribution, are special cases of the gamma distribution

$$G(x|\nu,\lambda) = \frac{\lambda^\nu}{\Gamma(\nu)} x^{\nu-1} \mathrm{e}^{-\lambda x}, \quad x > 0.$$

The parameter $\lambda > 0$ is a scale parameter, while the parameter $\nu > 0$ determines the shape of the distribution. With $\nu = 1$ we obtain the exponential distribution. The parameter ν is not restricted to natural numbers. With the special choice $\nu = f/2$ and $\lambda = 1/2$ we get the χ^2-distribution with f degrees of freedom (see Section 3.6.7).

The gamma distribution is used typically for the description of random variables that are restricted to positive values, as in the two cases just mentioned. The characteristic function is very simple:

$$\phi(t) = \left(1 - \frac{\mathrm{i}t}{\lambda}\right)^{-\nu}. \tag{3.50}$$

As usual, we obtain expected value, variance, moments about the origin, skewness and excess by differentiating $\phi(t)$:

$$\langle x \rangle = \frac{\nu}{\lambda},\ \mathrm{var}(x) = \frac{\nu}{\lambda^2},\ \mu_i = \frac{\Gamma(i+\nu)}{\lambda^i\,\Gamma(\nu)},\ \gamma_1 = \frac{2}{\sqrt{\nu}},\ \gamma_2 = \frac{6}{\nu}.$$

The maximum of the distribution is located at $x_{mod} = (\nu - 1)/\lambda$, $(\nu > 1)$.

The gamma distribution has the property of stability in the following sense: The sum of variates following gamma distributions with the *same*

[10] In the literature often the notation *number of degrees of freedom*, abbreviated by *n.d.f.* or *NDF* is used.

scaling parameter λ, but different shape parameters ν_i is again gamma distributed, with the shape parameter ν,

$$\nu = \sum \nu_i .$$

This result is obtained by multiplying the characteristic functions (3.50). It proves also the corresponding result (3.49) for the χ^2-distribution.

Example 3.35. Distribution of the mean value of decay times

Let us consider the sample mean $\overline{x} = \sum x_i/N$, of exponentially distributed variates x_i. The characteristic function is (see 3.3.4)

$$\phi_x(t) = \frac{1}{1 - \mathrm{i}t/\lambda} .$$

Forming the N-fold product, and using the scaling rule for Fourier transforms (3.19), $\phi_{x/N}(t) = \phi_x(t/N)$, we arrive at the characteristic function of a gamma distribution with scaling parameter $N\lambda$ and shape parameter N:

$$\phi_{\overline{x}}(t) = \left(1 - \frac{\mathrm{i}t}{N\lambda}\right)^{-N} . \tag{3.51}$$

Thus the p.d.f. $f(\overline{x})$ is equal to $G(\overline{x}|N, N\lambda)$. Considering the limit for large N, we convince ourselves of the validity of the law of large numbers and the central limit theorem. From (3.51) we derive

$$\begin{aligned} \ln \phi_{\overline{x}}(t) &= -N \ln \left(1 - \frac{\mathrm{i}t}{N\lambda}\right) \\ &= -N \left[\left(-\frac{\mathrm{i}t}{N\lambda}\right) - \frac{1}{2}\left(-\frac{\mathrm{i}t}{N\lambda}\right)^2 + O(N^{-3})\right], \\ \phi_{\overline{x}}(t) &= \exp\left[\mathrm{i}\frac{1}{\lambda}t - \frac{1}{2}\frac{1}{N\lambda^2}t^2 + O\left(N^{-2}\right)\right]. \end{aligned}$$

When N is large, the term of order N^{-2} can be neglected and with the two remaining terms in the exponent we get the characteristic function of a normal distribution with mean $\mu = 1/\lambda = \langle x \rangle$ and variance $\sigma^2 = 1/(N\lambda^2) = \mathrm{var}(x)/N$, (see 3.3.4), in agreement with the central limit theorem. If N approaches infinity, only the first term remains and we obtain the characteristic function of a delta distribution $\delta(1/\lambda - \overline{x})$. This result is predicted by the law of large numbers (see Appendix A.1). This law states that, under certain conditions, with increasing sample size, the difference between the sample mean and the population mean approaches zero.

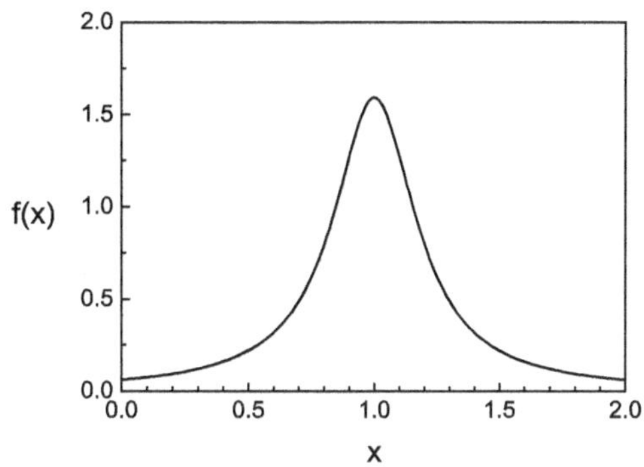

Fig. 3.19 Lorentz distribution with mean 1 and halfwidth $\Gamma/2 = 0.2$

3.6.9 *The Lorentz and the Cauchy Distributions*

The Lorentz distribution (Fig. 3.19)

$$f(x) = \frac{1}{\pi} \frac{\Gamma/2}{(x-a)^2 + (\Gamma/2)^2}$$

is symmetric with respect to $x = a$. Although it is bell-shaped like a Gaussian, it has, because of its long tails, no finite variance. This means that we cannot infer the location parameter a of the distribution[11] from the sample mean, even for arbitrary large samples. The Lorentz distribution describes resonance effects, where Γ represents the width of the resonance. In particle or nuclear physics, mass distributions of short-lived particles follow this p.d.f. which then is called *Breit–Wigner distribution*.

The Cauchy distribution corresponds to the special choice of the scale parameter $\Gamma = 2$. With the location parameter $a = 0$ it has the characteristic function $\phi(t) = \exp(-|t|)$, which obviously has no derivatives at $t = 0$, an other consequence of the nonexistence of moments. The characteristic function for the sample mean of N measurements, $\overline{x} = \sum_1^N x_i/N$, is found with the help of (3.19), (3.3.2) as

$$\phi_{\overline{x}}(t) = (\phi(t/N))^N = \phi(t).$$

The sample mean has the same distribution as the original population. It is therefore, as already stated above, not suited for the estimation of the location parameter.

[11]The first moment exists only as a Cauchy principal value and equals a.

3.6.10 *The Log-Normal Distribution*

The distribution of a variable $x > 0$ whose logarithm u is normally distributed

$$g(u) = \frac{1}{\sqrt{2\pi}s} \mathrm{e}^{-(u-u_0)^2/2s^2}$$

with mean u_0 and variance s^2 follows the log-normal distribution, see Fig. 3.20:

$$f(x) = \frac{1}{xs\sqrt{2\pi}} \mathrm{e}^{-(\ln x - u_0)^2/2s^2} .$$

This is, like the normal distribution, a two-parameter distribution where the parameters u_0, s^2, however, are not identical with the mean μ and variance σ^2, but the latter are given by

$$\begin{aligned} \mu &= \mathrm{e}^{u_0 + s^2/2} , \\ \sigma^2 &= (\mathrm{e}^{s^2} - 1)\mathrm{e}^{2u_0 + s^2} . \end{aligned} \tag{3.52}$$

Note that the distribution is declared only for positive x, while u_0 can also be negative.

The characteristic function cannot be written in closed form, but only as a power expansion. This means, the moments of order k about the origin are

$$\mu_k = \mathrm{e}^{ku_0 + \frac{1}{2}k^2 s^2} .$$

Other characteristic parameters are

$$\begin{aligned} \text{median}: &\quad x_{0.5} = \mathrm{e}^{u_0} , \\ \text{mode}: &\quad x_{mod} = \mathrm{e}^{u_0 - s^2} , \\ \text{skewness}: &\quad \gamma_1 = (\mathrm{e}^{s^2} + 2)\sqrt{\mathrm{e}^{s^2} - 1} , \\ \text{excess}: &\quad \gamma_2 = \mathrm{e}^{4s^2} + 2\mathrm{e}^{3s^2} + 3\mathrm{e}^{2s^2} - 6 . \end{aligned} \tag{3.53}$$

The distribution of a variate $x = \prod x_i$ which is the product of many variates x_i, each of which is positive and has a small variance, σ_i^2 compared to its mean squared μ^2, $\sigma_i^2 \ll \mu_i^2$, can be approximated by a log-normal distribution. This is a consequence of the central limit theorem (see 3.6.5). Writing

$$\ln x = \sum_{i=1}^{N} \ln x_i$$

we realize that $\ln x$ is normally distributed in the limit $N \to \infty$ if the summands fulfill the conditions required by the central limit theorem. Accordingly, x will be distributed by the log-normal distribution.

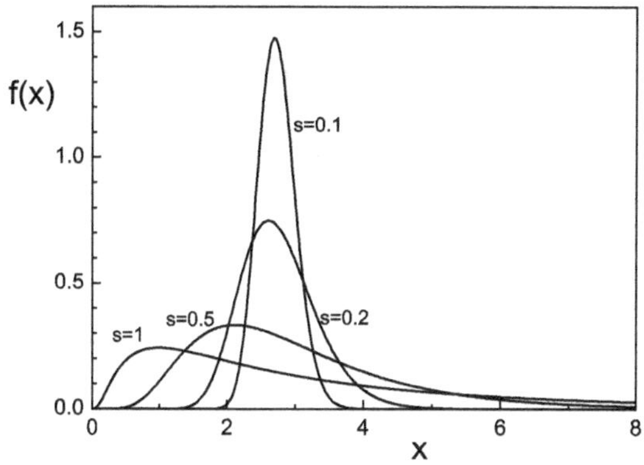

Fig. 3.20 Log-normal distribution with $u_0 = 1$ and different values of s

3.6.11 *Student's t Distribution*

This distribution, introduced by W. S. Gosset (pseudonym "Student") is frequently used to test the compatibility of a sample $x_i, ..., x_N$ drawn from a normal distribution with given mean μ but unknown variance. It describes the distribution of the so-called "studentized" variate t

$$t = \frac{\overline{x} - \mu}{s}. \tag{3.54}$$

The numerator is the difference between a sample mean and the mean of the Gaussian from which the sample of size N is drawn. It follows a normal distribution centered at zero. The denominator s is an estimate of the standard deviation of the distribution of the numerator, derived (3.15).

$$s^2 = \frac{1}{N(N-1)} \sum_{i=1}^{N} (x_i - \overline{x})^2.$$

The sum on the right-hand side, after division by the variance σ^2 of the Gaussian, follows a χ^2-distribution with $f = N - 1$ degrees of freedom, see (3.49). Dividing also the numerator of (3.54) by its standard deviation $\sigma/\sqrt{N}$, it follows a normal distribution of variance unity. Thus the variable t of the t distribution is the quotient of a normal variate and the square root of a χ^2 variate.

The analytical form of the p.d.f. can be found by the standard method used in Section 3.5.4. The result is

$$h(t|f) = \frac{\Gamma((f+1)/2)}{\Gamma(f/2)\sqrt{\pi f}} \left(1 + \frac{t^2}{f}\right)^{-\frac{f+1}{2}}.$$

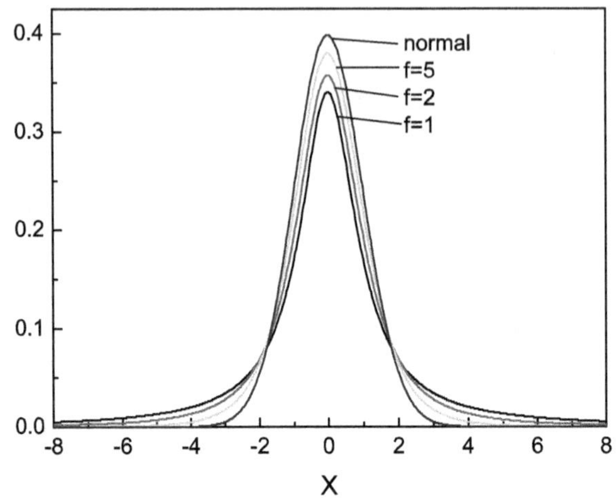

Fig. 3.21 Student's distributions for 1, 2, 5 degrees of freedom and a normal distribution.

The only parameter is f, the number of degrees of freedom. For $f = 1$ we recover the Cauchy distribution. For large f it approaches the normal distribution $\mathcal{N}(0,1)$ with variance equal to one. The distribution is symmetric, centered at zero, and bell shaped, but with longer tails than $\mathcal{N}(0,1)$. The even moments are

$$\mu_i = f^{\frac{i}{2}} \frac{1 \cdot 3 \cdots (i-1)}{(f-2)(f-4)\cdots(f-i)} .$$

They exist only for $i \leq f-1$. The variance for $f \geq 3$ is $\sigma^2 = f/(f-2)$, the excess for $f \geq 5$ is $\gamma_2 = 6/(f-4)$, disappearing for large f, in agreement with the fact that the distribution approaches the normal distribution.

The typical field of application for the t distribution is the derivation of tests or confidence intervals in cases where a sample is supposed to be taken from a normal distribution of unknown variance but known mean μ. Qualitatively, very large absolute values of t indicate that the sample mean is incompatible with μ. Sometimes the t distribution is used to approximate experimental distributions which differ from Gaussians because they have longer tails. In a way, the t distribution interpolates between the Cauchy (for $f = 1$) and the normal distribution (for $f \to \infty$).

3.6.12 *The Extreme Value Distributions*

The family of extreme value distributions is relevant for the following type of problem: Given a sample taken from a certain distribution, what can be said about the distribution of its maximal or minimal value? It is found that these distributions converge with increasing sample size to distributions of the types given below.

3.6.12.1 *The Weibull Distribution*

This distribution has been studied in connection with the lifetime of complex aggregates. It is a limiting distribution for the smallest member of a sample taken from a distribution limited from below. The p.d.f. is

$$f(x|a,p) = \frac{p}{a}\left(\frac{x}{a}\right)^{p-1} \exp\left(-\left(\frac{x}{a}\right)^{p}\right), \ x > 0 \tag{3.55}$$

with the positive scale and shape parameters a and p. The mode is

$$x_m = a\left(\frac{p-1}{p}\right)^{1/p} \quad \text{for } p \geq 1,$$

mean value and variance are

$$\mu = a\Gamma(1 + 1/p),$$
$$\sigma^2 = a^2\left(\Gamma(1+2/p) - \Gamma^2(1+1/p)\right).$$

The moments are

$$\mu_i = a^i\Gamma(1 + i/p).$$

For $p = 1$ we get an exponential distribution with decay constant $1/a$.

3.6.12.2 *The Fisher–Tippett Distribution*

Also this distribution with the p.d.f.

$$f_{\pm}(x|x_0, s) = \frac{1}{s}\exp\left(\pm\frac{x-x_0}{s} - \mathrm{e}^{\pm(x-x_0)/s}\right)$$

belongs to the family of extreme value distributions. It is sometimes called extreme value distribution (without further specification) or log-Weibull distribution.

If y is Weibull-distributed (3.55) with parameters a, p, the transformation to $x = -\ln y$ leads for x to a log-Weibull distribution with parameters $x_0 = -\ln a$ and $s = 1/p$. The first of these, the location parameter x_0, gives the position of the maximum, i.e. $x_{mod} = x_0$, and the parameter

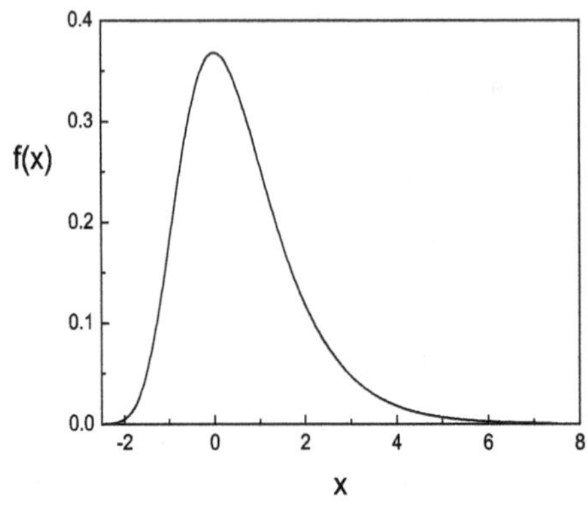

Fig. 3.22 Gumbel distribution.

$s > 0$ is a scale parameter. Mean value μ and variance σ^2 depend on these parameters through

$$\mu = x_0 \mp Cs \text{ , with Euler's constant } C = 0.5772\ldots \text{ ,}$$
$$\sigma^2 = s^2 \frac{\pi^2}{6} .$$

Mostly, the negative sign in the exponent is realized. Its normal form

$$f(x|0,1) = \exp\left(-x - \mathrm{e}^{-x}\right)$$

is also known as Gumbel's distribution and shown in Fig. 3.22.

Using mathematical properties of Euler's Γ function [25] one can derive the characteristic function in closed form:

$$\phi(t) = \Gamma(1 \pm \mathrm{i}st)\mathrm{e}^{\mathrm{i}x_0 t},$$

whose logarithmic derivatives give in turn the cumulants for this distribution:

$$\kappa_1 = x_0 \mp Cs, \ \kappa_{i \geq 2} = (\mp 1)^i (i-1)! s^i \zeta(i),$$

with Riemann's zeta function $\zeta(z) = \Sigma_{n=1}^{\infty} 1/n^z$. (see [25]). Skewness and excess are given by $\gamma_1 \approx 1.14$ and $\gamma_2 = 12/5$.

3.7 Mixed and Compound Distributions

In the statistical literature the terms mixed distribution and compound distribution are not clearly defined and distinguished. Sometimes the compound distribution is regarded as a specific mixed distribution.

3.7.1 *Superposition of Distributions*

The term *mixed distribution* is sometimes used for a superposition of distributions:

$$f(x) = \sum_{i=1}^{N} w_i f_i(x), \tag{3.56}$$

$$P(k) = \sum_{i=1}^{N} w_i P_i(k). \tag{3.57}$$

In physics applications superpositions occur, for instance, if a series of resonances or peaks is observed over a background. We have calculated the mean value and the variance of the superposition of two continuous distributions in Section 3.2. The relations in Section 3.2.3.4 can easily be extended to more then two components.

3.7.2 *Compound Distributions*

If a parameter of a distribution itself is randomly distributed, then we have a compound distribution. We can form different combinations of continuous and discrete distributions, but restrict ourselves to the case of a resulting continuous distribution:

$$f(x) = \int_{-\infty}^{\infty} h(x|\lambda) g(\lambda) d\lambda, \tag{3.58}$$

$$f(x) = \sum_k h(x|\lambda_k) P_k .$$

The relation (3.58) has the form of a convolution and is closely related to the marginalization of a two-dimensional distribution of x and λ. A compound distribution may also have the form of (3.56) or (3.57) where the weights are independent and randomly distributed.

Frequently we measure a statistical quantity with a detector that has a limited resolution. Then the probability to observe the value x' is a random distribution $R(x'|x)$ depending on the undistorted value x which itself is distributed according to a distribution $g(x)$. (In this context the notation is usually different from the one used in (3.58)) We have the convolution

$$f(x') = \int_{-\infty}^{\infty} R(x', x) g(x) dx .$$

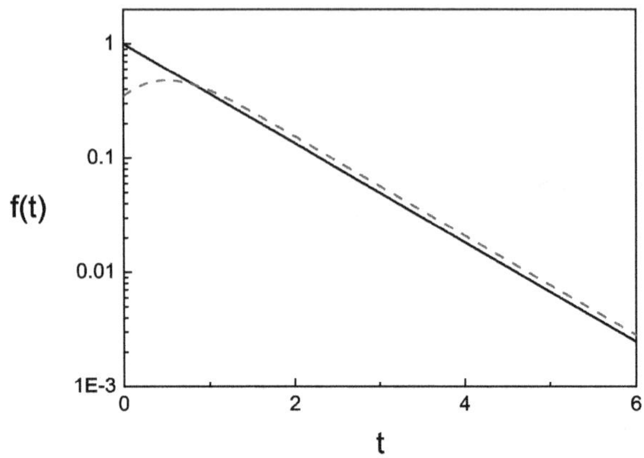

Fig. 3.23 Lifetime distribution, original (solid line) and measured with Gaussian resolution (dashed line).

Example 3.36. Measurement of a decay time distribution with Gaussian resolution

A muon stops in a scintillator and decays subsequently into an electron. The time t between the two corresponding light pulses follows an exponential distribution $\gamma e^{-\gamma t}$ with γ the muon decay constant. The observed value is t' with the response function

$$R(t',t) = \frac{1}{\sqrt{2\pi}\sigma} \exp\left(-\frac{(t'-t)^2}{2\sigma^2}\right).$$

The convolution integral

$$\begin{aligned} f(t') &= \frac{1}{\sqrt{2\pi}\sigma} \int_0^\infty \exp\left(-\frac{(t'-t)^2}{2\sigma^2}\right) \gamma e^{-\gamma t} dt \\ &= \gamma e^{-\gamma t'} \frac{1}{\sqrt{2\pi}\sigma} \int_0^\infty \exp\left(-\frac{(t'-t)^2}{2\sigma^2}\right) e^{\gamma(t'-t)} dt \\ &= \gamma e^{-\gamma t'} \frac{1}{\sqrt{2\pi}\sigma} \int_{-t'}^\infty \exp\left(-\frac{x^2}{2\sigma^2}\right) e^{-\gamma x} dx \end{aligned}$$

can be expressed with the help of the error function

$$\mathrm{erfc}(x) = 1 - \mathrm{erf}(x) = \frac{2}{\sqrt{\pi}} \int_x^\infty e^{-t^2} dt,$$

resulting in

$$f(t') = \frac{1}{2}\gamma \exp^{-\gamma t' - \frac{1}{2}\sigma^2\gamma^2} \operatorname{erfc}\left(\frac{-t' + \sigma^2\gamma}{\sqrt{2}\sigma}\right).$$

The result for $\gamma = 1$, $\sigma = 0.5$ is shown in Fig. 3.23. Except for small values of t, the observed time is shifted to larger values. In the asymptotic limit $t \to \infty$ the integral becomes a constant and the distribution $f_\infty(t')$ is exponentially decreasing with the same slope γ as the undistorted distribution:

$$f_\infty(t') \propto \gamma e^{-\gamma t'}.$$

This property applies also to all measurements where the resolution is a function of $t' - t$. This condition is usually realized.

3.7.3 *The Compound Poisson Distribution*

The *compound Poisson distribution* (CPD) describes the sum of a Poisson distributed number of independent and identical distributed weights. It applies if we weight events and sum up the weights. For example, when we measure the activity of a β source with a Geiger counter the probability that it fires, the detection probability may depend on the electron energy which varies from event to event. We can estimate the true number of decays by weighting each observation with the inverse of its detection probability. Sometimes weighting is used to measure the probability that an event belongs to a certain particle type. Weighted events play also a role in some Monte Carlo integration methods and in parameter inference (see Chapter 7, Section 7.3), if weighted observations are summed up in histogram bins.

The CPD also describes the sum $x = \sum_{i=1}^{N} n_i w_i$, if there is a given discrete weight distribution, w_i, $i = 1, 2, ..., N$ and where the numbers n_i are Poisson distributed. The equivalence of the two definitions of the CPD is shown in Appendix A.9.1. In Ref. [26] some properties of the compound Poisson distribution and the treatment of samples of weighted events is described. The CPD does not have a simple analytic expression. However, the cumulants and thus also the moments of the distribution can be calculated exactly.

Let us consider the definite case that on average λ_1 observations are obtained with probability ε_1 and λ_2 observations with probability ε_2. We correct the losses by weighting the observed numbers with $w_1 = 1/\varepsilon_1$ and $w_2 = 1/\varepsilon_2$. For the Poisson-distributed numbers k_1, k_2

$$\mathcal{P}_{\lambda_1}(k_1) = \frac{\lambda_1^{k_1}}{k_1!}\mathrm{e}^{-\lambda_1},$$
$$\mathcal{P}_{\lambda_2}(k_2) = \frac{\lambda_2^{k_2}}{k_2!}\mathrm{e}^{-\lambda_2},$$
$$k = w_1 k_1 + w_2 k_2,$$

The mean value μ of the variate k and its variance σ^2 are

$$\begin{aligned}\mu &= w_1\lambda_1 + w_2\lambda_2 \\ &= \lambda\langle w\rangle, \end{aligned} \tag{3.59}$$
$$\begin{aligned}\sigma^2 &= w_1^2\lambda_1 + w_2^2\lambda_2 \\ &= \lambda\langle w^2\rangle. \end{aligned} \tag{3.60}$$

with $\lambda = \lambda_1 + \lambda_2$, $\langle w\rangle = (w_1\lambda_1 + w_2\lambda_2)/\lambda$ and $\langle w^2\rangle = (w_1^2\lambda_1 + w_2^2\lambda_2)/\lambda$. We have used $\mathrm{var}(cx) = c^2\mathrm{var}(x)$.

According to (3.27), the cumulant κ_i of order i of the distribution of k is related to the cumulants $\kappa_i^{(1)}, \kappa_i^{(2)}$ of the corresponding distributions of k_1, k_2 through

$$\kappa_i = w_1^i\kappa_i^{(1)} + w_2^i\kappa_i^{(2)}. \tag{3.61}$$

With (3.26) we get also skewness γ_1 and excess γ_2:

$$\begin{aligned}\gamma_1 &= \frac{\kappa_3}{\kappa_2^{3/2}} = \frac{w_1^3\lambda_1 + w_2^3\lambda_2}{(w_1^2\lambda_1 + w_2^2\lambda_2)^{3/2}} \\ &= \frac{\langle w^3\rangle}{\lambda^{1/2}\langle w^2\rangle^{3/2}}, \end{aligned} \tag{3.62}$$
$$\begin{aligned}\gamma_2 &= \frac{\kappa_4}{\kappa_2^2} = \frac{w_1^4\lambda_1 + w_2^4\lambda_2}{(w_1^2\lambda_1 + w_2^2\lambda_2)^2} \\ &= \frac{\langle w^4\rangle}{\lambda\langle w^2\rangle^2}. \end{aligned} \tag{3.63}$$

The formulas can easily be generalized to more than two Poisson distributions and to a continuous weight distribution (see Appendix A.9.1). The relations (3.59), (3.60), (3.62), (3.63) remain valid.

In particular for a CPD with a weight distribution with variance $\mathrm{E}(w^2)$ and expected number of weights λ the variance of the sum of the weights is $\lambda\mathrm{E}(w^2)$ as indicated by (3.60).

For large values of λ the CPD can be approximated by a normal distribution or better by a *scaled Poisson distribution* (see Appendix A.9.1).

3.7.4 *Equivalent Number of Unweighted Poisson Events*

Now we consider n weights w_i where n follows a Poisson distribution with mean value λ and variance λ. The expected value of the sum is $\lambda \sum w_i$ and its variance is

$$\tilde{\lambda} = \left(\sum w_i\right)^2 / \sum(w_i^2). \qquad (3.64)$$

The relative uncertainty of the number n weighted observations is the same as that of $\tilde{n}$ unweighted events. The quantity is called *equivalent number of events.*

Chapter 4

Measurement Errors

4.1 General Considerations

4.1.1 *Introduction*

Measurements that are described by continuous variables are never arbitrarily precise but differ from the true value by an unavoidable random deviation. The expected size of this deviation is the measurement error or uncertainty. When we have no idea of the precision of a measurement, the measurement is rather useless. We need to know its accuracy to validate theoretical predictions, to combine measurements from different experiments and to provide information for its subsequent use in other scientific or engineering applications.

The correct determination and treatment of measurement errors is not always trivial. In principle, the evaluation of parameters and their uncertainties are part of the statistical problem of parameter inference, which we will treat in Chapters 6, 7 and 8 where we will come back to the estimation of errors.

There are uncertainties of different nature: *statistical errors* and *systematic errors*. Their definitions are not unambiguous, disagree from author to author and depend somewhat on the scientific discipline in which they are treated. We shall discuss systematic errors at the end of this chapter.

4.1.2 *Definition of the Measurement Error*

Measurements are either quantities read from a measurement device – we call them input quantities – or derived quantities, like the average of two or more input quantities, the slope of a street, or a rate which are computed from several input quantities. Let us first restrict ourselves to input quantities. An input quantity can be regarded as an *observation*, i.e. a random variable x drawn from a distribution centered around the true value μ of the quantity which we want to determine. The measurement process, including the experimental setup, determines the type of this distribution (Gauss, Poisson, etc.). Often this distribution is only approximately known. For the experimenter the true value μ is an unknown parameter of the distribution. The measurement x is an estimate of μ. The error δ is the the standard deviation or an estimate of it, defined as the root of the mean square difference $\Delta x = x - \mu$ between the true (μ) and the measured values (x), see Section 3.2:

$$\delta = \sqrt{\mathrm{E}(\Delta x^2)} = \sqrt{\mathrm{E}[(x-\mu)^2]} .$$

Sometimes the error depends on the true value μ, often linearly. For the Poisson distribution it is equal to $\sqrt{\mu}$ but it is estimated using x. One has to check whether the approximation $x \approx \mu$ is justified.

The error definition via the standard deviation permits to combine different error contributions. From the relation (3.12) that we have derived in Chapter 3 for the variance, follows that independent error contributions δ_i to the error δ of the measurement x are added quadratically,

$$\delta^2 = \sum \delta_i^2 .$$

Example 4.1. Combination of errors

The distance $s = x_2 - x_1$ between two points, x_1, x_2 is measured with a tape measure. The two positions have errors δ_1 and δ_2 and the tape, due to temperature effects, has an error $s\delta_3$ that is proportional to its length. The error δ of s is $\delta = \sqrt{\delta_1^2 + \delta_2^2 + s^2\delta_3^2}$.

In Section 8.1 we will, as mentioned above, also treat more complex cases, including asymmetric errors due to low event rates or other sources.

Apart from the definition of a measurement and its error by the estimated mean and standard deviation of the related distribution, there exist other conventions like maximal errors and confidence intervals. They are useful in specific situations but suffer from the crucial disadvantage that they are not suited for the combination of measurements or the determination of the errors of depending variables, i.e. error propagation.

4.1.3 *Declaration of Errors*

There are several ways to present measurements with their uncertainties. Some of the more frequent ones are given in the following examples:

$$\begin{aligned}
t &= (34.5 \pm 0.7)\ 10^{-3}\ \mathrm{s} \\
t &= 34.5\ 10^{-3}\ \mathrm{s} \pm 2\,\% \\
x &= 10.3^{+0.7}_{-0.3} \\
m_e &= (0.510\,999\,06 \pm 0.000\,000\,15)\ \mathrm{MeV}/c^2 \\
m_e &= 0.510\,999\,06\ (15)\ \mathrm{MeV}/c^2 \\
m_e &= 9.109\,389\,7\,10^{-31}\mathrm{kg}\ \pm 0.3\,ppm
\end{aligned}$$

The abbreviation *ppm* means *parts per million*. The third example shows how asymmetric errors are indicated.[1] The measurement and its error must have the same number of significant digits. Declarations like $x = 3.2 \pm 0.01$ or $x = 3.02 \pm 0.1$ are inconsistent.

A relatively crude declaration of the uncertainty is sufficient, one or two significant digits are adequate in any case, keeping in mind that often we do not know all sources of errors in detail or are unable to estimate their influence on the result with high accuracy.[2] This fact also justifies in most cases the approximations which we have to apply in the following.

4.2 Linear Error Calculus

We assume that the error distribution is approximately symmetric with respect to μ and that a possible error dependence on μ across the error interval is negligible. These conditions are violated for example for the Poisson distribution with low counting numbers.

4.2.1 *Linear Error Propagation*

We now want to investigate how a measurement error propagates into quantities which are functions of the measurement. We consider a measurement $x_m \pm \delta x$, a function $y(x)$ and are interested in the estimate $y_m = \langle y \rangle$ and its error δy. We expand y around x_m:

$$y = y(x_m) + y'(x_m)\Delta x + \frac{1}{2!}y''(x_m)(\Delta x)^2 + \cdots .$$

[1] The treatment of asymmetric errors will be postponed to Chapter 8.

[2] There are exceptions to this rule in hypothesis testing (see Chapter 10).

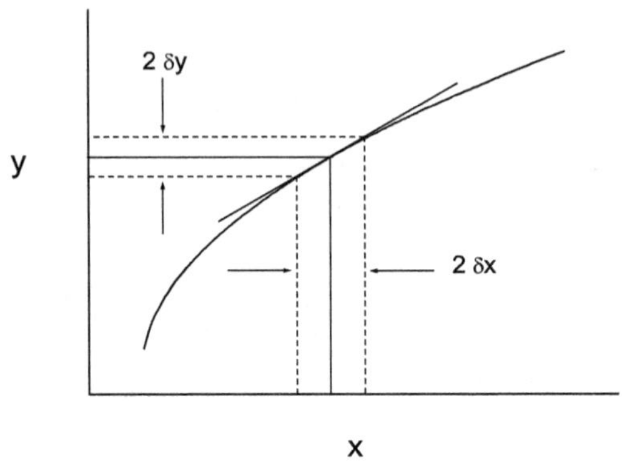

Fig. 4.1 Linear error propagation.

We neglect cubic and higher order terms. According to the assumed symmetry, the expected value of $\Delta x = x - x_m$ is zero, and that of $(\Delta x)^2$ equals $(\delta x)^2$. We get $y_m = y(x_m)$ and

$$\begin{aligned}(\delta y)^2 &= \langle (y - y_m)^2 \rangle \\ &\approx \langle (y(x_m) + y'(x_m)\Delta x - y_m)^2 \rangle \\ &= y'^2(x_m)\langle (\Delta x)^2 \rangle \\ &= y'^2(x_m)(\delta x)^2 , \\ \delta y &= |y'(x_m)|\delta x .\end{aligned}$$

This result can also be read off directly from Fig. 4.1.

Examples of the linear propagation of errors for some simple functions are compiled below:

$$\begin{aligned} &\text{Function :} && \text{Relation between errors :} \\ y &= ax^n \Rightarrow && \frac{\delta y}{|y|} = \frac{|n|\delta x}{|x|} , \\ y &= a\ln(bx) \Rightarrow && \delta y = \frac{|a|\delta x}{|x|} , \\ y &= ae^{bx} \Rightarrow && \frac{\delta y}{|y|} = |b|\delta x , \\ y &= \tan x \Rightarrow && \frac{\delta y}{|y|} = \frac{\delta x}{|\cos x \sin x|} . \end{aligned}$$

4.2.2 Error of a Function of Several Measured Quantities

Most physical measurements depend on several input quantities and their uncertainties. For example, a velocity measurement $v = s/t$ based on the measurements of length and time has an associated error which obviously depends on the errors of both s and t.

Let us first consider a function $y(x_1, x_2)$ of only two measured quantities with values x_{1m}, x_{2m} and errors $\delta x_1, \delta x_2$. With the Taylor expansion

$$y = y(x_{1m}, x_{2m}) + \frac{\partial y}{\partial x_1}(x_{1m}, x_{2m})\Delta x_1 + \frac{\partial y}{\partial x_2}(x_{1m}, x_{2m})\Delta x_2 + \cdots$$

we get as above to lowest order:

$$\begin{aligned} y_m &= \langle y(x_1, x_2) \rangle \\ &= y(x_{1m}, x_{2m}) \end{aligned}$$

and

$$\begin{aligned} (\delta y)^2 &= \langle (\Delta y)^2 \rangle \\ &= \left(\frac{\partial y}{\partial x_1}\right)^2 \langle (\Delta x_1)^2 \rangle + \left(\frac{\partial y}{\partial x_2}\right)^2 \langle (\Delta x_2)^2 \rangle + 2\left(\frac{\partial y}{\partial x_1}\right)\left(\frac{\partial y}{\partial x_2}\right) \langle \Delta x_1 \Delta x_2 \rangle \\ &= \left(\frac{\partial y}{\partial x_1}\right)^2 (\delta x_1)^2 + \left(\frac{\partial y}{\partial x_2}\right)^2 (\delta x_2)^2 + 2\left(\frac{\partial y}{\partial x_1}\right)\left(\frac{\partial y}{\partial x_2}\right) R_{12}\delta x_1 \delta x_2, \quad (4.1) \end{aligned}$$

with the correlation coefficient

$$R_{12} = \frac{\langle \Delta x_1 \Delta x_2 \rangle}{\delta x_1 \delta x_2}.$$

In most cases the quantities x_1 and x_2 are uncorrelated. Then the relation (4.1) simplifies with $R_{12} = 0$ to

$$(\delta y)^2 = (\frac{\partial y}{\partial x_1})^2 (\delta x_1)^2 + (\frac{\partial y}{\partial x_2})^2 (\delta x_2)^2.$$

If the function is a product of independent quantities, it is convenient to use relative errors as indicated in the following example:

$$z = x^n y^m,$$

$$\left(\frac{\delta z}{z}\right)^2 = \left(n\frac{\delta x}{x}\right)^2 + \left(m\frac{\delta y}{y}\right)^2.$$

It is not difficult to generalize our results to functions $y(x_1, .., x_N)$ of N measured quantities. To simplify the notation, we introduce the covariance matrix V,

$$\mathsf{V} = \begin{pmatrix} \langle \Delta x_1 \Delta x_1 \rangle, & \langle \Delta x_1 \Delta x_2 \rangle, & \ldots & \langle \Delta x_1 \Delta x_n \rangle \\ \langle \Delta x_2 \Delta x_1 \rangle, & \langle \Delta x_2 \Delta x_2 \rangle, & \ldots & \langle \Delta x_2 \Delta x_n \rangle \\ \vdots & \vdots & & \vdots \\ \langle \Delta x_n \Delta x_1 \rangle, & \langle \Delta x_n \Delta x_2 \rangle, & \ldots & \langle \Delta x_n \Delta x_n \rangle \end{pmatrix},$$

which, in this context, is also called error matrix. The covariance matrix is positive definite and symmetric by definition. It is related to the correlation matrix R through $V_{ij} = R_{ij}\delta x_i \delta x_j$. The error δy of the dependent variable y is then given in linear approximation by

$$(\delta y)^2 = \sum_{i,j=1}^{N} \left(\frac{\partial y}{\partial x_i} \frac{\partial y}{\partial x_j} V_{ij} \right) .$$

4.2.2.1 *Error Ellipsoids*

For two variables with normally distributed errors following (3.46) the error distribution is

$$\mathcal{N}(\Delta x_1, \Delta x_2) = \frac{1}{2\pi\delta_1\delta_2\sqrt{1-\rho^2}} \exp\left(-\frac{1}{2} \frac{\frac{(\Delta x_1)^2}{\delta_1^2} - 2\rho\frac{\Delta x_1 \Delta x_2}{\delta_1\delta_2} + \frac{(\Delta x_2)^2}{\delta_2^2}}{1-\rho^2} \right) . \tag{4.2}$$

The covariance matrix is

$$\mathsf{V} = \begin{pmatrix} \delta_1^2, & \rho\delta_1\delta_2 \\ \rho\delta_1\delta_2, & \delta_2^2 \end{pmatrix} .$$

Two-dimensional Gaussian error distributions like (4.2) (see Section 3.6.5) have the property that the curves of constant probability density are ellipses. Instead of $n\sigma$ error intervals in one dimension, we obtain $n\sigma$ error ellipses. The curve of constant probability density with density down by a factor of $\exp(-n^2/2)$ relative to the maximal density is the $n\sigma$ error ellipse.

For the error distribution in the form of (4.2) the error ellipse is

$$\frac{\frac{(\Delta x_1)^2}{\delta_1^2} - 2\rho\frac{\Delta x_1 \Delta x_2}{\delta_1\delta_2} + \frac{(\Delta x_2)^2}{\delta_2^2}}{1-\rho^2} = n^2 .$$

For uncorrelated errors the one standard deviation error ellipse is simply

$$\frac{(\Delta x_1)^2}{\delta_1^2} + \frac{(\Delta x_2)^2}{\delta_2^2} = 1 .$$

4.2.3 *Several Functions of Several Measured Quantities*

When we fix a straight line by two measured points in the plane, we are normally interested in its slope and its intercept with a given axis. The errors of these two quantities are usually correlated. The correlations often have to be known in subsequent calculations, e.g. of the crossing point with a second straight line.

In the general case we are dealing with K functions $y_k(x_1, .., x_N)$ of N variables with given measurement values x_i and error matrix V. The symmetric error matrix E related to the values y_k is

$$\langle \Delta y_k \Delta y_l \rangle = \sum_{i,j=1}^{N} \left(\frac{\partial y_k}{\partial x_i} \frac{\partial y_l}{\partial x_j} \langle \Delta x_i \Delta x_j \rangle \right), \tag{4.3}$$

$$E_{kl} = \sum_{i,j=1}^{N} \frac{\partial y_k}{\partial x_i} \frac{\partial y_l}{\partial x_j} V_{ij}.$$

Using the matrix

$$D_{ki} = \frac{\partial y_k}{\partial x_i},$$

we get a compact representation:

$$E_{kl} = \sum_{i,j=1}^{N} D_{ki} D_{lj} V_{ij}, \tag{4.4}$$

$$\mathsf{E} = \mathsf{D}\mathsf{V}\mathsf{D}^T. \tag{4.5}$$

4.2.4 *Examples*

The following examples represent some standard cases of error propagation.

Example 4.2. Error propagation: velocity of a sprinter

Given are $s = (100.0 \pm 0.1)\,\mathrm{m}$, $t = (10.00 \pm 0.02)\,\mathrm{s}$. The error of the velocity is *deltav*:

$$\left(\frac{\delta v}{v}\right)^2 = \left(\frac{\delta t}{t}\right)^2 + \left(\frac{\delta s}{s}\right)^2,$$

$$\frac{\delta v}{v} = \sqrt{\left(\frac{0.02}{10}\right)^2 + \left(\frac{0.1}{100}\right)^2} = 2.2\ 10^{-3}.$$

Example 4.3. Error propagation: area of a rectangular table

Given are the sides a, b with a reading error δ_1 and a relative scaling error δ_2, caused by a possible extension or shrinkage of the measuring tape. We want to calculate the error δF of the area $F = ab$. We find

$$(\delta_a)^2 = (\delta_1)^2 + (a\delta_2)^2,$$

$$(\delta_b)^2 = (\delta_1)^2 + (b\delta_2)^2,$$

$$V_{ab} = ab(\delta_2)^2,$$

$$(\delta F)^2 = b^2(\delta_a)^2 + a^2(\delta_b)^2 + 2abV_{ab},$$

$$\left(\frac{\delta F}{F}\right)^2 = (\delta_1)^2\left(\frac{1}{a^2} + \frac{1}{b^2}\right) + 2(\delta_2)^2.$$

For large areas, the contribution of the reading error is negligible compared to that of the scaling error.

Example 4.4. Straight line through two measured points

Given are two measured points $(x_1, y_1 \pm \delta y_1)$, $(x_2, y_2 \pm \delta y_2)$ of the straight line $y = mx + b$, where only the ordinate y possesses an error. We want to find the error matrix for the intercept

$$b = (x_2 y_1 - x_1 y_2)/(x_2 - x_1)$$

and the slope

$$m = (y_2 - y_1)/(x_2 - x_1).$$

According to (4.3) we calculate the errors

$$(\delta m)^2 = \frac{(\delta y_2)^2 + (\delta y_1)^2}{(x_2 - x_1)^2},$$

$$(\delta b)^2 = \frac{x_2^2(\delta y_1)^2 + x_1^2(\delta y_2)^2}{(x_2 - x_1)^2},$$

$$E_{12} = \langle \Delta m \Delta b \rangle = -\frac{x_2(\delta y_1)^2 + x_1(\delta y_2)^2}{(x_2 - x_1)^2}.$$

The error matrix E for m and b is therefore

$$\mathsf{E} = \frac{1}{(x_2 - x_1)^2}\begin{pmatrix} (\delta y_1)^2 + (\delta y_2)^2, & -x_2(\delta y_1)^2 - x_1(\delta y_2)^2 \\ -x_2(\delta y_1)^2 - x_1(\delta y_2)^2, & x_2^2(\delta y_1)^2 + x_1^2(\delta y_2)^2 \end{pmatrix}.$$

The correlation matrix element R_{12} is

$$\begin{aligned} R_{12} &= \frac{E_{12}}{\delta m\, \delta b}, \\ &= -\frac{x_2(\delta y_1)^2 + x_1(\delta y_2)^2}{\left\{\left[(\delta y_2)^2 + (\delta y_1)^2\right]\left[x_2^2(\delta y_1)^2 + x_1^2(\delta y_2)^2\right]\right\}^{1/2}}. \end{aligned} \tag{4.6}$$

For the special case $\delta y_1 = \delta y_2 = \delta y$ the results simplify to

$$(\delta m)^2 = \frac{2}{(x_1 - x_2)^2}(\delta y)^2,$$
$$(\delta b)^2 = \frac{(x_1^2 + x_2^2)}{(x_1 - x_2)^2}(\delta y)^2,$$
$$E_{12} = -\frac{(x_1 + x_2)}{(x_1 - x_2)^2}(\delta y)^2,$$
$$R_{12} = -\frac{x_1 + x_2}{\sqrt{2(x_1^2 + x_2^2)}}.$$

Remark: As seen from (4.6), for a suitable choice of the abscissa coordinates ($x \to x - x_s$) the correlation disappears. To achieve this, we take as origin the "center of gravity" x_s of the x-values x_i, weighted with the inverse squared errors of the ordinates, $1/(\delta y_i)^2$:

$$x_s = \sum \frac{x_i}{(\delta y_i)^2} / \sum \frac{1}{(\delta y_i)^2}.$$

4.2.5 *Averaging Uncorrelated Measurements*

Important measurements are usually performed by various experiments in parallel, or are repeated several times. The combination of the results from various measurements should be performed in such a way that it leads to the optimal accuracy. Under these conditions, we can calculate a so-called *weighted* mean, with an error smaller than that of any of the contributing measurements. We assume that the individual measurements are independent.

Remember: In this chapter we assume that the errors are small enough to neglect a dependence of the error on the value of the measured quantity within the range of the error.

As an example let us consider two measurements with measured values x_1, x_2 and errors δ_1, δ_2. With the relations given in Section 3.2.3, we find for the error squared δ^2 of a weighted sum

$$x = w_1 x_1 + w_2 x_2,$$
$$\delta^2 = w_1^2 \delta_1^2 + w_2^2 \delta_2^2.$$

Now we chose the weights in such a way that the error of the weighted sum is minimal, i.e. we seek for the minimum of δ^2 under the condition

$w_1 + w_2 = 1$. The result is

$$w_i = \frac{1/\delta_i^2}{1/\delta_1^2 + 1/\delta_2^2}$$

and for the combined error we get

$$\frac{1}{\delta^2} = \frac{1}{\delta_1^2} + \frac{1}{\delta_2^2} .$$

Generalization to N measurements yields

$$x = \sum_{i=1}^{N} \frac{x_i}{\delta_i^2} / \sum_{i=1}^{N} \frac{1}{\delta_i^2} , \tag{4.7}$$

$$\frac{1}{\delta^2} = \sum_{i=1}^{N} \frac{1}{\delta_i^2} . \tag{4.8}$$

Remark that for a non-linear function y(x) and two measurements x_1, x_2 the weighted mean values $< x >$ and $< y >$ satisfy only approximately $< y >= f(< x >)$.

When all measurements have the same error, all the weights are equal to $w_i = 1/N$, and we get the normal arithmetic mean, with the corresponding reduction of the error by the factor $1/\sqrt{N}$.

Remark: If the original raw data of different experiments are available, then we have the possibility to improve the averaging process compared to the simple use of the relations 4.7 and 4.8. When, for example, in two rate measurements of 1 and 2 hours duration, 2, respectively 12 events are observed, then the combined rate is $(2 + 12)/(1\,\mathrm{h} + 3\,\mathrm{h}) = 3.5\,\mathrm{h}^{-1}$, with an error $\pm 0.9\,\mathrm{h}^{-1}$. Averaging according to (4.7) would lead to too low a value of $(3.2 \pm 1.2)\,\mathrm{h}^{-1}$, due to the above mentioned problem of small rates and asymmetric errors. The optimal procedure is in any case the addition of the log-likelihoods which will be discussed in Chapter 8. It will correspond to the addition of the original data, as done here.

Example 4.5. Sum of weighted Poisson numbers

Particles are detected in three detectors with efficiencies $\varepsilon_1 = 0.7$, $\varepsilon_2 = 0.5$, $\varepsilon_3 = 0.9$. The observed event counts are $n_1 = 10$, $n_2 = 12$, $n_3 = 8$. A background contribution is estimated in a separate counting experiment as $b = 9$ with a reduction factor of $r = 2$. The estimate $\hat{n}$ for total number of particles which traverse the detectors is $\hat{n} = \sum n_i/\varepsilon_i - b/r = 43$. From linear error propagation we obtain the uncertainty $\delta_n = 9$. A more precise calculation based on the Poisson distribution of the equivalent number of events would yield asymmetric errors, $\hat{n} = 43^{+10}_{-8}$.

4.2.6 *Averaging Correlated Measurements*

In Section 4.2.5, we derived the expression for the weighted mean of independent measurements of one and the same quantity. This is a special case of a more general result for a sample of N measurements of the same quantity which differ not only in their variances, but are also correlated, and therefore not statistically independent.

We choose the weights for a weighted mean such that the variance of the combined value is minimal, in much the same way as in Section4.2.5 for uncorrelated measurements. For simplicity, we restrict ourselves to two measurements $x_{1,2}$. The weighted sum x is

$$x = w_1 x_1 + w_2 x_2, \quad \text{with } w_1 + w_2 = 1.$$

To calculate δ_x^2, we have to take into account the correlation terms:

$$\delta_x^2 \equiv \mathrm{var}(x) = w_1^2 V_{11} + w_2^2 V_{22} + 2 w_1 w_2 V_{12}.$$

We have used the symmetry of the covariance matrix, $V_{12} = V_{21}$. The minimum of $\delta_x^2$is achieved for

$$\begin{aligned} w_1 &= \frac{V_{22} - V_{12}}{V_{11} + V_{22} - 2V_{12}}, \\ w_2 &= \frac{V_{11} - V_{12}}{V_{11} + V_{22} - 2V_{12}}. \end{aligned} \tag{4.9}$$

The uncorrelated weighted mean corresponds to $V_{12} = 0$. Contrary to this case, where the expression for the minimal value of δ_x^2 is particularly simple, it is not as transparent in the correlated case.

The case of N correlated measurements leads to the following expression for the weights:

$$w_i = \frac{\sum_{j=1}^{N} C_{ij}}{\sum_{ij=1}^{N} C_{ij}},$$

where C is the weight matrix, the inverse of V.

The weighted mean and its error, derived by error propagation, are:

$$x = \sum_{i=1}^{N} w_i x_i = \frac{\sum_{ij=1}^{N} C_{ij} x_i}{\sum_{ij=1}^{N} C_{ij}}, \tag{4.10}$$

$$\delta^2 = \frac{\sum_{ijkl=1}^{N} C_{ij} C_{kl} V_{ik}}{\left(\sum_{ij=1}^{N} C_{ij}\right)^2} = \frac{1}{\sum_{ij=1}^{N} C_{ij}}. \tag{4.11}$$

Example 4.6. Average of measurements with common off-set error Several experiments (i) determine the energy E_i^* of an excited nuclear state by measuring its transition energy E_i with the uncertainty δ_i to the ground state with energy E_0. They take the value of E_0 from the same table which quotes an uncertainty of δ_0 of the ground state energy. Thus the results $E_i^* = E_i + E_0$ are correlated. The covariance matrix is

$$V_{ij} = \langle(\Delta_i + \Delta_0)(\Delta_j + \Delta_0)\rangle = \delta_i^2 \delta_{ij} + \delta_0^2.$$

Here δ_{ij} is the Kronecker symbol. V is the sum of a diagonal matrix and a matrix where all elements are identical, namely equal to δ_0^2. In this special situation the variance $\mathrm{var}(E^*) \equiv \delta^2$ of the combined result $E^* = \sum w_i E_i^*$ is

$$\begin{aligned}\delta^2 &= \sum_i w_i^2 V_{ii} + \sum_{i \neq j} w_i w_j V_{ij} \\ &= \sum w_i^2 \delta_i^2 + \left(\sum w_i\right)^2 \delta_0^2.\end{aligned}$$

Since the second sum is unity, the second term is unimportant when we minimize δ^2, with respect to the weights and we get the same result (4.7) for the weighted mean E^* as in the uncorrelated case. For its error we find, as could have been expected,

$$\delta^2 = \left(\sum \frac{1}{\delta_i^2}\right)^{-1} + \delta_0^2.$$

The following example is a warning that the naive application of linear error propagation can lead to false results.

Example 4.7. Average of correlated cross section measurements, Peelle's pertinent puzzle

The results of a cross section measurements is ξ_1 with uncertainties due to the event count, δ_{10}, and to the beam flux. The latter leads to an error $\delta_f \xi$ which is proportional to the cross section ξ. The two contributions are independent and thus the estimated error squared in the Gaussian approximation is $\delta_1^2 = \delta_{10}^2 + \delta_f^2 \xi_1^2$. A second measurement ξ_2 with different statistics but the same uncertainty on the flux has an uncertainty $\delta_2^2 = \delta_{20}^2 + \delta_f^2 \xi_2^2$. Combining the two measurements we have to take into account the correlation of the errors. In the literature [51] the following covariance matrix is discussed:

$$\mathsf{V} = \begin{pmatrix} \delta_{10}^2 + \delta_f^2 \xi_1^2 & \delta_f^2 \xi_1 \xi_2 \\ \delta_f^2 \xi_1 \xi_2 & \delta_{20}^2 + \delta_f^2 \xi_2^2 \end{pmatrix}.$$

It can lead to the strange result that the least square estimate $\hat{\xi}$ of the two cross sections is located outside the range defined by the individual results [52] , e.g. $\hat{\xi} < \xi_1, \xi_2$. This anomaly is known as *Peelle's Pertinent Puzzle* [53]. Its reason is that the normalization error is proportional to the true cross section and not to the observed one and thus has to be the same for the two measurements, i.e. in first approximation proportional to the estimate $\hat{\xi}$ of the true cross section. The correct covariance matrix is

$$\mathsf{V} = \begin{pmatrix} \delta_{10}^2 + \delta_f^2 \hat{\xi}^2 & \delta_f^2 \hat{\xi}^2 \\ \delta_f^2 \hat{\xi}^2 & \delta_{20}^2 + \delta_f^2 \hat{\xi}^2 \end{pmatrix}. \tag{4.12}$$

Since the best estimate of ξ cannot depend on the common scaling error it is given by the weighted mean

$$\hat{\xi} = \frac{\delta_{10}^{-2} \xi_1 + \delta_{20}^{-2} \xi_2}{\delta_{10}^{-2} + \delta_{20}^{-2}}. \tag{4.13}$$

The error δ is obtained by the usual linear error propagation,

$$\delta^2 = \frac{1}{\delta_{10}^{-2} + \delta_{20}^{-2}} + \delta_f^2 \hat{\xi}^2. \tag{4.14}$$

Proof: The weighted mean for $\hat{\xi}$ is defined as the combination

$$\hat{\xi} = w_1 \xi_1 + w_2 \xi_2$$

which, under the condition $w_1 + w_2 = 1$, has minimal variance (see Section 4.2):

$$\mathrm{var}(\hat{\xi}) = w_1^2 V_{11} + w_2^2 V_{22} + 2 w_1 w_2 V_{12} = min.$$

With (4.12), this can be written as

$$\mathrm{var}(\widehat{\xi}) = w_1^2\delta_{10}^2 + (1-w_1)^2\delta_{20}^2 + \delta_f^2\widehat{\xi}^2 .$$

Setting the derivative with respect to w_1 to zero, we get the usual result

$$w_1 = \frac{\delta_{10}^{-2}}{\delta_{10}^{-2} + \delta_{20}^{-2}} , \; w_2 = 1 - w_1 ,$$

$$\delta^2 = min[\mathrm{var}(\widehat{\xi})] = \frac{\delta_{10}^{-2}}{(\delta_{10}^{-2} + \delta_{20}^{-2})^2} + \frac{\delta_{20}^{-2}}{(\delta_{10}^{-2} + \delta_{20}^{-2})^2} + \delta_f^2\widehat{\xi}^2 ,$$

in agreement with relations (4.13), (4.14).

4.2.7 *Error of the Sum of Weighted Poisson Numbers*

In the evaluation of event numbers, the events are often counted with different weights, in order to take into account, for instance, a varying acceptance of the detector. Weighting is also important in Monte Carlo simulations (see 5.2.6) especially when combined with parameter estimation Section 7.3. For N different weights w_i, $i = 1, ..., N$ and n_i events with weight w_i the weighted number of events is

$$s = \sum_{i=1}^{N} n_i w_i .$$

As n_i is Poisson distributed, its uncertainty is $\sqrt{n_i}$. By error propagation we obtain for the error of the sum $\delta_s^2 = \sum n_i w_i^2$. If individual events are weighted, $n_i = 1$, we get

$$\delta_s^2 = \sum_{i=1}^{N} w_i^2 . \tag{4.15}$$

The sum of the weights follows a Compound Poisson distribution which is described in Section 3.7.3. The result (4.15) corresponds to (3.60).

4.2.8 *Error of the Product of Many Measurements*

If $y = \prod x_i$ is a product of many positive, independent variables x_i with relative standard deviations σ_i/x_i of similar size, then $\ln y$ satisfies

$$\langle \ln y \rangle = \Sigma \langle \ln x_i \rangle ,$$

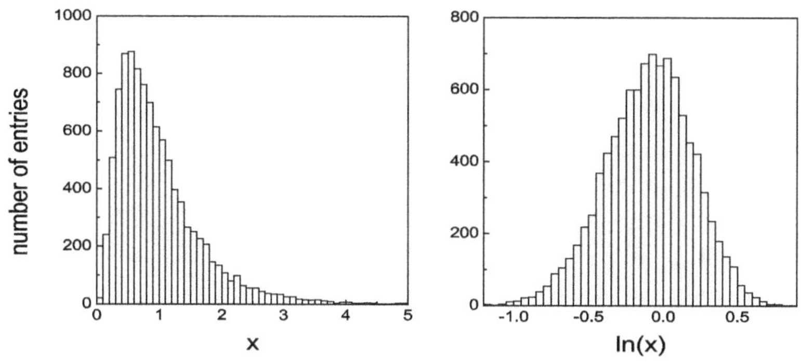

Fig. 4.2 Distribution of the product of 10 variates with mean 1 and standard deviation 0.2 (left) and of its logarithm (right).

and

$$\sigma^2_{\ln y} \approx \sum \sigma^2_{\ln x_i} .$$

Due to the central limit theorem, if applied to the logarithms of the variables, $\ln y$ is approximately normally distributed which means that y follows approximately the log-normal distribution, that has been introduced in Section 3.6.10). The variance of the distribution of y, obtained by error propagation is $\sigma^2 = y^2 \sum \sigma^2_{x_i}/x_i^2$, but as the distribution is asymmetric, the corresponding errors do not describe the uncertainty very well. It is recommended to simulate the distribution and to associate the mode to y and to quote asymmetric errors, see Chapters 5 and 8.

Example 4.8. Distribution of a product of measurements

We simulate the distribution of $y = \prod x_i$, of 10 measured quantities with mean equal to 1 and standard deviation of 0.2, all normally distributed. The result is very different from a Gaussian and is well described by the log-normal distribution as is shown in Fig. 4.2. The mean is compatible with $\langle y \rangle = 1$ and the standard deviation is 0.69, slightly larger than the prediction from simple error propagation of 0.63. These results remain valid if we replace the Gaussian errors by uniform ones with the same standard deviation. Thus details of the distributions of the input variables x_i are not important.

4.2.9 *Non-Linear Error Propagation*

Often we have to estimate of the error of a non-linear function y of several (random), possibly correlated variables. To simplify the discussion of this problem we restrict ourselves to two variable. The p.d.f. $f(x_1, x_2)$ be given.

We simply simulate the pair (x_1, x_2) N times, compute each time the value of y and extract from the distribution of the N y-values the mean value and the variance, or if necessary, asymmetric errors. (The simulation technique will be explained in the following chapter.) If the p.d.f. f is not known, we have to use our best guess.

A special case is the following. We have many identically and independent distributed (i.i.d.)observations $x_1, ..., x_N$ and a non-linear function $y(x)$. We could use the sample to estimate $< x >$, $var(x)$ and apply the standard error propagation. To take the non-linearity into account, the values $y_i) = y(x_i)$ should be computed instead and then the (y) sample can be used to compute the desired parameters for y.

4.3 Statistical Errors

4.3.1 *Errors Following a Known Statistical Distribution*

Relatively simple is the interpretation of measurements if *the distributions of the errors follow known statistical laws.* The corresponding uncertainties are called *statistical errors.* Examples are the measurement of counting rates (Poisson distribution), counter efficiency (binomial distribution) or of the lifetime of unstable particles (exponential distribution). Characteristic for statistical errors is that sequential measurements are uncorrelated and thus the precision of the combined results is improved by the repetition of the measurement. In these cases the error distribution is known up to a parameter – the true value μ. The error is computed replacing μ by the actually observed value x.

Example 4.9. Error of a Poisson number Recorded have been $N = 150$ decays. We set the rate and its error equal to $Z = N \pm \sqrt{N} = (150 \pm \sqrt{150}) \approx 150 \pm 12$.

Example 4.10. Digital measurement (uniform distribution) With a digital clock the time $t = 237\,\text{s}$ has been recorded. The error is $\delta t = 1/\sqrt{12}\ \text{s} \approx 0.3\,\text{s}$, thus $t = (237.0 \pm 0.3)\,\text{s}$.

Example 4.11. Efficiency of a detector (binomial distribution)

From $N_0 = 60$ particles which traverse a detector, $N = 45$ are registered. The efficiency is $\varepsilon = N/N_0 = 0.75$. The error derived from the binomial distribution is

$$\delta\varepsilon = \delta N/N_0 = \sqrt{\varepsilon(1-\varepsilon)/N_0} = \sqrt{0.75 \cdot 0.25/60} = 0.06 .$$

Example 4.12. Calorimetric energy measurement (normal distribution)

The energy of an high energy electron is measured by a scintillating fiber calorimeter by collecting light produced by the electromagnetic cascade in the scintillator of the device. From the calibration of the calorimeter with electrons of known energies E we know that the calorimeter response is well described by a Gaussian with mean proportional to E and variance proportional to E.

Many experimental signals follow to a very good approximation a normal distribution. This is due to the fact that they consist of the sum of many contributions and is a consequence of the central limit theorem.

In particle physics we derive parameters usually from a sample of events and thus take the average of many independent measurements. We have seen that the relative error of the mean from N i.i.d. measurements decreases with $1/\sqrt{N}$, see relation (3.13). This behavior is typical for statistical errors and applies to arbitrary parameters derived from a sample.

4.3.2 *Errors Determined from a Sample of Measurements*

An often used method for the estimation of errors is to repeat a measurement several times and to estimate the error from the fluctuation of the results.

In the simplest case, for instance in calibration procedures, the true value x_t of the measured quantity x is known, and the measurement is only performed in order to get information about the accuracy of the measurement. An estimate of the average error δx of x from N measurements is in this case the empirical standard deviation

$$(\delta x)^2 = \frac{1}{N}\sum_{i=1}^{N}(x_i - x_t)^2 .$$

We have to require that the fluctuations are purely statistical and that correlated systematic variations are absent, i.e. the data have to be independent from each other. The relative uncertainty of the error estimate follows the $1/\sqrt{N}$ law. It will be studied below. For example with 100 repetitions of the measurement, the uncertainty of the error itself is reasonably small, i.e. about 10 % but depends on the distribution of x.

When the true value is unknown, we can approximate it by the sample mean $\overline{x} = \frac{1}{N}\sum_{i=1}^{N} x_i$ and use the following recipe:

$$(\delta x)^2 = \frac{1}{N-1}\sum_{i=1}^{N}(x_i - \overline{x})^2 . \tag{4.16}$$

In the denominator of the formula used to determine the mean quadratic deviation $(\delta x)^2$ of a single measurement figures $N-1$ instead of N. This is plausible because, when we compute the empirical mean value $\overline{x}$, the measurements x_i enter and thus they are expected to be in average nearer to their mean value than to the true value. In particular the division by N would produce the value $\delta x = 0$ for $N = 1$, while the division by $N-1$ yields an indefinite result. The derivation of (4.16) follows from (3.15). The quantity $(\delta x)^2$ in (4.16) is sometimes called *empirical variance*. We have met it already in Section 3.2.3 of the previous chapter.

Frequently, we want to find the error for measurements x_i corresponding to different true values μ_i that are constrained by physical or mathematical laws and where the true values are estimated to $\hat{x}_i$ by a parameter fit (to be explained in subsequent chapters). The expression (4.16) then is generalized to

$$(\delta x)^2 = \frac{1}{N-Z}\sum_{i=1}^{N}(x_i - \hat{x}_i)^2 . \tag{4.17}$$

where Z is the number of parameters that have been adjusted using the data. When we compare the data of a sample to the sample mean we have $Z = 1$ parameter, namely $\bar{x}$, see relation (4.16). When we compare coordinates to the values of a straight line fit then we have $Z = 2$ free parameters to be adjusted from the data, for instance, the slope and the intercept of the line with the ordinate axis. Again, the denominator $N - Z$ is intuitively plausible, since for $N = Z$ we have 2 points lying exactly on the straight line which is determined by them, so also the numerator is zero and the result then is indefinite.

An simple application of (4.17) is the error determination from a data sample.

$$(\delta\overline{x})^2 = (\delta x)^2/N, \\ = \frac{1}{N(N-1)}\sum_{i=1}^{N}(x_i - \overline{x})^2. \tag{4.18}$$

Example 4.13. Error of the mean value of 5 measurements

In the following table five measurements are displayed.

measurements x_i	quadratic deviations $(x_i - \overline{x})^2$
2.22	0.0009
2.25	0.0000
2.30	0.0025
2.21	0.0016
2.27	0.0004
$\sum x_i = 11.25$	$\sum(x_i - \overline{x})^2 = 0.0054$
$\overline{x} = 2.25$	$(\delta x)^2 = \sum(x_i - \overline{x})^2/4 = 0.0013$

The average is $\overline{x} = 2.25 \pm 0.02$. We have used that the error of the mean value is smaller by the factor $\sqrt{5}$ than that of a single measurement, $\delta\overline{x} = \delta x/\sqrt{5}$. With only 5 repetitions the precision of the error estimate is rather poor.

Our recipe yields $\delta\overline{x} \sim 1/\sqrt{N}$, i.e. the error becomes arbitrarily small if the number of the measurements approaches infinity. The validity of the $1/\sqrt{N}$ behavior relies on the assumption that the fluctuations are purely statistical and that correlated systematic variations are absent, i.e. the data have to be independent of each other. When we measure repeatedly the period of a pendulum, then the accuracy of the measurements can be deduced from the variations of the results only if the clock is not stopped systematically too early or too late, and if the clock is not running too fast or too slow. Our experience tells us that some correlation between the different measurements usually cannot be avoided completely and thus there is a lower limit for $\delta\overline{x}$. To obtain a reliable estimate of the uncertainty, we have to take care that the systematic uncertainties are small compared to the statistical error $\delta\overline{x}$.

4.3.3 *Error of the Empirical Variance*

Sometimes we are interested in the variance of an empirical distribution and in its uncertainty. In the same category falls the problem to estimate *the error of the error* of a parameter which is determined from a series of measurements. For example, we may need to know the resolution of a meter or the width of a spectral line and the related accuracy. It is also of interest to know how often a calibration measurement has to be performed to estimate the corresponding error with sufficient accuracy. In these situations the variance s^2 itself is the result of the investigation to which we would like to associate an uncertainty.

The variance of $(x-\mu)^2$ for a given distribution is easily calculated using the formulas of Section 3.2.3. We omit the details of the calculation and quote the result which is related of the second and fourth central moments.

$$\begin{aligned}\mathrm{var}[(x-\mu)^2] &= \left\langle \left[(x-\mu)^2 - \sigma^2\right]^2 \right\rangle \\ &= \mu_4' - \sigma^4 .\end{aligned}$$

We now assume that our sample is large, $N \gg 1$, and replace the distribution moments μ_n' by the empirical central moments m_n',

$$m_n' = \frac{1}{N}\sum (x_i - \overline{x})^n .$$

The moment $s^2 = m_2'$ is an estimate for σ^2. For N events in the sample, we get for the uncertainty δs^2 of s^2

$$(\delta s^2)^2 = \frac{m_4' - m_2'^2}{N}$$

and by error propagation we obtain

$$\begin{aligned}\frac{\delta s}{s} &= \frac{1}{2}\frac{\delta s^2}{s^2} , \\ &= \frac{1}{2\sqrt{N}}\frac{\sqrt{m_4' - s^4}}{s^2} .\end{aligned}$$

If the type of distribution is known, we can use the relations between the moments. Thus, for the normal distribution we have $\mu'_4 = 3\sigma^4$ (see Section 3.6.5), and get $\delta s/s = 1/\sqrt{2N}$ which also follows from the variance $2N$ of the χ^2-distribution. This relation sometimes is applied to arbitrary distributions. It then often underestimates the uncertainty. The error can also be obtained with the bootstrap method (see Section 12.2.2).

4.4 Systematic Errors

The errors assigned to measurements serve primarily the purpose to verify or reject theoretical predictions and to establish new discoveries. Sometimes a significance of four or five standard deviations is required to accept a new finding, for example a new particle that manifests itself through a bump in a mass distribution. Obviously our reasoning here is based on the assumption that the errors are approximately normally distributed which is the case for the statistical error derived from the number of events that have been involved. If background has to be subtracted which usually is extrapolated from regions left and right of the bump location, then the distribution of the background has to be estimated. In this way additional uncertainties are introduced, which are summarized in a *systematic error*. Nearly every measurement is subject to systematic errors, typically associated with auxiliary parameters related to the measuring apparatus, or to model assumptions. Their evaluation is especially important in high precision measurements like those of the magnetic dipole moment of the muon or of the CP violation constants in the neutral kaon system.

The result of a measurement is typically presented in the form $x = 2.34 \pm 0.06 = 2.34 \pm 0.05(stat.) \pm 0.03(syst.)$.

The main reason for the separate quotation of the two uncertainties is that the systematic uncertainties are usually less well known than the purely statistical errors. Thus, for example, excluding a prediction by say a 4 standard deviation measurement where the errors are dominantly of systematic type is certainly less convincing than if the result is purely statistical. Furthermore the separate quotation is informative for subsequent experiments; it helps to design an experiment in such a way that the systematic errors are reduced or avoided such that the precision of a measurement can be improved.

4.4.1 *Definition and Examples*

Systematic errors are at least partially based on assumptions made by the experimenter, are model dependent or follow unknown distributions. This leads to correlations between repeated measurements because the assumptions entering into their evaluations are common to all measurements. Therefore, contrary to statistical errors, the relative error of mean value of repeated measurements, suffering from systematic errors, violates the $1/\sqrt{N}$ law.

A systematic error arises for instance if we measure a distance with a tape-measure which may have expanded or shrunk due to temperature effects. Corrections can be applied and the corresponding uncertainty can be estimated roughly from the estimated range of temperature variations and the known expansion coefficient of the tape material if it is made out of metal. It may also be guessed from previous experience.

Systematic errors occur also when an auxiliary parameter is taken from a technical data sheet where the given uncertainty is usually not of the type "statistical". It may happen that we have to derive a parameter from two or three observations following an unknown distribution. For instance, the current of a magnet may have been measured at the beginning and at the end of an experiment. The variation of the current introduces an error for the momentum measurement of charged particles. The estimate of the uncertainty from only two measurements obeying an unknown distribution of the magnet variations will be rather vague and thus the error is classified as systematic.

A relatively large systematic error has affected the measurement of the mass of the Z^0 particle by the four experiments at the LEP collider. It was due to the uncertainty in the beam energy and has led to sizable correlations between the four results.

Typical systematic uncertainties in particle physics are the following:

(1) Uncertainties in the experimental conditions (Calibration uncertainties for example of a calorimeter or the magnetic field, unknown beam conditions, unknown geometrical acceptance, badly known detector resolutions, temperature and pressure dependence of the performance of gaseous tracking detectors.),
(2) unknown background behavior,
(3) limited quality of the Monte Carlo simulation due to technical approximations,

(4) uncertainties in the theoretical model used in the simulation (approximations in radiative or QCD corrections, poorly known parton densities),
(5) systematic uncertainties caused by the elimination of nuisance parameters (see Section 7.8),
(6) uncertainties in auxiliary parameters taken from data sheets or previous experiments.

Contrary to some authors [27] we classify uncertainties from a limited number of Monte Carlo events as statistical.

4.4.2 *How to Avoid, Detect and Estimate Systematic Errors*

Some systematic errors are difficult to retrieve.[3] If, for instance, in the data acquisition system the deadtime is underestimated, all results may look perfectly all right. In order to detect and to estimate systematic errors, experience, common sense, and intuition is needed. A general advice is to try to suppress them as far as possible already by an appropriate design of the experiment and to include the possibility of control measurements, like regular calibration. Since correlation of repeated measurements is characteristic for the presence of systematic errors, observed correlations of results with parameters related to the systematic effects provide the possibility to estimate and reduce the latter. In the pendulum example, where the frequency is determined from a time measurement for a given number of periods, systematic contribution to the error due to a possible unknown bias in the stopping procedure can be estimated by studying the result as a function of the number of periods and it can be reduced by increasing the measurement time. In particle physics experiments where usually only a fraction of events is accepted by some filtering procedure, it is advisable to record also a fraction of those events that are normally rejected (downscaling) and to try to understand their nature. Some systematic effects are related to the beam intensity, thus a variation of the beam intensity helps to study them.

How can we detect systematic errors caused for instance by background subtraction or efficiency corrections at the stage of data analysis? Clearly, a thorough comparison of the collected data with the simulation in as many

[3]For example, the LEP experiments had to discover that monitoring the beam energy required a magnet model which takes into account leakage currents from nearby passing trains and tidal effects.

different distributions as possible is the primary method. All effects that can be simulated are necessarily understood.

Often kinematical or geometrical constraints can be used to retrieve systematic shifts and to estimate the uncertainties. A trivial example is the comparison of the sum of measured angles of a triangle with the value 180^0 which is common in surveying. In the experiments of particle physics we can apply among other laws the constraints provided by energy and momentum conservation. When we adjust curves, e.g. a straight line to measured points, the deviations of the points from the line permit us to check the goodness of the fit, and if the fit is poor, we might reject the presumed parametrization or revise the error assignment. (Goodness-of-fit tests will be treated in Chapter 10.) Biases in the momentum measurement can be detected by comparing the locations and widths of mass peaks to the nominal values of known particles.

A widely used method is also the investigation of the results as a function of the selection criteria. A correlation of the interesting parameter with the value of a cut-off parameter in a certain variable is a clear indication for the presence of systematic errors. It is evident though that the systematic errors then have to be much larger than the normal statistical fluctuations in order to be detected. Obviously, we want to discriminate also systematic errors which are of the same order of magnitude as the statistical ones, preferably much smaller. Therefore we have to investigate samples, where the systematic effects are artificially enhanced. If we suspect rate dependent distortion effects as those connected with dead times, it is recommended to analyze a control sample with considerably enhanced rate. When we eliminate a background reaction by a selection criterion, we should investigate its importance in the region which has been excluded, where it is supposed to be abundant.

Frequently made mistakes are:

(1) From the fact that the data are consistent with the absence of systematic errors, it is supposed that they do not exist. This leads always to underestimation of systematic errors.
(2) The changes of the results found by changing the selection criteria are directly converted into systematic errors. This in most cases leads to overestimates because the variations are partially due to the normal statistical fluctuations.

There is no simple recipe for the estimation of systematic uncertainties. Let us consider again the problem of background subtraction under an interesting physics signal. If we know nothing about the background, we cannot exclude with absolute certainty that the whole signal is faked by the background. We should exploit all possibilities to reduce the background by looking into many different distributions to derive efficient kinematic cuts. In the end we have to use plausible extrapolations of the background shape based on experience and common sense.

4.4.3 *Treatment of Systematic Errors*

As mentioned, the systematic and the statistical contributions to the measurement error should be declared separately.

In many experiments there appears a quite large number – typically a dozen or so – of such systematic uncertainties. When we combine systematic errors (see Section 8.1), we can often profit from the central limit theorem (see Section 3.6.5), provided that they are all of the same order of magnitude and that the contributions to the measurement are additive. The distribution of the sum of variables suffering from systematic uncertainties approaches a normal distribution, with variance equal to the sum of variances of the contributing distributions. In this case tails in the distributions of the individual systematic errors are less disturbing.

Sometimes a systematic effect affects several experiments which measure the same quantity. For example, as mentioned above, the measurements of the mass of the Z^0 particle by the four experiments at the $e^+ - e^-$ - collider LEP suffered from a common systematic uncertainty of the beam energy. When we combine the results the correlation has to be taken into account.

Sometimes systematic errors are combined linearly. There is no justification for such a procedure.

Interesting discussions of systematic error can be found in Refs. [27,28]. In Ref. [29], a very detailed and competent study of systematic errors as met in particle physics experiments is presented.

In Ref. [27], purely statistical uncertainties related to detector effects or secondary measurements are called *class 1 systematic errors*, but the author states that a classification of these uncertainties as statistical errors would be more informative. He subdivides further the real systematic errors following his definition of systematic errors (which coincides with ours), into systematic errors related to experimental effects (*class 2*) and those depending on theoretical models (*class 3*). This distinction makes sense, because our possibilities to reduce, detect and estimate class 2 and class 3 errors are very different.

4.4.4 *Averaging Measurements with Systematic Errors*

The combination of measurements with systematic errors proceeds in the same way as for measurements with purely random errors. We form the weighted sum where the weights are computed from the full error and we associate to it again a statistical and a systematic error that we calculate by simple error propagation.

To avoid complicated indexes, we write the result of a measurement as $x \pm \delta$, $x \pm a \pm b$, where a stands for the statistical and b for the systematic error. For N measurements, $x_i \pm \delta_i$, $x_i \pm a_i \pm b_i$ we have as before

$$x = \sum_{j=1}^{N} w_i x_i ,$$

$$w_i = \frac{1/\delta_i^2}{\sum_{i=1}^{N} 1/\delta_i^2} .$$

The statistical and the systematic errors are

$$a^2 = \sum_{i=1}^{N} w_i^2 a_i^2 ,$$

$$b^2 = \sum_{i=1}^{N} w_i^2 b_i^2 .$$

Now we consider correlated errors. As always, we assume that the statistical errors are not correlated with the systematic errors. Then we have, apart from the combined covariance matrix V, statistical and systematic covariance matrices A and B which add up to V, $V_{ij} = A_{ij} + B_{ij}$. The formulas (4.10) and (4.11) remain valid and if we split the error into its statistical and its systematic part we get:

$$a^2 = \frac{\sum_{ijkl=1}^{N} C_{ij} C_{kl} A_{ik}}{\left(\sum_{ij=1}^{N} C_{ij}\right)^2} ,$$

$$b^2 = \frac{\sum_{ijkl=1}^{N} C_{ij} C_{kl} B_{ik}}{\left(\sum_{ij=1}^{N} C_{ij}\right)^2} .$$

Example 4.14. Average of Z^0 mass measurements

In four experiments at the LEP storage rings the mass of the Z^0 particle has been measured. The results in unit of GeV are summarized in the first four lines of the following table:

experiment	mass x	error δ	stat. error a	syst. error b
OPAL	91.1852	0.0030	0.0023	0.0018
DELPHI	91.1863	0.0028	0.0023	0.0016
L3	91.1898	0.0031	0.0024	0.0018
ALEPH	91.1885	0.0031	0.0024	0.0018
mean	91.1871	0.0023	0.0016	0.0017

The estimated covariance matrices in MeV^2 are:

$$\mathsf{V} = \begin{pmatrix} 30^2 & 16^2 & 16^2 & 16^2 \\ 16^2 & 28^2 & 16^2 & 16^2 \\ 16^2 & 16^2 & 31^2 & 16^2 \\ 16^2 & 16^2 & 16^2 & 31^2 \end{pmatrix},$$

$$\mathsf{A} = \begin{pmatrix} 23^2 & 0 & 0 & 0 \\ 0 & 23^2 & 0 & 0 \\ 0 & 0 & 24^2 & 0 \\ 0 & 0 & 0 & 24^2 \end{pmatrix}, \quad \mathsf{B} = \begin{pmatrix} 18^2 & 16^2 & 16^2 & 16^2 \\ 16^2 & 16^2 & 16^2 & 16^2 \\ 16^2 & 16^2 & 18^2 & 16^2 \\ 16^2 & 16^2 & 16^2 & 18^2 \end{pmatrix}.$$

The covariance matrices are estimates derived from numbers given in Ref. [31]. The systematic errors are almost completely correlated. The weight matrix $\mathsf{C} = \mathsf{V}^{-1}$ is:

$$\mathsf{C} = \begin{pmatrix} 1.32 & -0.29 & -0.29 & -0.29 \\ -0.29 & 1.54 & -0.29 & -0.29 \\ -0.29 & -0.29 & 1.22 & -0.29 \\ -0.29 & -0.29 & -0.29 & 1.22 \end{pmatrix} \cdot 10^{-3}.$$

We insert these numbers into (4.10) and (4.11) and obtain the results displayed in the last line of the table. Remark, had we neglected the correlation, the uncertainty would have been only 1.6 MeV compared to the correct number 2.3 MeV. The results do not exactly agree with the numbers $m(Z^0) = 91.1876 \pm 0.0021$ quoted in Ref. [31] where theoretical corrections have been applied to the Z^0 mass.

4.5 Biased Measurements

We have required that our measurement values x_i are undistorted (unbiased). We have used this property in the discussion of error propagation. Anyway, it is rather plausible that we should always avoid biased measurements, because averaging measurements with a *common*, e.g. *correlated* bias would produce a result with the same bias. The average from infinitely many measurements would thus be different from the true parameter value but the associated error would be infinitely small. However a closer look at the problem reveals that to require that independent measurements be unbiased, is not justified. When we average measurements, the measurements x_i are weighted with $1/\delta_i^2$, their inverse squared errors, as we have seen above. *To be consistent, it is therefore required that the quantities* x_i/δ_i^2 *are unbiased!* Of course, we explicitly excluded the possibility of errors which depend on the measurement values, but since this requirement is violated so often, we present an example as a warning.

Example 4.15. Bias in averaging measurements

Let us assume that several measurements of a constant x_0 produce unbiased results x_i with errors $\delta_i \sim x_i$ which are proportional to the measurements. This could be, for instance, measurements of particle lifetimes, where the relative error is determined by the number of recorded decays and thus the absolute error is set proportional to the observed mean life. When we compute the weighted mean x over many such measurements

$$\begin{aligned} x &= \sum \frac{x_i}{\delta_i^2} \Big/ \sum \frac{1}{\delta_i^2} \\ &= \sum \frac{1}{x_i} \Big/ \sum \frac{1}{x_i^2} \\ &\approx \langle 1/x \rangle \, / \, \langle 1/x^2 \rangle \end{aligned}$$

the expected value is shifted systematically to lower values. This is easily seen from a Taylor expansion of the expected values:

$$\begin{aligned} \langle x - x_0 \rangle &= \frac{\langle 1/x \rangle}{\langle 1/x^2 \rangle} - x_0 , \\ \left\langle \frac{1}{x} \right\rangle &= \frac{1}{x_0} \left(1 - \left\langle \frac{\Delta x}{x_0} \right\rangle + \left\langle \frac{\Delta x^2}{x_0^2} \right\rangle + \cdots \right) \\ &\approx \frac{1}{x_0} (1 + \frac{\delta^2}{x_0^2}) , \end{aligned}$$

$$\begin{aligned}
\left\langle\frac{1}{x^2}\right\rangle &= \frac{1}{x_0^2}\left(1-2\left\langle\frac{\Delta x}{x_0}\right\rangle+3\left\langle\frac{\Delta x^2}{x_0^2}\right\rangle+\cdots\right) \\
&\approx \frac{1}{x_0^2}(1+3\frac{\delta^2}{x_0^2}), \\
\langle x-x_0\rangle &\approx x_0\frac{1+\delta^2/x_0^2}{1+3\delta^2/x_0^2}-x_0 \\
&\approx x_0(1-2\delta^2/x_0^2)-x_0, \\
\frac{\langle x-x_0\rangle}{x_0} &\approx -2\frac{\delta^2}{x_0^2}.
\end{aligned}$$

Here δ^2 is the expectation of the error squared in an individual measurement. For a measurement error δ/x_0 of 20% we obtain a sizable final bias of 8% for the asymptotic result of infinitely many contributions.

4.6 Confidence Intervals

Table 4.1 Confidence levels for different values of the standard deviation σ.

	Dimension			
Deviation	1	2	3	4
$1\ \sigma$	0.683	0.393	0.199	0.090
$2\ \sigma$	0.954	0.865	0.739	0.594
$3\ \sigma$	0.997	0.989	0.971	0.939
$4\ \sigma$	1.	1.	0.999	0.997

Under the condition that the error distribution is a one-dimensional Gaussian, with a width independent of the expected value, the error intervals of many repeated measurements will cover the true parameter value in 68.3 % of the cases, because for any true value μ the probability to observe x inside one standard deviation interval is

$$\frac{1}{\sqrt{2\pi}\delta}\int_{-\delta}^{\delta}\exp\left[-\frac{(x-\mu)^2}{2\delta^2}\right]\mathrm{d}x \approx 0.683.$$

The region $[x-\delta, x+\delta]$ is called a confidence interval[4] with the confidence level (α) of 68.3 %, or, in physicists' jargon, a 1σ confidence interval. Thus

[4]We will discuss confidence intervals in more detail in Chapter 8.

Table 4.2 Error limits in units of the standard deviation σ for several confidence levels.

Confidence	Dimension			
level	1	2	3	4
0.50	0.67	1.18	1.54	1.83
0.90	1.65	2.14	2.50	2.79
0.95	1.96	2.45	2.79	3.08
0.99	2.58	3.03	3.37	3.64

in about one third of the cases our standard error intervals, under the above assumption of normality, will not contain the true value. Often a higher safety is desired, for instance 90%, 95%, or even 99%. The respective limits can be calculated, provided the probability distribution is known with sufficient accuracy. For the normal distribution, we present some limits in units of the standard deviation in Table 4.2. The numerical values can be taken from tables of the χ^2-distribution function.

For distributions of several variates, the probability to find *all* variables inside their error limits is strongly decreasing with the number of variables. Some probabilities for Gaussian errors are given in Table 4.1. In three dimensions only 20 % of the observations are found in the 1σ ellipsoid. Figure 4.3 shows confidence ellipses and probabilities for two variables.

Example 4.16. Confidence level for the mean of normally distributed variates

Let us consider a sample of N measurements $x_1, \dots, x_N$ which are supposed to be normally distributed with *unknown* mean μ but *known* variance σ^2. The sample mean $\overline{x}$ is also normally distributed with standard deviation $\delta_N = \sigma/\sqrt{N}$. The 1σ confidence interval $[\overline{x} - \delta_N, \overline{x} + \delta_N]$ covers, as we have discussed above, the true value μ in 68.3 % of the cases. We can, with the help of Table 4.2, also find a 99 % confidence level, i.e. $[\overline{x} - 2.58\delta_N, \overline{x} + 2.58\delta_N]$.

We have to keep in mind that the Gaussian confidence limits do not or only approximately apply to other distributions. Error distributions often have tails which are not well understood. Then it is impossible to derive

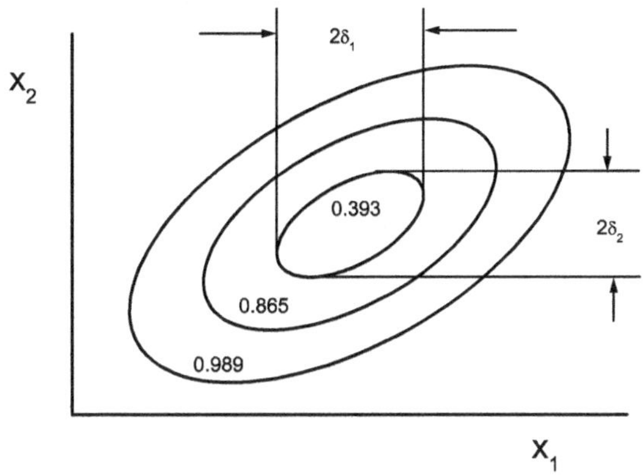

Fig. 4.3 Confidence ellipses for 1, 2 and 3 standard deviations and corresponding probabilities.

reliable confidence limits with high confidence levels. The same is true when systematic errors play a role, for example due to background and acceptance which usually are not known with great accuracy. Then for a given confidence level much wider intervals than in the above case are required. We come back to our previous example but now we assume that the error has to be estimated from the sample itself, according to (4.16), (4.18):

$$\delta_N^2 = \sum_{i=1}^{N}(x_i - \overline{x})^2/[N(N-1)].$$

To compute the confidence level for a given interval in units of the standard deviation, we now have to switch to the Student's distribution $h_f(t)$ (see Section 3.6.11). The variate t, given by $(\overline{x} - \mu)/\delta_N$, can be shown to be distributed according to $h_f(t)$ with $f = N - 1$ degrees of freedom. The confidence level for a given number of standard deviations will now be lower, because of the tails of Student's distribution. Instead of quoting this number, we give in Table 4.3 the factor k by which we have to increase the interval length, to get the same confidence level as in the Gaussian case. To clarify its meaning, let us look at two special cases: For 68.3% confidence and $N = 3$ we require a 1.32 standard deviation interval and for 99% confidence and $N = 10$ a $1.26 \times 2.58 = 3.25$ standard deviation interval. As expected, the discrepancies are largest for small samples and high confidence levels. In the limit when N approaches infinity, the factor k has to become equal to one.

Table 4.3 Values of the factor k for the Student's t-distribution as a function of the confidence levels α and sample size N.

N	68.3%	99%
3	1.32	3.85
10	1.06	1.26
20	1.03	1.11
∞	1.00	1.00

Chapter 5

Monte Carlo Simulation

5.1 Introduction

The possibility to simulate stochastic processes and of numerical modeling on the computer simplifies extraordinarily the solution of many problems in science and engineering. The deeper reason for this is characterized quite aptly by the German saying "Probieren geht über studieren" (Trying beats studying). Monte Carlo methods replace intellectual by computational effort which, however, is realized by the computer.

A few simple examples will demonstrate the advantages, but also the limits of this method. The first two of them are purely mathematical integration problems which could be solved also by classical numerical methods, but show the conceptual simplicity of the statistical approach.

Example 5.1. Area of a circle of diameter d

We should keep in mind that without the knowledge of the quantity π the problem requires some mathematics but even a child can solve this problem experimentally. It may inscribe a circle into a square with edge length d, and sprinkles confetti with uniform density over it. The fraction of confetti confined inside the circle provides the area of the circle in units of the square area. Digital computers have no problem in "sprinkling confetti" homogeneously over given regions.

Example 5.2. Volume of the intersection of a cone and a torus

We solve the problem simply by scattering points homogeneously inside a cuboid containing the intersection. The fraction of points inside both bodies is a measure for the ratio of the intersection volume to that of the cuboid.

In the following three examples we consider the influence of the measurement process on the quantity to be determined.

Example 5.3. Correction of decay times

The decay time of unstable particles is measured with a digital clock which is stopped at a certain maximal time. How can we determine the mean lifetime of the particles? The measured decay times are distorted by both the limited resolution as well as by the finite measurement time, and have to be corrected. The correction can be determined by a simulation of the whole measurement process. (We will come back to details below.)

Example 5.4. Efficiency of particle detection

Charged particles passing a scintillating fiber produce photons. A fraction of the photons is reflected at the surface of the fiber, and, after many reflections, eventually produces a signal in a photomultiplier. The photon yield per crossing particle has to be known as a function of several parameters like track length of the particle inside the fiber, its angle of incidence, fiber length and curvature, surface parameters of the fiber etc.. Here a numerical solution using classical integration methods would be extremely involved and an experimental calibration would require a large number of measurements. Here, and in many similar situations, a Monte Carlo simulation is the only sensible approach.

Example 5.5. Measurement of a cross section in a collider experiment

Particle experiments often consist of millions of detector elements which have to measure the trajectories of sometimes thousands of particles and the energies deposited in an enormous number of calorimeter cells. To measure a specific cross section, the corresponding events have to be selected, acceptance losses have to be corrected, and unavoidable background has to be estimated. This can only be achieved by sophisticated Monte Carlo simulations which often require a large amount of computing time. These simulations consist of two distinct parts, namely the generation of the particle reaction (event generation) which contains the interesting physics, and the simulation of the detector response. The computing time needed for the event generation is negligible compared to that required for the detector simulation. As a consequence one tries to avoid the repetition of the detector simulation and takes, if possible, modifications of the physical process into account by re-weighting events.

Example 5.6. Reaction rates of gas mixtures

A vessel contains different molecules with translational and rotational movements according to the given temperature. The molecules scatter at the walls or interact with each other and possibly transform into other molecules by chemical processes, depending on their energy. To be determined is the composition of the gas after a certain time. The process can be simulated only for a limited number of molecules because the computation of the particle trajectories and the reactions is rather time consuming.

All examples lead finally to integration problems. In the first three examples also numerical integration, even exact analytical methods, could have been used. For the Examples 5.4 and 5.6, however, this is hardly possible, since the number of variables is too large. Furthermore, the mathematical formulation of the problems becomes rather involved.

Monte Carlo simulation does not require a profound mathematical expertise. Due to its simplicity and transparency, mistakes can be avoided. It is true, though, that the results are subject to statistical fluctuations which, however, may be kept small enough in most cases thanks to the fast computers available nowadays. For the simulation of chemical reactions, however, (Example 5.6) we reach the limits of computing power quite soon, even with super-computers. For instance, the simulation of the movement of the molecules of one mole of gas is impossible. Most problems can however be solved with a reasonably small number of variables.

Nowadays, even statistical problems are often solved through Monte Carlo simulations. In some large experiments the error estimation for parameters determined in a complex analysis is so involved that it is easier to simulate the experiment, including the analysis, several times, and to derive the errors quasi experimentally from the distribution of the resulting parameter values. The relative statistical fluctuations of the simulated sample can be scaled down with the square root of the sample size. The computing speed can be increased using bootstrap samples, see Section 12.2.2.

In the following section we will treat the simulation of the basic univariate distributions which are needed for the generation of more complex processes. The generalization to several dimensions is not difficult. Then we continue with a short summary on Monte Carlo integration methods.

5.2 Generation of Statistical Distributions

As mentioned above, computers provide random numbers that are uniformly distributed in the interval $[0, 1]$. The uniform distribution serves as basis for the generation of all other distributions. In the following we will introduce some frequently used methods to generate random numbers with desired distributions.

Some of the simpler methods have been introduced already in Chapter 3, Sections 3.6.4 and 3.6.5: By a linear transformation we can generate uniform distributions of any location and width. The sum of two uniformly distributed random numbers follows a triangular distribution. The addition of only five such numbers produces a quite good approximation of a Gaussian variate.

Since our computers work deterministically, they cannot produce numbers that are really random, but they can be programmed to deliver for practically any application sufficiently unordered numbers, *pseudo random numbers* which approximate random numbers to a very good accuracy.

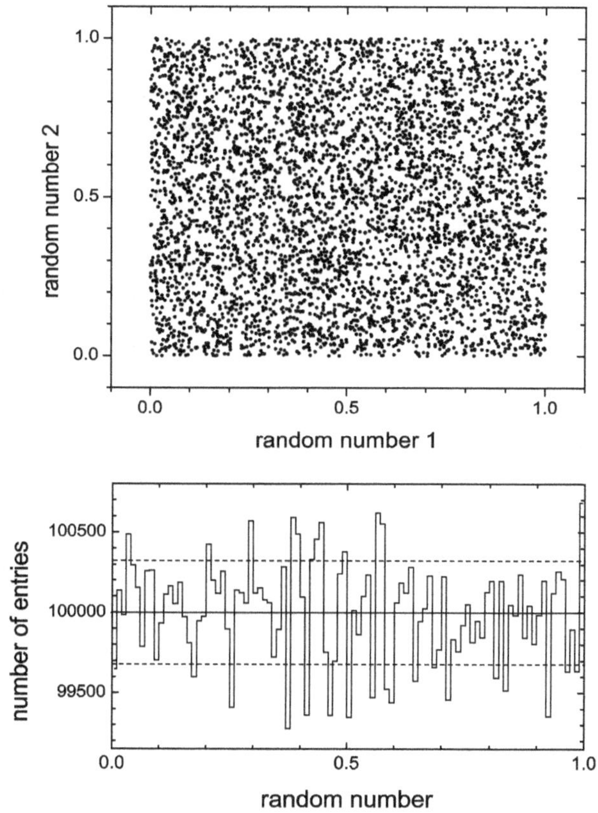

Fig. 5.1 Correlation plot of consecutive random numbers (top) and frequency of random numbers (bottom).

5.2.1 *Computer Generated Pseudo Random Numbers*

The computer delivers pseudo random numbers in the interval between zero and one. Because of the finite number of digits used to represent data in a computer, these are discrete, rational numbers which due to the usual floating point accuracy in double precision can take about 10^{15} different values, and follow a fixed, reproducible sequence which, however, appears as stochastic to the user. More refined algorithms can avoid a repetition of the same sequence after a finite number of calls. The *Mersenne twister*, one of the fastest random number generators, invented in 1997 by M. Matsomoto and T. Nishimura has the enormous period of 2^{19937} which never can be exhausted and is shown to be uniformly distributed in 623 dimensions. In all generators, the user has the possibility to set some starting value, called *seed*, and thus to repeat exactly the same sequence or to interrupt a simulation and to continue with the sequence in order to generate statistically independent samples.

In the following we will speak of random numbers also when we mean pseudo random numbers.

There are many algorithms for the generation of random numbers. A simple generator is the following. The principle is quite simple: One performs an arithmetic operation and uses only the insignificant digits of the resulting number. How this works is shown by the prescription

$$x_{i+1} = n^{-1} \operatorname{mod}(\lambda x_i; n),$$

producing from the old random number x_i a new one between zero and one. The parameters λ and n fulfill the condition $\lambda \gg n$. With the values $x_1 = 0.7123$, $\lambda = 4158$, $n = 1$ we get, for instance, the number

$$x_2 = \operatorname{mod}(2961.7434; 1) = 0.7434.$$

The apparent "randomness" is due to the cutting off the significant digits by the *mod* operation.

This random number generator is far from being perfect, as can be shown experimentally by investigation of the correlations of consecutive random numbers. The generators installed in the commonly used program libraries are almost always sufficiently good. Nevertheless it is advisable to check their quality before starting important calculations. Possible problems with random number generators are that they have a shorter than expected repetition period, correlations of successive values and lack of uniformity. Though intuitively attractive, randomly mixing the results of different random number generators does not improve the overall quality.

In Fig. 5.1 the values of two consecutive random numbers from a PC are plotted against each other. Obvious correlations and clustering cannot be detected. The histogram of a projection is well compatible with a uniform distribution. A quantitative judgment of the quality of random number generators can be derived with goodness-of-fit tests (see Chapter 10).

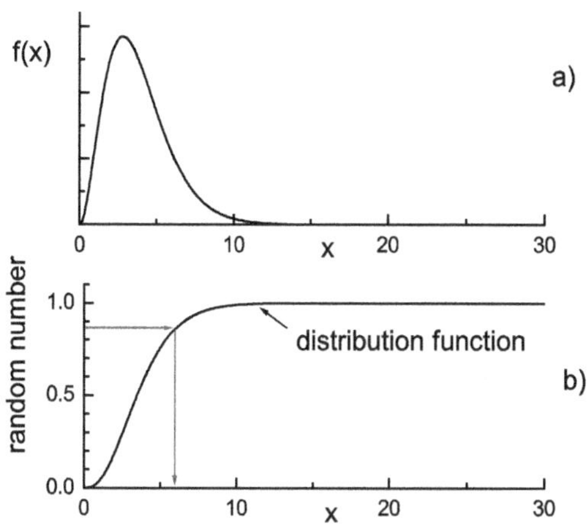

Fig. 5.2 The p.d.f (top) follows from the distribution function as indicated by the arrows.

5.2.2 *Generation of Distributions by Variable Transformation*

5.2.2.1 *Continuous Variables*

With the restrictions discussed above, we can generate with the computer random numbers obeying the uniform distribution

$$u(r) = 1 \text{ for } 0 \le r \le 1 .$$

In the following we use the notations u for the uniform distribution and r for a uniformly distributed variate in the interval $[0, 1]$. Other univariate distributions $f(x)$ are obtained by variable transformations $r(x)$ with r a monotone function of x (see Chapter 3):

$$\begin{aligned} f(x)\mathrm{d}x &= u(r)\mathrm{d}r , \\ \int_{-\infty}^{x} f(x')\mathrm{d}x' &= \int_{0}^{r(x)} u(r')\mathrm{d}r' = r(x) , \\ F(x) &= r , \\ x(r) &= F^{-1}(r) . \end{aligned}$$

The variable x is calculated from the inverse function F^{-1} where $F(x)$ is the distribution function which is set equal to r. For an analytic solution the p.d.f. has to be analytically integrable and the distribution function must have an inverse in analytic form.

The procedure is explained graphically in Fig. 5.2: A random number r between zero and one is chosen on the ordinate. The distribution function (or rather its inverse) then delivers the respective value of the random variable x.

In this way it is possible to generate the following distributions by simple variable transformation from the uniform distribution:

- Linear distribution:

$$f(x) = 2x \quad 0 \leq x \leq 1,$$
$$x(r) = \sqrt{r}.$$

- Power-law distribution:

$$f(x) = (n+1)x^n \quad 0 \leq x \leq 1,\ n > -1,$$
$$x(r) = r^{1/(n+1)}.$$

- Exponential distribution (Section 3.6.6) :

$$f(x) = \gamma e^{-\gamma x},$$
$$x(r) = -\frac{1}{\gamma}\ln(1-r).$$

- Normal distribution (Section 3.6.5): Two independent normally distributed random numbers x, y are obtained from two uniformly distributed random numbers r_1, r_2, see (3.36), (3.37).

$$f(x,y) = \frac{1}{2\pi}\exp\left[-\frac{x^2+y^2}{2}\right],$$
$$x(r_1,r_2) = \sqrt{-2\ln(1-r_1)}\cos(2\pi r_2),$$
$$y(r_1,r_2) = \sqrt{-2\ln(1-r_1)}\sin(2\pi r_2).$$

- Breit-Wigner distribution (Sect 3.6.9):

$$f(x) = \frac{1}{\pi\Gamma/2}\frac{(\Gamma/2)^2}{x^2+(\Gamma/2)^2},$$
$$x(r) = \frac{\Gamma}{2}\tan\left[\pi(r-\frac{1}{2})\right].$$

- Log-Weibull (Fisher–Tippett) distribution (3.6.12.1)

$$f(x) = \exp(-x - \mathrm{e}^{-x}),$$
$$x(r) = -\ln(-\ln r).$$

The expression $1-r$ can be replaced by r in the formulas. More general versions of these distributions are obtained by translation and/or scaling operations. A triangular distribution can be constructed as a superposition of two linear distributions. Correlated normal distributed random numbers are obtained by scaling x and y differently and subsequently rotating the coordinate frame. How to generate superpositions of distributions will be explained in Section 5.2.5.

5.2.2.2 *Uniform Angular, Circular and Spherical Distributions*

Very often the generation of a uniform angular distribution is required. The azimuthal angle φ is given by

$$\varphi = 2\pi r .$$

To obtain a spatially isotropic distribution, we have also to generate the polar angle θ. As we have discussed in Section 3.5.8, its cosine is uniformly distributed in the interval $[-1, 1]$. Therefore

$$\begin{aligned} \cos\theta &= (2r_1 - 1) , \\ \theta &= \arccos(2r_1 - 1) , \\ \varphi &= 2\pi r_2 . \end{aligned}$$

A uniform distribution inside a circle of radius R_0 is generated by

$$\begin{aligned} R &= R_0\sqrt{r_1} , \\ \varphi &= 2\pi r_2 . \end{aligned}$$

A uniform distribution inside a sphere of radius R_0 is obtained similarly from

$$\begin{aligned} R &= R_0 r_1^{1/3} , \\ \theta &= \arccos(2r_2 - 1) , \\ \varphi &= 2\pi r_3 . \end{aligned}$$

5.2.2.3 *Discrete Distributions*

The generation of random numbers drawn from discrete distributions is performed in a completely analogous fashion. We demonstrate the method with a simple example: We generate random numbers k following a Poisson distribution (see Section 3.6.3) $\mathcal{P}_{4.6}(k)$ with expected value 4.6 which is displayed in Fig. 5.3. By summation of the bins starting from the left (integration), we obtain the distribution function $S(k) = \Sigma_{i=0}^{i=k}\mathcal{P}_{4.6}(i)$ shown

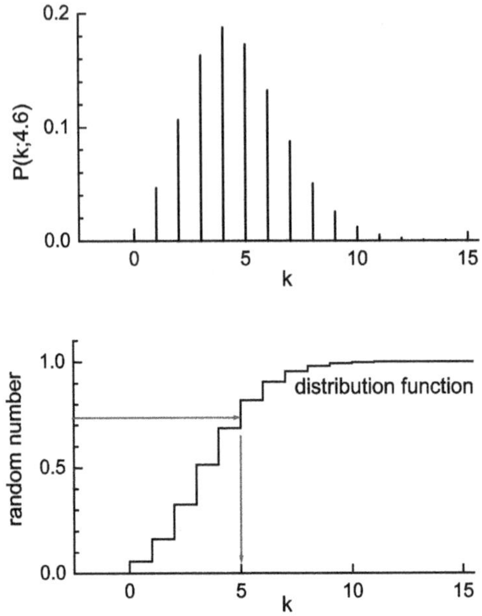

Fig. 5.3 Generation of a Poisson distributed random number.

in the figure. To a uniformly distributed random number r we attach the value k which corresponds to the minimal $S(k)$ fulfilling $S > r$. The numbers k follow the desired distribution.

5.2.2.4 *Histograms*

A similar method is applied when an empirical distribution given in the form of a histogram has to be simulated. The random number r determines the bin j. The remainder $r - S(j-1)$ is used for the interpolation inside the bin interval. Often the bins are small enough to justify a uniform distribution for this interpolation. A linear approximation does not require much additional effort.

For two-dimensional histograms h_{ij} we first produce a projection,

$$g_i = \sum_j h_{ij} \,,$$

normalize it to one, and generate at first i, and then for given i in the same way j.

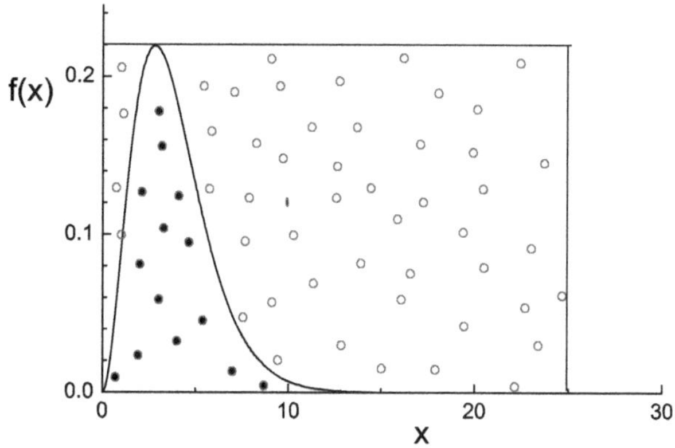

Fig. 5.4 Random selection method. The projection of the points located below the curve follow the desired distribution.

5.2.3 *Simple Rejection Sampling*

In the majority of cases it is not possible to find and invert the distribution function analytically. As an example for a non-analytic approach, we consider the generation of photons following the Planck black-body radiation law. The appropriately scaled frequency x obeys the distribution

$$f(x) = c\frac{x^3}{\mathrm{e}^x - 1} \tag{5.1}$$

with the normalization constant c. This function is shown in Fig. 5.4 for $c = 1$, i.e. not normalized. We restrict ourselves to frequencies below a given maximal frequency x_{max}.

A simple method to generate this distribution $f(x)$ is to choose two uniformly distributed random numbers, where r_1 is restricted to the interval (x_{min}, x_{max}) and r_2 to $(0, f_{max})$. This pair of numbers $P(r_1, r_2)$ corresponds to a point inside the rectangle shown in Fig. 5.4. We generate points and those lying above the curve $f(x)$ are rejected. The density of the remaining r_1 values follows the desired p.d.f. $f(x)$.

A disadvantage of this method is that it requires several randomly distributed pairs to select one random number following the distribution. In our example the ratio of successes to trials is about 1:10. For generating photons up to arbitrary large frequencies the method cannot be applied at all.

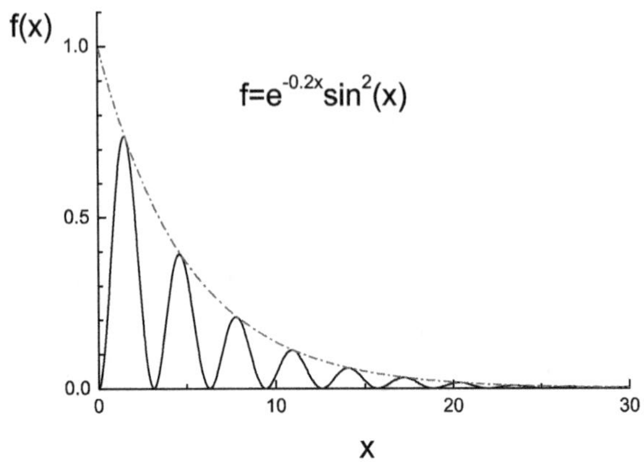

Fig. 5.5 Majorant (dashed) used for importance sampling.

5.2.4 *Importance Sampling*

An improved selection method, called importance sampling, is the following: We look for an appropriate function, a majorant, $m(x)$, with the properties

- $m \geq f$ for all x,
- $x = M^{-1}(r)$, i.e. the indefinite integral $M(x) = \int_{-\infty}^{x} m(x')\mathrm{d}x'$ is invertible.

If it exists (see Fig. 5.5), we generate x according to $m(x)$ and, in a second step, for given x drop randomly the fraction $[m(x) - f(x)]/f(x)$ of the events. This means, for each event (i.e. each generated x) a second, this time uniform random number between zero and $m(x)$ is generated, and if it is larger than $f(x)$, the event is abandoned. The advantage is, that for $m(x)$ being not much different from $f(x)$ in most cases, the generation of one event requires only two random numbers. Moreover, in this way it is possible to generate also distributions which extend to infinity, as for instance the Planck distribution, and many other distributions.

We illustrate the method with a simple example (Fig. 5.5):

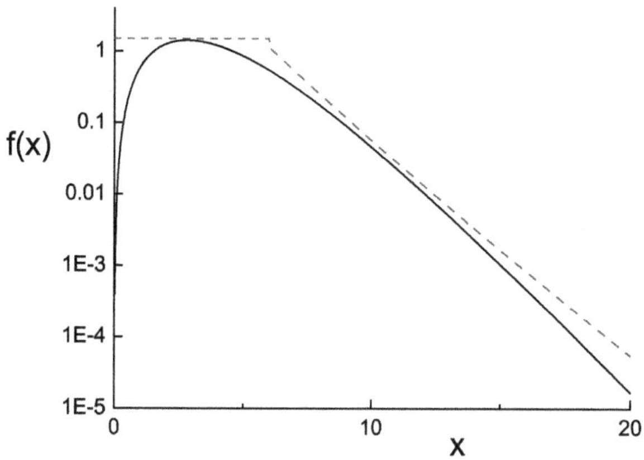

Fig. 5.6 Planck spectrum with majorant.

Example 5.7. Importance sampling

To generate

$$f(x) = c(\mathrm{e}^{-0.2x}\sin^2 x) \quad \text{for} \quad 0 < x < \infty$$

with the majorant

$$m(x) = c\,\mathrm{e}^{-0.2x},$$

we normalize $m(x)$ and calculate its distribution function

$$\begin{aligned} r &= \int_0^x 0.2\mathrm{e}^{-0.2x'}\mathrm{d}x' \\ &= 1 - \mathrm{e}^{-0.2x}. \end{aligned}$$

Thus the variate transformation from the uniformly distributed random number r_1 to x is

$$x = -\frac{1}{0.2}\ln(1 - r_1).$$

We draw a second uniform random number r_2, also between zero and one, and test whether $r_2\, m(x)$ exceeds the desired p.d.f. $f(x)$. If this is the case, the event is rejected:

$$\begin{aligned} \text{for } r_2\, m(x) < \sin^2 x \quad &\rightarrow \text{ keep } x, \\ \text{for } r_2\, m(x) > \sin^2 x \quad &\rightarrow \text{ drop } x. \end{aligned}$$

With this method a uniform distribution of random points below the majorant curve is generated, while only those points are kept which lie below the p.d.f. to be generated. On average about 4 random numbers per event are needed in this example, because the test has a positive result in about half of the cases.

If an appropriate continuous, analytical majorant function cannot be found, often a piecewise constant function (step function) is chosen.

Example 5.8. Generation of the Planck distribution

Here a piecewise defined majorant is useful. We consider again the Planck distribution (5.1), and define the majorant in the following way: For small values $x < x_1$ we chose a constant majorant $m_1(x) = 6\,c$. For larger values $x > x_1$ the second majorant $m_2(x)$ should be integrable with invertible integral function. Due to the x^3-term, the Planck distribution decreases somewhat more slowly than e^{-x}. Therefore we chose for m_2 an exponential factor with x substituted by $x^{1-\varepsilon}$. With the arbitrary choice $\varepsilon = 0.1$ we take

$$m_2(x) = 200\, c\, x^{-0.1} \mathrm{e}^{-x^{0.9}} .$$

The factor $x^{-0.1}$ does not influence the asymptotic behavior significantly but permits the analytical integration:

$$\begin{aligned} M_2(x) &= \int_{x_1}^{x} m_2(x') \mathrm{d}x' , \\ &= \frac{200c}{0.9} \left[\mathrm{e}^{-x_1^{0.9}} - \mathrm{e}^{-x^{0.9}} \right] . \end{aligned}$$

This function can be easily solved for x, therefore it is possible to generate m_2 via a uniformly distributed random number. Omitting further details of the calculation, we show in Fig. 5.6 the Planck distribution with the two majorant pieces in logarithmic scales.

5.2.5 *Treatment of Additive Probability Densities*

Quite often the p.d.f. to be considered is a sum of several terms. Let us restrict ourselves to the simplest case with two terms,

$$f(x) = f_1(x) + f_2(x),$$

with

$$S_1 = \int_{-\infty}^{\infty} f_1(x)\mathrm{d}x,$$
$$S_2 = \int_{-\infty}^{\infty} f_2(x)\mathrm{d}x,$$
$$S_1 + S_2 = 1.$$

Now we choose a uniformly distributed random number r_1. If r_1 is smaller (greater) than S_1, x is generated with a second random number r_2 according to f_1, $(_2)$.[1]

The generalization to more than two terms is trivial.

Example 5.9. Generation of an exponential distribution with constant background

In order to generate the p.d.f.

$$f(x) = \varepsilon \frac{\lambda \mathrm{e}^{-\lambda x}}{1-\mathrm{e}^{-\lambda a}} + (1-\varepsilon)\frac{1}{a} \quad \text{für} \quad 0 < x < a,$$

we chose for $r < \varepsilon$

$$x = \frac{-1}{\lambda} \ln\left(1 - \frac{1-\mathrm{e}^{-\lambda a}}{\varepsilon} r\right),$$

and for $r > \varepsilon$

$$x = a\frac{r-\varepsilon}{1-\varepsilon}.$$

We need only one random number per event. The direct way to use the inverse of the distribution function $F(x)$ would not have been successful, since it cannot be given in analytic form.

The separation into additive terms is always recommended, even when the individual terms cannot be handled by simple variate transformations as in the example above.

[1]The function $f(x)$ can also be generated with a single random number. To find the solution is left to the reader.

5.2.6 *Weighting Events*

In Section 3.6.3 we have discussed some statistical properties of weighted events and realized that the relative statistical error of a sum of N weighted events can be much larger than the Poisson value $1/\sqrt{N}$, especially when the individual weights are very different. Thus we will usually refrain from weighting. However, there are situations where it is not only convenient but essential to work with weighted events. If a large sample of events has already been generated and stored and the p.d.f. has to be changed afterwords, it is of course much more economical to re-weight the stored events than to generate new ones because the simulation of reactions in highly complex detectors is quite computer time consuming. Furthermore, for small changes the weights are close to one and will not much increase the statistical error per simulated event. As we will see later, parameter inference based on a comparison of data with a Monte Carlo simulation usually requires re-weighting anyway.

An event with weight w stands for w identical events with weight 1. When interpreting the results of a simulation, i.e. calculating errors, one has to take into account the distribution of a sum of weights, see last example in Section 4.2.4. There we showed that

$$\mathrm{var}\left(\sum w_i\right) = \sum w_i^2 .$$

Relevant is the relative error of a sum of weights:

$$\frac{\delta\left(\sum w_i\right)}{\sum w_i} = \frac{\sqrt{\sum w_i^2}}{\sum w_i} .$$

To simulate a distribution

$$f(x): \quad \text{with} \quad x_a < x < x_b$$

with weighted events is especially simple: We generate events x_i that are uniformly distributed in the interval $[x_a, x_b]$ and weight each event with $w_i = f(x_i)$.

In the Example 5.7 we could have generated events following the majorant $m(x)$, weighting them with $\sin^2 x$. The weights would then be $w_i = f(x_i)/m(x_i)$.

When we have generated events following a p.d.f. $f(x|\theta)$, depending on a parameter θ, and if we are interested in the distribution $f'(x|\theta)$, we have only to re-weight the events by f'/f.

5.2.7 *Markov Chain Monte Carlo*

5.2.7.1 *Introduction*

The purpose of this section is to give a rough idea of Markov chain Monte Carlo and its application. The interested reader is referred to the extensive literature which is nicely summarized in [32]. The generation of high dimensional distributions is difficult with the methods that we have described above. Markov chain Monte Carlo (MCMC) is able to generate samples of distributions with hundred or thousand dimensions. It has become popular in thermodynamics where statistical distributions are simulated to compute macroscopic mean values and especially to study phase transitions. It has also been applied for the approximation of functions on discrete lattices. The method is used mainly in theoretical physics to sample multidimensional distributions. It is also applied to optimize artificial neural networks.

Characteristic of a Markov chain is that a random variable x is modified stochastically in discrete steps, its value at step i depending only on its value at the previous step $i-1$. Values of older steps are forgotten: $P(x_i|x_{i-1}, x_{i-2}, ..., x_1) = P(x_i|x_{i-1})$. A typical example of a Markov chain is random walk. Of interest are Markov chains that converge to an equilibrium distribution, like random walk in a fixed volume. MCMC generates a Markov chain that has as its equilibrium distribution the desired distribution. Continuing with the chain once the equilibrium has been reached, it produces further variates of the distribution. To satisfy this requirement, the chain has to satisfy certain conditions which are fulfilled for instance for the so-called Metropolis algorithm, which we will use below. There exist also several other sampling methods.

5.2.7.2 *Thermodynamical Model, Metropolis Algorithm*

In thermodynamics the molecules of an arbitrary initial state always approach – if there is no external intervention – a stationary equilibrium distribution. Transitions then obey the *principle of detailed balance.* In a simple model with atoms or molecules in only two possible states in the stationary case, the rate of transitions from state 1 to state 2 has to be equal to the reverse rate from 2 to 1. For occupation numbers N_1, N_2 of the respective states and transition rates per molecule and time W_{12}, respectively W_{21}, we have the equation of balance

$$N_1 W_{12} = N_2 W_{21} .$$

For instance, for atoms with an excited state, where the occupation numbers are very different, the equilibrium corresponds to a Boltzmann distribution, $N_1/N_2 = \mathrm{e}^{-\Delta E/kT}$, with ΔE being the excitation energy, k the Boltzmann constant and T the absolute temperature. When the stationary state is not yet reached, e.g. the number N_1 is smaller than in the equilibrium, there will be less transitions to state 2 and more to state 1 on average than in equilibrium. The occupation number of state 1 will therefore increase until the equilibrium is reached. Since transitions are stochastic, even in equilibrium the occupation numbers will fluctuate around their nominal values.

If now, instead of discrete states, we consider systems that are characterized by a continuous variable x, the occupation numbers are to be replaced by a density distribution $f(x)$ where x is multi-dimensional. It represents the total of all energies of all molecules. As above, for a stationary system we have

$$f(x)W(x \to x') = f(x')W(x' \to x).$$

As probability $P(x \to x')$ for a transition from state x to a state x' we choose the Boltzmann acceptance function

$$\begin{aligned} P(x \to x') &= \frac{W(x \to x')}{W(x \to x') + W(x' \to x)} \\ &= \frac{f(x')}{f(x) + f(x')}. \end{aligned}$$

In an ideal gas and in many other systems the transition regards only one or two molecules and we need only consider the effect of the change of those. Then the evaluation of the transition probability is rather simple. Now we simulate the stochastic changes of the states with the computer, by choosing a molecule at random and change its state with the probability $P(x \to x')$ into a also randomly chosen state x'. The choice of the initial distribution for x is relevant for the speed of convergence but not for the asymptotic result.

This mechanism has been introduced by Metropolis *et al.* [33] with a different acceptance function in 1953. It is well suited for the calculation of mean values and fluctuations of parameters of thermodynamical or quantum statistical distributions. The process continues after the equilibrium is reached and the desired quantity is computed periodically. This process simulates a periodic measurement, for instance of the energy of a gas with small number of molecules in a heat bath. Measurements performed shortly

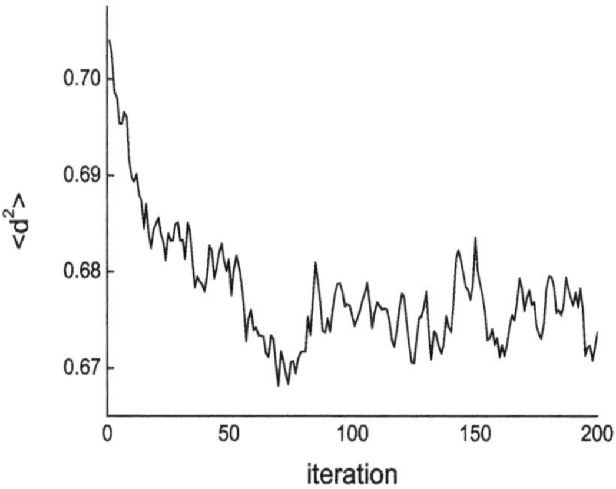

Fig. 5.7 Mean distance of spheres as a function of the number of iterations.

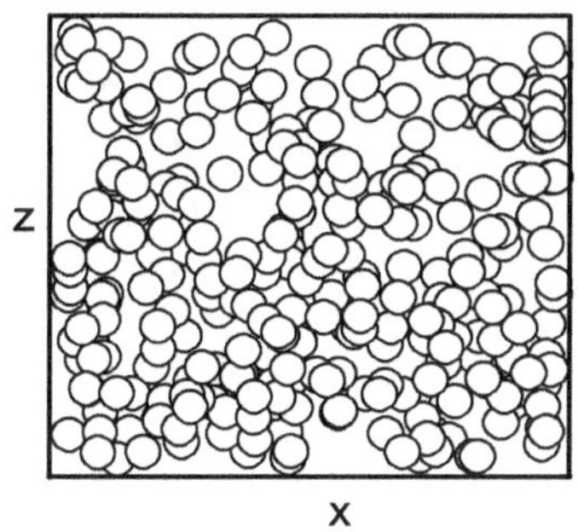

Fig. 5.8 Solid spheres in a box. The plot is a projection onto the x-z plane.

one after the other will be correlated. The same is true for sequentially probed quantities of the MCMC sampling. For the calculation of statistical fluctuations the effect of correlations has to be taken into account. It can be estimated by varying the number if moves between subsequent measurements.

Example 5.10. Mean distance of gas molecules

We consider an atomic gas enclosed in a cubic box located in the gravitational field of the earth. The N atoms are treated as hard balls with given radius R. Initially the atoms are arranged on a regular lattice.

The p.d.f. is zero for overlapping atoms, and proportional to $e^{-\alpha z}$, where z is the vertical coordinate of a given atom. The exponential factor corresponds to the Boltzmann distribution for the potential energy in the gravitational field. An atom is chosen randomly. Its position may be (x, y, z). A second position inside the box is randomly selected by means of three uniformly distributed random numbers. If within a distance of less than $2R$ an other atom is found, the move is rejected and we repeat the selection of a possible new location. If the position search with the coordinates (x', y', z') is successful, we form the ratio $w = e^{-az'}/(e^{-az'} + e^{-\alpha z})$. The position is changed if the condition $r < w$ is fulfilled, with a further random number r. Periodically, the quantity being studied, here the mean distance between atoms, is calculated. It is displayed in Fig. 5.7 as a function of the iteration number. Its mean value converges to an asymptotic value after a number of moves which is large compared to the number of atoms. Figure 5.8 shows the position of atoms projected to the x-z plane, for 300 out of 1000 considered atoms, after 20000 moves. Also the statistical fluctuations can be found and, eventually, re-calculated for a modified number of atoms according to the $1/\sqrt{N}$-factor.

5.3 Evaluation of Integrals and Process Simulation

The generation of distributions has always the aim, finally, to evaluate integrals. There the integration consists in simply counting the sample elements (the events), for instance, when we determine the acceptance or efficiency of a detector.

The integration methods follow very closely those treated above for the generation of distributions. To simplify the discussion, we will consider mainly one-dimensional integrals. The generalization to higher dimensions, where the advantages of the Monte Carlo method become even more pronounced than for one-dimensional integration, does not impose difficulties.

Monte Carlo integration is especially simple and has the additional advantage that the accuracy of the integrals can be determined by the usual methods of statistical error estimation. Error estimation is often quite involved with the conventional numerical integration methods.

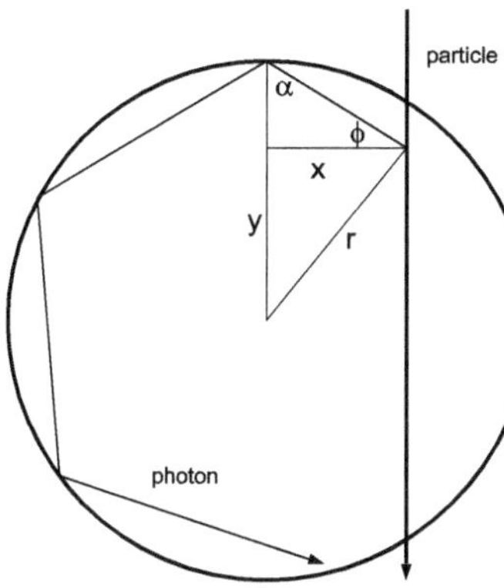

Fig. 5.9 Geometry of photon radiation in a scintillating fiber.

5.3.1 *Simple Random Selection Method*

Integrals with the integrand changing sign are subdivided into integrals over intervals with only positive or only negative integrand. Hence it is sufficient to consider only the case

$$I = \int_{x_a}^{x_b} y(x)\,\mathrm{d}x \quad \text{with } y > 0. \tag{5.2}$$

As in the analogous case when we generate a distribution, we produce points which are distributed randomly and uniformly in a rectangle covering the integrand function. An estimate $\widehat{I}$ for the area I is obtained from the ratio of successes – this are the points falling below the function $y(x)$ – to the number of trials N_0, multiplied by the area I_0 of the rectangle:

$$\widehat{I} = I_0 \frac{N}{N_0}.$$

To evaluate the uncertainty of this estimate, we refer to the binomial distribution, in which we approximate the probability of success ε by the experimental value $\varepsilon = N/N_0$:

$$\delta N = \sqrt{N_0 \varepsilon (1-\varepsilon)},$$
$$\frac{\delta \widehat{I}}{\widehat{I}} = \frac{\delta N}{N} = \sqrt{\frac{1-\varepsilon}{N}}. \tag{5.3}$$

As expected, the accuracy raises with the square root of the number of successes and with ε. The smaller the deviation of the curve from the rectangle, the less will be the uncertainty.

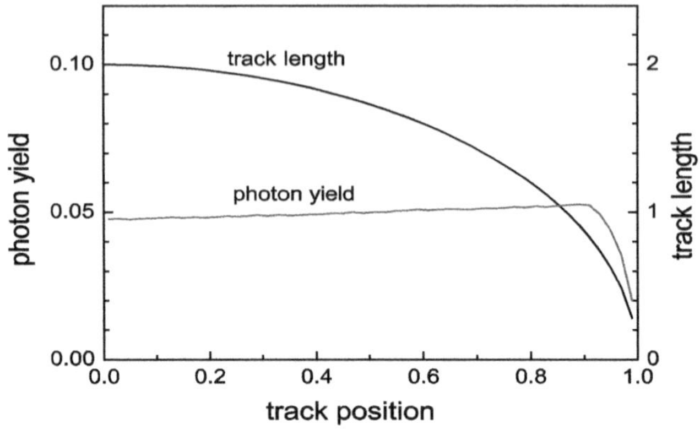

Fig. 5.10 Photon yield as a function of track position.

Example 5.11. Photon yield for a particle crossing a scintillating fiber

Ionizing particles are crossing a scintillating fiber with circular cross section. The trajectories are perpendicular to the fiber axis (Fig. 5.9), and generate photons with a spatially isotropic angular distribution (see 5.2.2.2). Photons hitting the fiber surface will be reflected if the angle with respect to the surface normal is larger than $\beta_0 = 60^o$. For smaller angles they will be lost. We want to know, how the number of captured photons depends on the distance x of the trajectory from the fiber axis. To evaluate the acceptance, we perform the following steps:

- Set the fiber radius $R = 1$, create a photon at x, y uniformly distributed in the square $0 < x,\ y < 1$,
- calculate $r^2 = x^2 + y^2$, if $r^2 > 1$ reject the event,
- chose the azimuthal angle φ for the photon direction, with respect to an axis parallel to the fiber direction in the point x, y, $0 < \varphi < 2\pi$, φ uniformly distributed,
- calculate the projected angle α (see Fig. 5.9) using $\sin\alpha = r\sin\varphi$,
- choose a polar angle ϑ for the photon direction, $0 < \cos(\vartheta) < 1$, $\cos(\vartheta)$ uniformly distributed,
- calculate the angle β of the photon with respect to the surface normal of the fiber, $\cos\beta = \sin\vartheta\cos\alpha$,

- for $\beta < \beta_0$ reject the event,
- store x for the successful trials in a histogram and normalize to the total number of trials.

The efficiency is normalized such that particles crossing the fiber at $x = 0$ produce exactly 1 photon.

Figure 5.10 shows the result of the simulation. For large values of x the track length is small, but the photon capture efficiency is large, therefore the yield increases with x almost until the fiber edge.

5.3.2 *Improved Selection Method*

5.3.2.1 *Reducing the Reference Area*

We can gain in accuracy by reducing the area in which the random points are distributed, in the same way as above by introduction of a majorant function for thr generation of distributions, Fig. 5.5. The relative error is proportional to the square root of the inefficiency (5.3).

We come back to the first example of this chapter:

Example 5.12. Determination of π

The area of a circle with radius 1 is π. For N_0 uniformly distributed trials in a circumscribed square of area 4 (Fig. 5.11) the number of successes N is on average

$$\langle N \rangle = \frac{\pi}{4} N_0 .$$

An estimate $\widehat{\pi}$ for π is

$$\widehat{\pi} = \frac{4N}{N_0} ,$$

$$\frac{\delta \widehat{\pi}}{\pi} = \frac{\sqrt{1 - \pi/4}}{\sqrt{N_0 \pi/4}} ,$$

$$\approx 0.52 \frac{1}{\sqrt{N_0}} .$$

Choosing a circumscribed octagon as the reference area, the error is reduced by about a factor two. A further improvement is possible by inscribing another polygon inside the circle and considering only the area between the polygons.

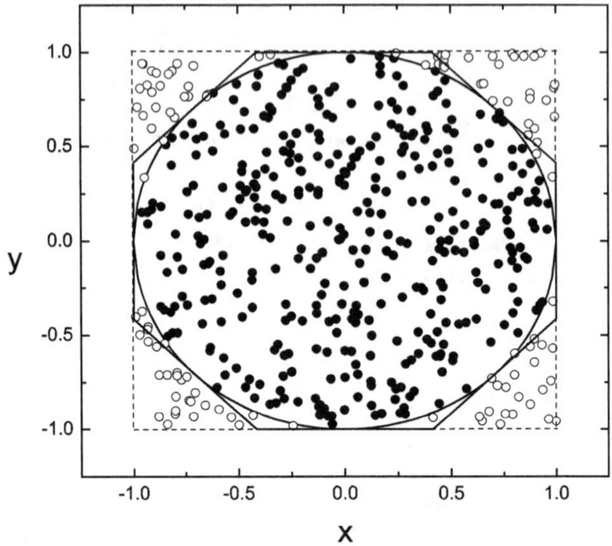

Fig. 5.11 Estimation of the number π.

5.3.2.2 *Importance Sampling*

If there exists a majorant $m(x)$ for the function $y(x)$ to be integrated,

$$I = \int_{x_a}^{x_b} y(x)\mathrm{d}x, \tag{5.4}$$

with the property that the indefinite integral $M(x)$,

$$M(x) = \int_{x_a}^{x} m(x')\mathrm{d}x',$$

can be inverted, we generate N_0 x-values according to the distribution $m(x)$. For each x_i a further random number y_i in the interval $0 < y < m(x_i)$ is generated. Again, as for the simulation of distributions, points lying above $y(x_i)$ are rejected. The number N of the remaining events provides the integral

$$\widehat{I} = M(x_b)\frac{N}{N_0}.$$

5.3.3 *Weighting Method*

5.3.3.1 *Simple Weighting*

We generate N random numbers x_i in the interval $x_a < x < x_b$ and average over the function values:

$$\overline{y} = \sum_{i=1}^{N} y(x_i)/N.$$

An estimate for the integral (5.4) is given by

$$\widehat{I} = (x_b - x_a)\overline{y}.$$

This method corresponds to the usual numerical integration, with the peculiarity that the supporting points on the abscissa are not chosen regularly but are distributed at random. This alone cannot be an advantage, and indeed the Monte Carlo integration in one and two dimensions for a given number of supporting points is less efficient than conventional methods. It is, however, superior to other methods for multi-dimensional integrations. Already in three dimensions it competes favorably in many cases.

To estimate the accuracy, we apply the usual statistical error estimation. We consider the numbers $y_i = y(x_i)$ as N stochastic measurements of $\overline{y}$. The expected mean squared error of $\overline{y}$ is then given by (4.18):

$$(\delta\overline{y})^2 = \frac{1}{N(N-1)} \sum (y_i - \overline{y})^2 .$$

The relative errors of $\overline{y}$ and $\widehat{I}$ are the same,

$$\begin{aligned} \left(\frac{\delta\widehat{I}}{\widehat{I}}\right)^2 &= \left(\frac{\delta\overline{y}}{\overline{y}}\right)^2, \\ &= \frac{\sum (y_i - \overline{y})^2}{N(N-1)\overline{y}^2}. \end{aligned} \tag{5.5}$$

The numerator is an estimate of the variance of the y distribution. The accuracy is the better, the smaller the fluctuations of the function around its mean value are.

5.3.3.2 *Subtraction method*

The accuracy can be improved through a reduction of the fluctuations of the integrand.

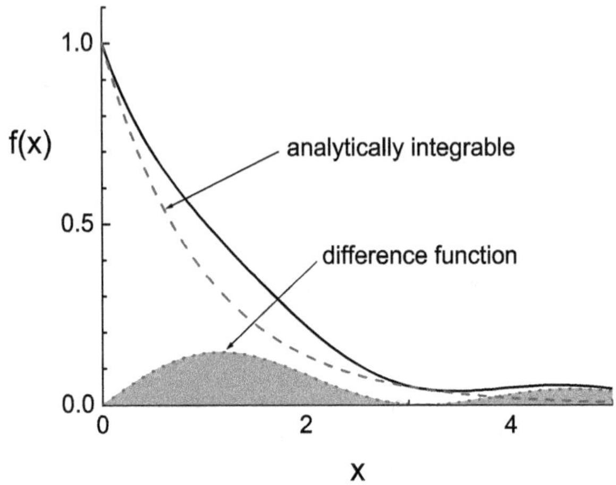

Fig. 5.12 Monte Carlo integration of the difference between the function to be integrated and an integrable function.

If we find a function $\widetilde{y}(x)$ which is integrable analytically and does not differ too much from the original integrand $y(x)$ we cast the integral into the form

$$\int_{x_a}^{x_b} y(x)\mathrm{d}x = \int_{x_a}^{x_b} \widetilde{y}(x)\mathrm{d}x + \int_{x_a}^{x_b} \left(y(x) - \widetilde{y}(x)\right)\mathrm{d}x.$$

We now have to evaluate by Monte Carlo only the second term with relatively small fluctuations (Fig. 5.12).

5.3.4 *Reduction to Expected Values*

In many cases it makes sense to factorize the integrand $y(x) = f(x)y_1(x)$ into a factor $f(x)$ corresponding to a p.d.f. normalized inside the integration interval which is easy to generate, and a second factor $y_1(x)$. To be effective, the method requires that f is close to y. Our integral has now the form of an expected value:

$$\begin{aligned}\int_{x_a}^{x_b} y(x)\mathrm{d}x &= \int_{x_a}^{x_b} f(x)y_1(x)\mathrm{d}x \\ &= \langle y_1 \rangle .\end{aligned}$$

We generate values x_i distributed according to $f(x)$ and obtain from these an estimate for the integral I:

$$\widehat{I} = \frac{\sum_i y_1(x_i)}{N},$$

$$\left(\frac{\delta \widehat{I}}{\widehat{I}}\right)^2 = \frac{\sum \left[y_1(x_i) - \overline{y}_1\right]^2}{N(N-1)\overline{y}_1^2}.$$

The estimate is again the better, the less the y_1-values are fluctuating, i.e. the more similar the functions y and f are. The error estimate is analogous to (5.5).

5.3.5 *Stratified Sampling*

In stratified sampling the domain of integration is partitioned into sub-domains. Over each of these we integrate separately. The advantage is that the distribution in each sub-domain is more uniform and thus the fluctuations of the random variables are smaller and the statistical error is reduced. This method is somewhat antithetical to the basic idea of the simple Monte Carlo method, since it produces a more uniform (equidistant) distribution of the supporting points and requires some effort to combine the errors from the different contributions. Thus we recommend it only if the integrand shows very strong variations.

5.4 General Remarks

Often we need to solve integrals over different domains but always with the same integrand. In these cases the Monte Carlo approach is particularly advantageous. We store all single simulated values (usually called "events") and are able to select events afterwards according to the chosen domain, and obtain the integral with relatively small computing expense by summation. Similarly a change of event weights is possible without repeating the generation of the events.

Let us illustrate this feature with a mechanical example: If, for instance, we want to obtain the tensor of inertia for a complex mass distribution like a car, we distribute points randomly within the body and store their coordinates together with the respective mass densities. With these data it is easy to calculate by summations the mass, the center of mass and the moments of inertia with respect to arbitrary axes. If desired, parts of the body can be eliminated simply by rejecting the corresponding points in the sums and different materials can be considered by changing the density.

In thermodynamic systems we are often interested in several mean values, like the mean free path length, mean kinetic or potential energy, velocities etc.. Once a statistical ensemble has been generated, all these quantities are easily obtained, while with the usual integration methods, one has to repeat each time the full integration.

Even more obvious are these advantages in acceptance calculations. Big experiments in particle physics and other areas have to be simulated as completely and realistically as allowed by the available computing power. The acceptance of a given system of particle detectors for a certain class of events is found in two steps: first, a sample of interesting events is generated and the particles produced are traced through the detecting apparatus. The hits in various detectors together with other relevant information (momenta, particle identities) are stored in data banks. In a second step the desired acceptance for a class of events is found by simulating the selection procedure and counting the fraction of events which are retained. Arbitrary changes in the selection procedure are readily implemented without the need to simulate large event samples more than once.

Finally, we want to stress again how easy it is to estimate the errors of Monte Carlo integration. It is almost identical[2] to the error estimation for the experimental data. We usually will generate a number of Monte Carlo reactions which is large enough to neglect their statistical error compared to the experimental error. In other words, the number of Monte Carlo events should be large compared to the number of experimental events. Usually a factor of ten is sufficient.

[2]The Monte Carlo errors are usually described by the binomial distribution, those of the experimental data by the Poisson distribution.

Chapter 6

Estimation I

6.1 Introduction

We now leave the probability calculus and its simple applications and turn to the field of statistics. More precisely, in this chapter we are concerned with inferential statistics.

While the probability calculus, starting from distributions, predicts properties of random samples, in statistics, given a data sample, we look for a theoretical description of the population from which it has been derived by some random process. In the simplest case, the sample consists of one, but usually of several, observations, randomly drawn from a parent population. If not specified differently, we assume that the population is a collection of elements which all follow the same discrete or continuous distribution. Frequently, the sample consists of data collected in a measurement sequence. We either want to differentiate between several hypotheses, or to determine unknown parameters of a given theory and to quantify the result.

Let us look at two simple examples:

(1) At a table we find eight playing cards: two *kings*, three *ladies*, one *ten*, one *eight* and one *seven*. Do the cards belong to a set of Canasta cards or to a set of Skat cards?
(2) Five decays of a specific unstable particle, a muon, have been observed and the times between production and decay have been measured. We want to determine the mean lifetime τ of muon particles and the uncertainty of the result.

In our first example we would favor the Skat game because none of the cards *two* to *six* is present which, however, are part of Canasta card sets.

Assuming that the cards have been taken at random from a complete card set, we can summarize the available information in the following way: The probability to observe no card with value below *seven* in eight cards of a Canasta game is $L_C = (5/13)^8 = 4.8 \times 10^{-4}$ whereas it is $L_S = 1$ for a Skat game. We call these quantities *likelihoods*.[1] The likelihood indicates how well a given hypothesis is supported by the observation, but the likelihood alone is not sufficient for a decision in favor of one or another hypothesis. Additional considerations may play an important role. When the cards are located in a Swiss youth hostel, we would consider the hypothesis *Skat* more sceptically than when the cards are found in a pub at Hamburg. We therefore would weight our hypotheses with *prior probabilities* which quantify this additional piece of information. The prior probability or prior is the probability that is assigned to a hypothesis independent of the data. Prior probabilities are usually vague and often there is no bases to justify them. As a consequence, results depending on priors are usually model dependent. We will avoid to introduce prior probabilities and stay with likelihoods. Sometimes this is not possible. Then the results have to be interpreted conditional on the validity of the applied prior probabilities.

In our second example the number of hypotheses is infinite. To each value of the unknown parameter, i.e. to each different mean lifetime τ, corresponds a different prediction that is given by the exponential decay distribution $f(t|\tau) = \exp(-t/\tau)/\tau$. The probability to obtain the observed times is infinitely small but proportional to $f(t_i|\tau)$ for each of the times t_i, $i = 1, ..., 5$. We are intended – as in the previous example – to select τ such that the likelihood L_τ is maximal. The result is, as expected, that the parameter estimate (point estimate) τ is equal to the mean value of the observed times. We can also, as we will see below, estimate the uncertainty (interval estimate) δ_τ of the estimate. Also here, both, the point estimate and the interval estimate depend on a prior probability of τ that is difficult to quantify.

We shall address these problems in the following sections.

6.2 Inference with Given Prior

If prior information is available, it is possible by means of Bayes' theorem, to derive from a given sample probabilities for hypotheses or parameters.

[1]The term *likelihood* was first used by Sir Ronald Aylmer Fisher (1890-1962), British biologist and statistician.

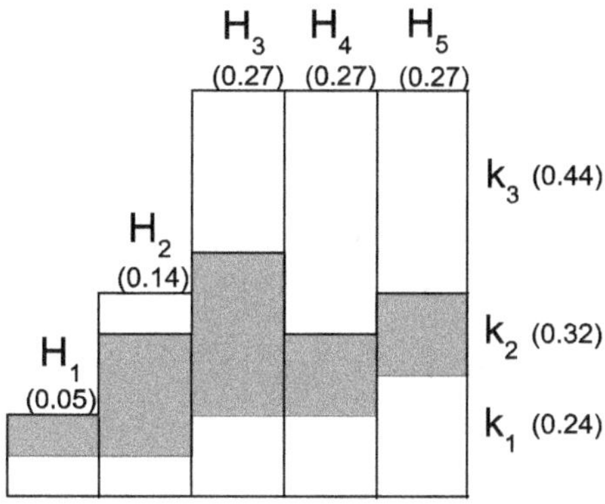

Fig. 6.1 Quantitative Venn diagram. The areas indicate the probabilities for certain combinations of hypotheses H_i and discrete events of type k_j. The marginal probabilities are given in brackets.

6.2.1 *Discrete Hypotheses*

In Chapter 1 we had shown that conditional probabilities fulfill the following relation (*Bayes' theorem*):

$$P\{A \cap B\} = P\{A|B\}P\{B\} = P\{B|A\}P\{A\}. \tag{6.1}$$

The probability $P\{A \cap B\}$, that both the properties A *and* B apply is equal to the probability $P\{B\}$, to find property B multiplied by the conditional probability $P\{A|B\}$ to find A, when B is realized. This is the first part of the relation above. The second part is analogous.

We apply this relation to a discrete random variable k and hypotheses H_i. The index denoting the hypothesis is interpreted as a random variable.[2]

We assume that the probability $P\{k|H_i\}$ to observe k is given for a finite number of mutually exclusive hypotheses H_i. Then we have

$$P\{k|H_i\}P\{H_i\} = P\{H_i|k\}P\{k\},$$
$$P\{H_i|k\} = \frac{P\{k|H_i\}P\{H_i\}}{P\{k\}}. \tag{6.2}$$

Here $P\{H_i\}$ is the assumed probability for the validity of hypothesis i before the observation happens, it is the a priori probability.

In Fig. 6.1 we illustrate relation (6.2) in form of a so called Venn diagram where in the present example 3 out of the 5 hypotheses have the same prior.

[2]In this case this is a categorical variable which denotes a certain class.

Each hypothesis bin is divided into 3 regions with areas proportional to the probabilities to observe $k = k_1$, $k = k_2$ and $k = k_3$, respectively. For example, when the observation is $k = k_2$ (gray) then the gray areas provide the relative probabilities of the validity of the corresponding hypotheses. In our example hypothesis H_3 is the most probable, H_1 the most unlikely.

The computation of $P\{k\}$ which is the marginal distribution of k, i.e. the probability of a certain observation, summed over all hypotheses, yields:

$$P\{k\} = \sum_i P\{k|H_i\}P\{H_i\}.$$

As required, $P\{H_i|k\}$ is normalized in such a way that the probability that any of the hypotheses is fulfilled is equal to one. We get

$$P\{H_i|k\} = \frac{P\{k|H_i\}P\{H_i\}}{\sum_j P\{k|H_j\}P\{H_j\}}. \tag{6.3}$$

In words: The probability for the validity of hypothesis H_i after the measurement k is equal to the prior $P\{H_i\}$ of H_i multiplied with the probability to observe k if H_i applies and divided by a normalization factor. When we are only interested in the relative probabilities of two different hypotheses H_i and H_j for an observation k, we have:

$$\frac{P\{H_i|k\}}{P\{H_j|k\}} = \frac{P\{k|H_i\}P\{H_i\}}{P\{k|H_j\}P\{H_j\}}.$$

Example 6.1. Bayes' theorem: Pion or kaon decay?

A muon has been detected. Does it originate from a pion or from a kaon decay? The decay probabilities inside the detector are known and are $P\{\mu|\pi\} = 0.02$ and $P\{\mu|K\} = 0.10$, respectively. The ratio of pions and kaons in the beam is $P\{\pi\} : P\{K\} = 3 : 1$. With these numbers we obtain:

$$\frac{P\{K|\mu\}}{P\{\pi|\mu\}} = \frac{0.10 \times 1}{0.02 \times 3} = \frac{5}{3},$$

$$\frac{P\{K|\mu\}}{P\{K|\mu\} + P\{\pi|\mu\}} = \frac{0.10 \times 1}{0.02 \times 3 + 0.10 \times 1} = 0.625.$$

The kaon hypothesis is more likely than the pion hypothesis. Its probability is 0.625.

6.2.2 *Continuous Parameters*

Now we extend our considerations to the case where the hypothesis index is replaced by a continuous parameter θ, i.e. we have an infinite number of hypotheses. Instead of probabilities we obtain probability densities. Bayes' theorem now reads

$$f(x,\theta) = f_x(x|\theta)\pi_\theta(\theta) = f_\theta(\theta|x)\pi_x(x) \tag{6.4}$$

which is just the relation 3.5.1.1 of Section 3.5, where f_x, f_θ are conditional distribution densities and $\pi_x(x)$, $\pi_\theta(\theta)$ are the marginal distributions of $f(x,\theta)$. We interpret the second equality

$$f_\theta(\theta|x) = \frac{f_x(x|\theta)\pi_\theta(\theta)}{\pi_x(x)} .$$

as a function of θ given a measurement x. To get rid of $\pi(x)$ which does not depend on θ we use the fact that the probability has to be normalized,

$$f_\theta(\theta|x) = \frac{f_x(x|\theta)\pi_\theta(\theta)}{\int_{-\infty}^{\infty} f_x(x|\theta)\pi_\theta(\theta)\mathrm{d}\theta} . \tag{6.5}$$

The term $f_x(x|\theta)$ is large for values of θ that lead to a large probability to produce the observation x. It is the likelihood function $L(\theta|x)$. The p.d.f. of θ on the left hand side is proportional to the likelihood function of θ and to the prior probability density of θ.

Example 6.2. Time of a decay with exponential prior

A detector with finite resolution registers at time t the decay of a K meson. The time resolution corresponds to a Gaussian with variance σ^2. We are interested in the time θ at which the decay occurred. The mean lifetime τ of kaons is known. The probability density for the parameter θ before the measurement, the prior, is $\pi(\theta) = e^{-\theta/\tau}/\tau$, $\theta \geq 0$. The probability density for t with θ fixed is the Gaussian. Applying (6.5) we obtain the probability density $f(\theta) = f(\theta|t)$ of the parameter θ,

$$f(\theta) = \frac{\mathrm{e}^{-(t-\theta)^2/(2\sigma^2)}\mathrm{e}^{-\theta/\tau}}{\int_0^{\infty} \mathrm{e}^{-(t-\theta)^2/(2\sigma^2)}\mathrm{e}^{-\theta/\tau}\mathrm{d}\theta} ,$$

which is displayed in Fig. 6.2. As a consequence of the exponential prior it is visibly shifted to the left with respect to the observation.

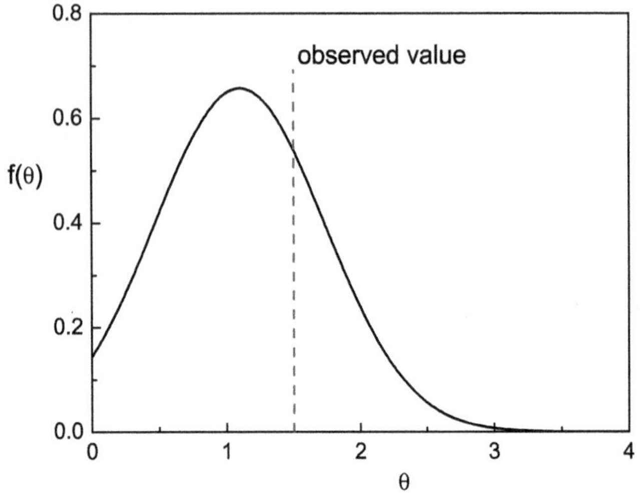

Fig. 6.2 Fit with known prior: Probability density for the true decay time. The maximum of the distribution is located at $\theta = 1$, the observed time is 1.5.

If the value of the probability density $f_x(x|\theta)$ in (6.5) varies much more rapidly with θ than the prior – this is the case when the observation restricts the parameter drastically – then to a good approximation the prior can be regarded as constant in the interesting region. We then have

$$f_\theta(\theta|x) \approx \frac{f_x(x|\theta)}{\int_{-\infty}^{\infty} f_x(x|\theta)\mathrm{d}\theta} .$$

In this approximation the probability density f_θ of the parameter corresponds to the normalized likelihood function.

In practice, f_θ often follows to a good approximation a normal distribution. The value $\hat{\theta}$ where f_θ is maximal then is the estimate of θ and the values where f_θ has decreased by the factor $\mathrm{e}^{1/2}$ define a standard deviation error interval and thus fix the uncertainty of the estimate $\hat{\theta}$.

6.3 Likelihood and the Likelihood Ratio

Usually we do not know the prior or our ideas about it are rather vague.

Example 6.3. Likelihood ratio: $V + A$ or $V - A$ reaction?

An experiment is performed to measure the energy E of muons produced in the decay of the tau lepton, $\tau^- \rightarrow \mu^- \nu_\tau \bar{\nu}_\mu$, to determine whether the decay corresponds to a $V - A$ or a $V + A$ matrix element. We know the corresponding normalized decay distributions $f_-(E)$ and $f_+(E)$. For a single observation E' we can compute the likelihood ratio $R_L = L_+/L_-$ of the likelihoods $L_+ = f_+(E')$, $L_- = f_-(E')$. But how should we choose the prior densities for the two alternative hypotheses? In this example it would not make sense to quantify the prejudices for the two hypothesis and to compute the resulting probabilities. One would rather publish only the ratio R_L.

In the absence of prior information the likelihood ratio is the only element which we have, to judge the relative virtues of alternative hypotheses. According to a lemma of J. Neyman[3] and E. Pearson[4] there is no other more powerful quantity to discriminate between competing hypotheses. (see Chapter 10).

We summarize the findings.

Definition: The likelihood L_i of a hypothesis H_i, to which corresponds a probability density $f_i(x) \equiv f(x|H_i)$ or a discrete probability distribution $W_i(k) \equiv P\{k|H_i\}$, when the observation x, k, respectively, has been realized, is equal to

$$L_i \equiv L(H_i|x) = f_i(x)$$

and

$$L_i \equiv L(H_i|k) = W_i(k),$$

respectively. For a continuous parameter θ and a parameter dependent p.d.f. $f(x|\theta)$ or a discrete probability distribution $W(k|\theta)$ and observations x, k, the corresponding likelihoods are

$$L(\theta) \equiv L(\theta|x) = f(x|\theta),$$
$$L(\theta) \equiv L(\theta|k) = W(k|\theta).$$

[3] Jercy Neyman (1894-1981) mathematician and statistician
[4] Egon Pearson (1895-1980) statistician

While the likelihood is related to the validity of a hypothesis given an observation, the p.d.f. is related to the probability to observe a variate for a given hypothesis. In our notation, the quantity which is considered as fixed is placed behind the bar while the random variable is located left of it. When both quantities are fixed the function values of both the likelihood and the p.d.f. are equal. To attribute a likelihood makes sense only if alternative hypotheses, either discrete or differing by parameters, can apply. If the likelihood depends on one or several continuous parameters, we have a *likelihood function*.

Remark: The likelihood function is not a probability density of the parameter. It has no differential element like $\mathrm{d}\theta$ involved and does not obey the laws of probability. To distinguish it from probability, R. A. Fisher had invented the name likelihood. Multiplied by a prior and normalized, a probability density of the parameter is obtained. Statisticians call this *inverse probability* or *probability of causes* to emphasize that compared to the *direct probability* where the parameter is known and the chances of an event are described, we are in the inverse position where we have observed the event and want to associate probabilities to the various causes that could have led to the observation.

As already stated above, the likelihood of a certain hypothesis is large if the observation is probable for this hypothesis. It measures how strongly a hypothesis is supported by the data. If an observation is very unlikely the validity of the hypothesis is doubtful – however this classification applies only when there is an alternative hypothesis with larger likelihood. Relevant are only ratios of likelihoods.

Usually experiments provide a sample of N independent observations x_i which all follow independently the same p.d.f. $f(x|\theta)$ which depends on the unknown parameter θ. The combined p.d.f. $\tilde{f}$ then is equal to the product of the N simple p.d.f.s,

$$\tilde{f}(x_1, \ldots, x_N|\theta) = \prod_{i=1}^{N} f(x_i|\theta) .$$

For discrete variates we have the corresponding relation,

$$\tilde{W}(k_1, \ldots, k_N|\theta) = \prod_{i=1}^{N} W(k_i|\theta) .$$

For all values of θ the function $\tilde{f}$ evaluated for the sample $x_1, \ldots, x_N$ is equal to the *likelihood* $\tilde{L}$,

$$\begin{aligned}\tilde{L}(\theta) &\equiv \tilde{L}(\theta|x_1,x_2,\ldots,x_N)\\ &= \tilde{f}(x_1,x_2,\ldots,x_N|\theta)\\ &= \prod_{i=1}^{N} f(x_i|\theta)\\ &= \prod_{i=1}^{N} L(\theta|x_i).\end{aligned}$$

The same relation also holds for discrete variates:

$$\begin{aligned}\tilde{L}(\theta) &\equiv \tilde{L}(\theta|k_1,\ldots,k_N)\\ &= \prod_{i=1}^{N} W(k_i|\theta)\\ &= \prod_{i=1}^{N} L(\theta|k_i).\end{aligned}$$

When we have a sample of independent observations, it is convenient to consider the logarithm of the likelihood. It is called *log-likelihood*. It is equal to

$$\ln \tilde{L}(\theta) = \sum_{i=1}^{N} \ln\left[f(x_i|\theta)\right]$$

for continuous variates. A corresponding relation holds for discrete variates.

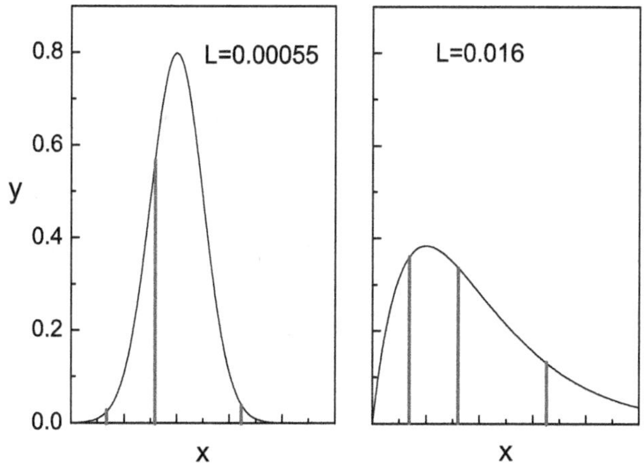

Fig. 6.3 Likelihood of three observations and two hypotheses with different p.d.f.s.

Figure 6.3 illustrates the notion of likelihood in a concrete case of two hypotheses which predict different p.d.f.s of the variate x. For a sample of three observations we present the values of the likelihood, i.e. the products of the three corresponding p.d.f. values. The broad p.d.f. in the right hand picture matches better. Its likelihood is about thirty times higher than that of the left hand hypothesis.

So far we have considered the likelihood of samples of i.i.d. variates. Also the case where two independent experiments A, B measure the same quantity x is of considerable interest. The combined likelihood L is just the product of the individual likelihoods. From

$$f(x_1, x_2|\theta) = f_A(x_1|\theta) f_B(x_2|\theta),$$
$$L(\theta) = f(x_1, x_2|\theta),$$

follows

$$L = L_A L_B,$$
$$\ln L = \ln L_A + \ln L_B.$$

Example 6.4. Likelihood ratio of Poisson frequencies

We observe 5 decays and want to compute the relative probabilities for three hypotheses. Prediction H_1 assumes a Poisson distribution with expectation value 2, H_2 and H_3 have expectation values 9 and 20, respectively. The likelihoods following from the Poisson distribution $\mathcal{P}_\lambda(k)$ are:

$$\begin{array}{lclcl} L_1 & = & \mathcal{P}_2(5) & \approx & 0.036, \\ L_2 & = & \mathcal{P}_9(5) & \approx & 0.061, \\ L_3 & = & \mathcal{P}_{20}(5) & \approx & 0.00005. \end{array}$$

We can form different likelihood ratios. If we are interested for example in hypothesis 2, then the quotient $L_2/(L_1 + L_2 + L_3) \approx 0.63$ is relevant. If we observe in a second measurement in the same time interval 8 decays, we obtain:

$$\begin{array}{lclclcl} L_1 & = & \mathcal{P}_2(5)\mathcal{P}_2(8) & = & \mathcal{P}_4(13) & \approx & 6.4 \cdot 10^{-3}, \\ L_2 & = & \mathcal{P}_9(5)\mathcal{P}_9(8) & = & \mathcal{P}_{18}(13) & \approx & 5.1 \cdot 10^{-2}, \\ L_3 & = & \mathcal{P}_{20}(5)\mathcal{P}_{20}(8) & = & \mathcal{P}_{40}(13) & \approx & 6.1 \cdot 10^{-7}. \end{array}$$

The likelihood ratio $L_2/(L_1 + L_2 + L_3) \approx 0.89$ (for H_1 and H_3 correspondingly 0.11 and 10^{-5}) now is much more significant. The fact that all values

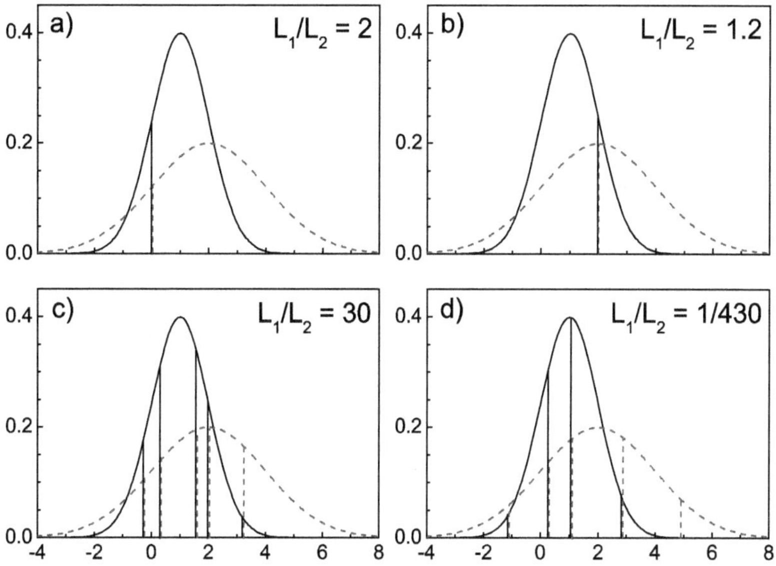

Fig. 6.4 Likelihood ratio for two normal distributions. Top: 1 observation, bottom: 5 observations.

L_i are small is unimportant because one of the three hypotheses has to be valid.

We now turn to hypotheses with probability densities.

Example 6.5. Likelihood ratio of normal distributions

We compare samples drawn from one out of two alternative normal distributions with different expectation values and variances (Fig. 6.4).

$$f_1 = \frac{1}{\sqrt{2\pi}1} e^{-(x-1)^2/2} ,$$

$$f_2 = \frac{1}{\sqrt{2\pi}2} e^{-(x-2)^2/8} .$$

(a) Initially the sample consists of a single observation at $x = 0$, for both cases one standard deviation off the mean values of the two distributions

(Fig. 6.4a):

$$\frac{L_1}{L_2} = 2\frac{e^{-1/2}}{e^{-4/8}} = 2.$$

(b) Now we place the observation at $x = 2$, the maximum of the second distribution (Fig. 6.4b):

$$\frac{L_1}{L_2} = 2\frac{e^{-1/2}}{e^{-0}} = 1.2.$$

(c) We now consider five observations which have been taken from distribution f_1 (Fig. 6.4c) and distribution f_2, respectively (Fig. 6.4d). We obtain the likelihood ratios

$$\begin{aligned} L_1/L_2 &= 30 \qquad \text{(Fig. 6.4c)),} \\ L_1/L_2 &= 1/430 \ \text{(Fig. 6.4d).} \end{aligned}$$

It turns out that small distributions are easier to exclude than broad ones. On the other hand we get in case (b) a preference for distribution 1 even though the observation is located right at the center of distribution 2.

Example 6.6. Likelihood ratio for two decay time distributions

A sample of N decay times t_i has been recorded in the time interval $t_{\min} < t < t_{\max}$. The times are expected to follow either an exponential distribution $f_1(t) \sim e^{-t/\tau}$ (hypothesis 1) or a uniform distribution $f_2(t) = const.$ (hypothesis 2). How likely are H_1, H_2? First we have to normalize the p.d.f.s:

$$\begin{aligned} f_1(t) &= \frac{1}{\tau}\frac{e^{-t/\tau}}{e^{-t_{\min}/\tau} - e^{-t_{\max}/\tau}}, \\ f_2(t) &= \frac{1}{t_{\max} - t_{\min}}. \end{aligned}$$

The likelihoods are equal to the product of the p.d.f.s at the observations:

$$\begin{aligned} L_1 &= \left[\tau\left(e^{-t_{\min}/\tau} - e^{-t_{\max}/\tau}\right)\right]^{-N} \exp\left(-\sum_{i=1}^{N} t_i/\tau\right), \\ L_2 &= 1/(t_{\max} - t_{\min})^N. \end{aligned}$$

With $\bar{t} = \sum t_i/N$ the mean value of the times, we obtain the likelihood ratio

$$\frac{L_1}{L_2} = \left(\frac{t_{\max} - t_{\min}}{\tau(\mathrm{e}^{-t_{\min}/\tau} - \mathrm{e}^{-t_{\max}/\tau})}\right)^N \mathrm{e}^{-N\bar{t}/\tau}.$$

6.4 Prior Probability and Bayesian Inference

If we apply Bayes principle and introduce prior probabilities, we are able to infer probabilities for competing hypotheses and p.d.f.s for continuous parameters. We repeat the relevant relations (6.3) and (6.5) and replace the probability distributions by the corresponding likelihood functions,

$$P\{H_i|k\} = \frac{\pi(H_i)L\{H_i,\}}{\sum_j \pi(H_j)L\{H_j|k\}},$$

$$f_\theta(\theta|x) = \frac{\pi_\theta(\theta)L(\theta|x)}{\int_{-\infty}^{\infty} \pi_\theta(\theta)f_x(x|\theta)\mathrm{d}\theta}.$$

To distinguish the posterior distribution $f_\theta(\theta|x)$ from a genuine distribution that is independent of prior information, it is called credibility distribution.

The prior probability or prior density is based either on previous knowledge (experience), symmetry considerations, mathematical constructions (reference priors) or the *principle of indifference.* A detailed discussion of prior probability is presented in [30].

For discrete hypotheses follows from the *principle of indifference* that the prior probabilities are set equal for all hypotheses. A often cited example is the following: A ball is hidden in one out of ten equal boxes. Of course, one would attribute to all boxes the same probability to contain the ball. However, as seen in the card example, mentioned in the introduction, prior probabilities are not necessarily equal.

For distributions depending on parameters, the *principle of indifference* leads to uniform prior densities, often cited as uninformative. However, uniform prior probabilities are not invariant against variable transformations. A prior that is uniform in the lifetime τ is not uniform in the decay rate $\gamma = 1/\tau$. Experience tells us that in particle physics the distribution of decay rates are more uniform than those of the lifetimes. But in practice this consideration is irrelevant.

Often the likelihood function is so narrow, that the probability density is insensitive to the shape of acceptable prior densities, see Example 1.

Bayesian statistics is indispensable in decision theory. In sciences where a major requirement is to document results, it is less important.

6.5 The Maximum Likelihood Method for Parameter Inference

In the previous examples we have compared a sample with alternative hypotheses. Now we are interested in the estimation of the parameter of a parameter dependent distribution. As in the discrete case, in the absence of a given prior probability for the parameter, *the only available piece of information which allows us to judge different parameter values is the likelihood function.* A formal justification for this assertion is given by the *likelihood principle* (LP) which states that the likelihood function exhausts all the information contained in the observations related to the parameters. We shall discuss it in the following section. It is then plausible to choose as estimate of a parameter the location of the maximum of the likelihood function, the *maximum likelihood estimate, MLE.*

Remark that the MLE, as well as likelihood intervals, are invariant against transformations of the parameter. The likelihood is not a p.d.f. but a function of the parameter and therefore $L(\theta) = L'(\theta')$ for $\theta'(\theta)$. Thus a likelihood analysis estimating, for example, the mass of a particle will give the same result as that inferring the mass squared, and estimates of the decay rate γ and mean life $\tau = 1/\gamma$ will be consistent.

Here and in the following sections we assume that the likelihood function is continuous and differentiable and has exactly one maximum inside the valid range of the parameter. This condition is fulfilled in the majority of all cases.

In most cases it is impossible to compute the location of the maximum of the likelihood function analytically. To simplify the numerical computation, still linear approximations (e.g. linear regression) are used quite frequently. These methods find the solution by matrix operations and iteration. They are dispensable nowadays. With common PCs and maximum searching programs the maximum of a function of tens of parameters can be determined without problem, given enough observations to fix it.

6.5.1 *The Recipe for a Single Parameter*

We proceed according to the following recipe. Given a sample of N i.i.d. observations $\{x_1, \ldots, x_N\}$ from a p.d.f. $f(x|\theta)$ with unknown parameter θ,

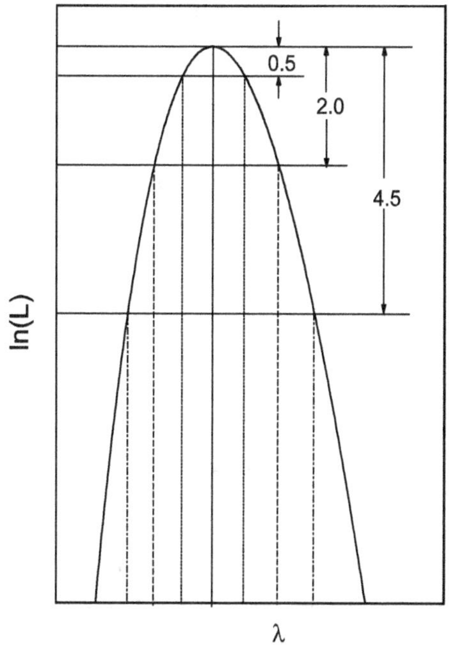

Fig. 6.5 Log-likelihood function and uncertainty limits for 1, 2, 3 standard deviations.

we form the likelihood or its logarithm, respectively, in the following way:

$$L(\theta) = \prod_{i=1}^{N} f(x_i|\theta), \tag{6.6}$$

$$\ln L(\theta) = \sum_{i=1}^{N} \ln f(x_i|\theta). \tag{6.7}$$

In most cases the likelihood function resembles a bell shaped Gaussian and $\ln L(\theta)$ approximately a downwards open parabola (see Fig. 6.5). This approximation is especially good for large samples.

To find the maximum of L ($\ln L$ and L have their maxima at the same location), we derive the log-likelihood[5] with respect to the parameter and set the derivative equal to zero. The value $\hat{\theta}$, that satisfies the equation

[5]The advantage of using the log-likelihood compared to the normal likelihood is that we do not need to differentiate a product but a sum which is more convenient.

which we obtain in this way is the MLE of θ:

$$\frac{\mathrm{d}\ln L}{\mathrm{d}\theta}|_{\hat{\theta}} = 0. \tag{6.8}$$

Since only likelihood ratios are important, factors in the likelihood or summands in the log-likelihood which are independent of the parameter of interest can be omitted. In most cases we are not able to compute analytically the location of the maximum of the likelihood, the point estimate, and have to resort to numerical methods.

The point estimate has to be accompanied by an error interval. Point estimate and error interval form an ensemble and cannot be discussed separately. Choosing as point estimate the value that maximizes the likelihood function it is natural to include inside the error limits parameter values with higher likelihood than all parameters that are excluded. This prescription leads to so-called likelihood ratio error intervals.

We will discuss the error interval estimation in a separate chapter, but fix the error limit already now by definition:

Definition: The limits of a standard error interval are located at the parameter values where the likelihood function has decreased from its maximum by a factor $\mathrm{e}^{1/2}$. For two and three standard error intervals the factors are e^2 and $\mathrm{e}^{4.5}$. This choice corresponds to differences for the log-likelihood of 0.5 for one, of 2 for two and of 4.5 for three standard error intervals as illustrated in Fig. 6.5. We assume that these limits exist inside the parameter range.

The reason for this definition is the following: As already mentioned, asymptotically, when the sample size N tends to infinity, under very general conditions the likelihood function approaches a Gaussian and becomes proportional to the probability density of the parameter (for a proof, see Appendix A.3). Then our error limit corresponds exactly to the standard deviation of the p.d.f., i.e. the square root of the variance of the Gaussian. We keep the definition also for non normally shaped likelihood functions and small sample sizes. Then we get usually asymmetric error limits. There are two reasons for this extension which contradicts the definition that we have chosen in Chapter 4: The necessity to describe asymmetric errors and the requirement that the point estimate and its error are invariant under a transformation of the parameter. We shall come back to the latter argument below.

6.5.2 Examples

Example 6.7. Maximum likelihood estimate (MLE) of the mean life of an unstable particle

Given be N decay times t_i of an unstable particle with unknown mean life τ. For an exponential decay time distribution

$$f(t|\gamma) = \gamma \mathrm{e}^{-\gamma t}$$

with $\gamma = 1/\tau$ the likelihood is

$$\begin{aligned} L &= \gamma^N \prod_{i=1}^{N} \mathrm{e}^{-\gamma t_i} \\ &= \gamma^N \mathrm{e}^{-\sum_{i=1}^{N} \gamma t_i}, \\ \ln L &= N \ln \gamma - \gamma \sum_{i=1}^{N} t_i . \end{aligned}$$

The estimate $\hat{\gamma}$ satisfies

$$\begin{aligned} \frac{\mathrm{d} \ln L}{\mathrm{d}\gamma}|_{\hat{\gamma}} &= 0, \\ 0 &= \frac{N}{\hat{\gamma}} - \sum_{i=1}^{N} t_i, \\ \hat{\tau} = \hat{\gamma}^{-1} &= \sum_{i=1}^{N} t_i/N = \bar{t}. \end{aligned}$$

Thus the estimate is just equal to the mean value $\bar{t}$ of the observed decay times. In practice, the full range up to infinitely large decay times is not always observable. If the measurement is restricted to an interval $0 < t < t_{\max}$, the p.d.f. changes, it has to be re-normalized:

$$\begin{aligned} f(t|\gamma) &= \frac{\gamma \mathrm{e}^{-\gamma t}}{1 - \mathrm{e}^{-\gamma t_{\max}}}, \\ \ln L &= N \left[\ln \gamma - \ln(1 - \mathrm{e}^{-\gamma t_{\max}}) \right] - \gamma \sum_{i=1}^{N} t_i . \end{aligned}$$

The maximum is now located at the estimate $\hat{\gamma}$, which fulfils the relation

$$\begin{aligned} 0 &= N \left(\frac{1}{\hat{\gamma}} - \frac{t_{\max} \mathrm{e}^{-\hat{\gamma} t_{\max}}}{1 - \mathrm{e}^{-\hat{\gamma} t_{\max}}} \right) - \sum_{i=1}^{N} t_i, \\ \hat{\tau} &= \bar{t} + \frac{t_{\max} \mathrm{e}^{-t_{\max}/\hat{\tau}}}{1 - \mathrm{e}^{-t_{\max}/\hat{\tau}}}, \end{aligned}$$

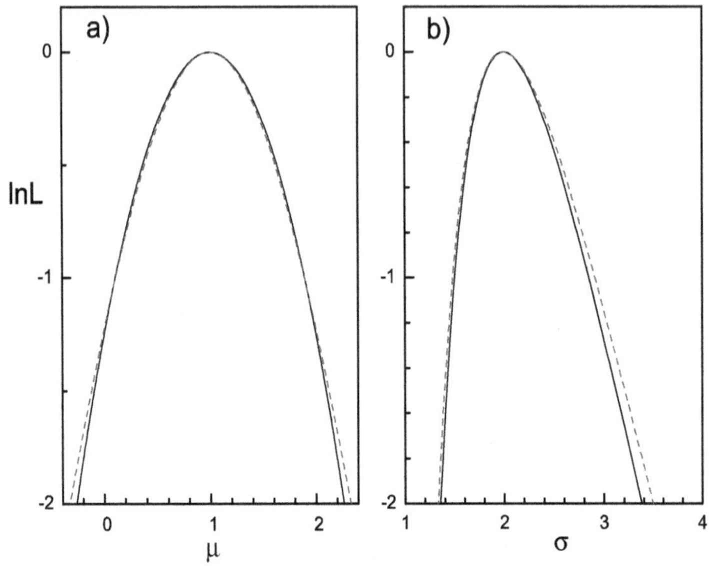

Fig. 6.6 Log-likelihood functions for the parameters of a normal distribution: (a) for the mean μ with know width (solid curve) and unknown width (dashed curve), (b) for the width σ with known mean (solid curve) and unknown mean (dashed curve). The curves represent expected likelihoods for 10 events.

which has to be evaluated numerically. If the time interval is not too short, $t_{\max} > \tau$, an iterative computation lends itself: The correction term at the right hand side is neglected in zeroth order. At the subsequent iterations we insert in this term the value τ of the previous iteration. We notice that the estimate again depends solely on the mean value $\bar{t}$ of the observed decay times. The quantity $\bar{t}$ is a *sufficient statistic.* We will explain this term in more detail later. The case with also a lower bound $t_{\min}$ of t can be reduced easily to the previous one by transforming the variable to $t' = t - t_{\min}$.

In the following examples we discuss the likelihood functions and the MLEs of the parameters of the normal distribution with mean μ and standard deviation σ evaluated for N normally distributed events.

Example 6.8. MLE of the mean value of a normal distribution with known width

Given are N observation x_i drawn from a normal distribution of known width σ and mean value μ to be estimated:

$$f(x|\mu) = \frac{1}{\sqrt{2\pi}\sigma} \exp\left[-\frac{(x-\mu)^2}{2\sigma^2}\right],$$

$$L(\mu) = \prod_{i=1}^{N} \frac{1}{\sqrt{2\pi}\sigma} \exp\left[-\frac{(x_i-\mu)^2}{2\sigma^2}\right],$$

$$\ln L(\mu) = -\sum_{i=1}^{N} \frac{(x_i-\mu)^2}{2\sigma^2} + const \tag{6.9}$$

$$= -N\frac{\overline{x^2} - 2\overline{x}\mu + \mu^2}{2\sigma^2} + const.$$

The log-likelihood function is a parabola. It is shown in Fig. 6.6a for $\sigma = 2$. Deriving it with respect to the unknown parameter μ and setting the result equal to zero, we get

$$\hat{\mu} = \overline{x}.$$

The likelihood estimate $\hat{\mu}$ for the mean of the normal distribution is equal to the arithmetic mean $\overline{x}$ of the sample. It is independent of σ, but σ determines the width of the likelihood function and the standard error $\delta_\mu = \sigma/\sqrt{N}$.

Example 6.9. MLE of the width of a normal distribution with given mean

Given are now N observations x_i which follow a normal distribution of unknown width σ to be estimated for known mean.

$$L(\sigma) = \prod_{i=1}^{N} \frac{1}{\sqrt{2\pi}\sigma} \exp\left(-\frac{(x_i-\mu)^2}{2\sigma^2}\right),$$

$$\ln L(\sigma) = -N\left(\frac{1}{2}\ln 2\pi + \ln\sigma\right) - \sum_{i=1}^{N} \frac{(x_i-\mu)^2}{2\sigma^2}$$

$$= -N\left[\ln\sigma + \frac{\overline{(x-\mu)^2}}{2\sigma^2}\right] + const.$$

The log-likelihood function for our numerical values is presented in Fig. 6.6b. Deriving it with respect to the parameter of interest and setting the result equal to zero we find

$$\hat{\sigma} = \sqrt{\overline{(x-\mu)^2}}.$$

Again we obtain a well known result. The mean square deviation of the sample values provides an estimate for the width of the normal distribution. This relation is the usual distribution-free estimate of the standard deviation if the expected value is known. The error bounds from the drop of the log-likelihood function by 1/2 become asymmetric. Solving the respective transcendental equation, neglecting higher orders in $1/N$, one finds

$$\delta_\sigma^{\pm} = \frac{\hat{\sigma}\sqrt{\frac{1}{2N}}}{1 \mp \sqrt{\frac{1}{2N}}}.$$

Example 6.10. MLE of the mean of a normal distribution with unknown width

The solution of this problem can be taken from Section 3.6.11 where we found that $t = (\overline{x} - \mu)/s$ with $s^2 = \sum(x_i - \overline{x})^2/[N(N-1)] = v^2/(N-1)$ follows the Student's distribution with $N-1$ degrees of freedom. Here v is the empirical width.

$$h(t|N-1) = \frac{\Gamma(N/2)}{\Gamma((N-1)/2)\sqrt{\pi(N-1)}} \left(1 + \frac{t^2}{N-1}\right)^{-\frac{N}{2}}.$$

The corresponding log-likelihood is

$$\ln L(\mu) = -\frac{N}{2} \ln\left[1 + \frac{(\overline{x}-\mu)^2}{v^2}\right]$$

with the maximum $\mu = \overline{x}$. It corresponds to the dashed curve in Fig. 6.6a. From the drop of $\ln L$ by 1/2 we get now for the standard error squared the expression

$$\delta_\mu^2 = (e^{1/N} - 1)v^2.$$

This becomes for large N, after expanding the exponential function, very similar to the expression for the standard error in case with known width, but with σ exchanged by v.

Example 6.11. MLE of the width of a normal distribution with unknown mean

Obviously, shifting a sample changes the mean value but not the true or the empirical variance $v^2 = \overline{(x - \overline{x})^2}$. Thus the empirical variance v^2 can only depend on σ and not on μ. Without going into the details of the calculation, we state that Nv^2/σ^2 follows a χ^2-distribution of $N-1$ degrees of freedom.

$$f(v^2|\sigma) = \frac{N}{\Gamma\left[(N-1)/2\right] 2\sigma^2} \left(\frac{Nv^2}{2\sigma^2}\right)^{(N-3)/2} \exp\left(-\frac{Nv^2}{2\sigma^2}\right)$$

with the log-likelihood

$$\ln L(\sigma) = -(N-1)\ln\sigma - \frac{Nv^2}{2\sigma^2}$$

corresponding to the dashed curve in Fig. 6.6b. (The numerical value of the true value of μ was chosen such that the maxima of the two curves are located at the same value in order to simplify the comparison.) The MLE is

$$\hat{\sigma}^2 = \frac{N}{N-1} v^2 ,$$

in agreement with our result (3.15). For the asymmetric error limits we find in analogy to example 6.9

$$\delta_\sigma^\pm = \frac{\hat{\sigma}\sqrt{\frac{1}{2(N-1)}}}{1 \mp \sqrt{\frac{1}{2(N-1)}}} .$$

6.5.3 *Likelihood Inference for Several Parameters*

We can extend our concept easily to several parameters λ_k which we combine to a vector $\boldsymbol{\lambda} = \{\lambda_1, \ldots, \lambda_K\}$.

$$L(\boldsymbol{\lambda}) = \prod_{i=1}^{N} f(x_i|\boldsymbol{\lambda}), \tag{6.10}$$

$$\ln L(\boldsymbol{\lambda}) = \sum_{i=1}^{N} \ln f(x_i|\boldsymbol{\lambda}). \tag{6.11}$$

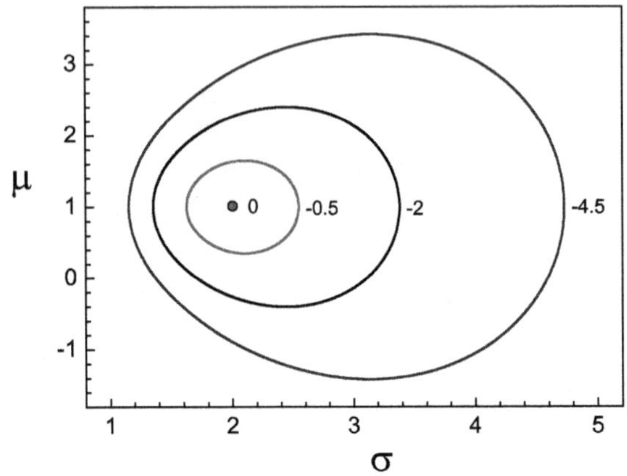

Fig. 6.7 MLE of the parameters of a normal distribution and lines of constant log-likelihood. The numbers indicate the values of the log-likelihood relative to the maximum.

To find the maximum of the likelihood function, we set the partial derivatives equal to zero. Those values $\hat{\lambda}_k$ which satisfy the system of equations obtained this way, are the MLEs $\hat{\lambda}_k$ of the parameters λ_k:

$$\frac{\partial \ln L}{\partial \lambda_k}|_{\hat{\lambda}_1,\ldots,\hat{\lambda}_K} = 0. \tag{6.12}$$

The error interval is now to be replaced by an error volume with its surface defined again by the drop of $\ln L$ by $1/2$:

$$\ln L(\widehat{\boldsymbol{\lambda}}) - \ln L(\boldsymbol{\lambda}) = 1/2.$$

We have to assume, that this defines a closed surface in the parameter space, in two dimensions just a closed contour, as shown in the next example.

Example 6.12. MLEs of the mean value and the width of a normal distribution

Given are N observations x_i which follow a normal distribution where now both the width and the mean value μ are unknown. As above, the log-likelihood is

$$\ln L(\mu, \sigma) = -N\left[\ln \sigma + \frac{\overline{(x-\mu)^2}}{2\sigma^2}\right] + const.$$

The derivation with respect to the parameters leads to the results:

$$\hat{\mu} = \frac{1}{N}\sum_{i=1}^{N} x_i = \overline{x},$$

$$\hat{\sigma}^2 = \frac{1}{N}\sum_{i=1}^{N}(x_i - \hat{\mu})^2 = \overline{(x-\overline{x})^2} = v^2.$$

The MLE and log-likelihood contours for a sample of 10 events with empirical mean values $\overline{x} = 1$ and $\overline{x^2} = 5$ are depicted in Fig. 6.7. The innermost line encloses the standard error area. If one of the parameters, for instance $\mu = \mu_1$ is given, the log-likelihood of the other parameter, here σ, is obtained by the cross section of the likelihood function at $\mu = \mu_1$.

Similarly, any other relation between μ and σ defines a curve in Fig. 6.7 along which a one-dimensional likelihood function is defined.

Remark: Frequently, we are interested only in one of the parameters, and we want to eliminate the others, the nuisance parameters. How to achieve this, will be discussed in Section 7.8. Generally, it is not allowed to use the MLE of a single parameter in the multi-parameter case separately, ignoring the other parameters. While in the previous example $\hat{\sigma}$ is the correct estimate of σ if $\hat{\mu}$ applies, the solution for the estimate and its likelihood function independent of μ has been given in example 6.11 and that of μ independent of σ in example 6.10.

Example 6.13. Determination of the axis of a given distribution of directions

(This example has been borrowed from the book of L. Lyons [7].) Given are the directions of N tracks by the unit vectors $\boldsymbol{e}_k$. The distribution of the direction cosines $\cos\alpha_i$ with respect to an axis $\boldsymbol{u}$ corresponds to

$$f(\cos\alpha) = \frac{3}{8}(1+\cos^2\alpha).$$

We search for the direction of the axis. The axis $\boldsymbol{u}(u_1, u_2, u_3)$ is parametrized by its components, the direction cosines u_k. (There are only two independent parameters u_1, u_2 because $u_3 = \sqrt{1-u_1^2-u_2^2}$ depends on u_1 and u_2.) The log-likelihood function is

$$\ln L = \sum_{i=1}^{N} \ln(1+\cos^2\alpha_i),$$

where the values $\cos\alpha_i = \boldsymbol{u}\cdot\boldsymbol{e}_i$ depend on the parameters of interest, the direction cosines. Maximizing $\ln L$, yields the parameters u_1, u_2. We omit the calculation.

Example 6.14. Likelihood analysis for a signal with a linear background

We want to fit a normal distribution with a linear background to a given sample. (The procedure for a background described by a higher order polynomial is analogous.) The p.d.f. is

$$f(x) = \theta_1 x + \theta_2 + \theta_3 \mathcal{N}(x|\mu,\sigma).$$

Here $\mathcal{N}$ is the normal distribution with unknown mean μ and standard deviation σ. The other parameters are not independent because f has to be normalized in the given interval $x_{\rm min} < x < x_{\rm max}$. Thus we can eliminate one parameter. Assuming that the normal distribution is negligible outside the interval, the norm D is:

$$D = \frac{1}{2}\theta_1(x_{\rm max}^2 - x_{\rm min}^2) + \theta_2(x_{\rm max} - x_{\rm min}) + \theta_3.$$

The normalized p.d.f. is therefore

$$f(x) = \frac{\theta_1' x + \theta_2' + \mathrm{N}(x|\mu,\sigma)}{\frac{1}{2}\theta_1'(x_{\rm max}^2 - x_{\rm min}^2) + \theta_2'(x_{\rm max} - x_{\rm min}) + 1},$$

with the new parameters $\theta_1' = \theta_1/\theta_3$ and $\theta_2' = \theta_2/\theta_3$. The likelihood function is obtained in the usual way by inserting the observations of the sample into $\ln L = \sum \ln f(x_i|\theta_1', \theta_2', \mu, \sigma)$. Maximizing this expression, we obtain the four parameters and from those the fraction of signal events $S = \theta_3/D$:

$$S = \left[1 + \frac{1}{2}\theta_1'(x_{\rm max}^2 - x_{\rm min}^2) + \theta_2'(x_{\rm max} - x_{\rm min})\right]^{-1}.$$

6.5.4 *Visualization of Complicated Likelihood Functions*

If the likelihood function deviates considerably from a normal distribution in the vicinity of its maximum, e.g. contains several significant maxima, then it is not appropriate to parametrize it by the maximum and error limits. In this situation the full function or a likelihood map should be

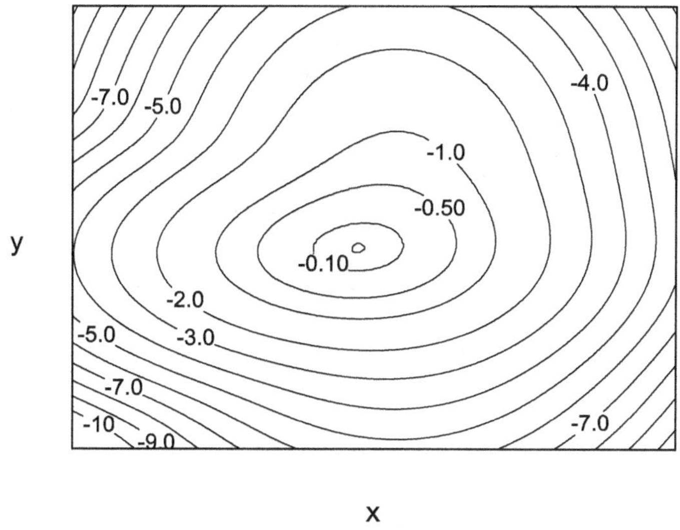

Fig. 6.8 Likelihood contours.

presented. Such a map is shown in Fig. 6.8. The presentation reflects very well which combinations of the parameters are supported by the data. Under certain conditions, with more than two parameters, several projections have to be considered.

6.5.5 *Combining Measurements*

The likelihood method makes it possible to combine experimental results in a extremely simple and at the same time optimal way. We have just to add the log-likelihoods. The experimental data can originate from completely heterogeneous experiments because no assumptions about the p.d.f.s of the individual experiments enter, except that they are independent of each other.

6.6 Likelihood and Information

6.6.1 *Sufficiency*

In a previous example, we have seen that the likelihood function for a sample of exponentially distributed decay times is a function only of the sample mean. In fact, in many cases, the i.i.d. individual elements of a sample

$\{x_1, \ldots, x_N\}$ can be combined to fewer quantities, ideally to a single one without affecting the estimation of the interesting parameters. The set of these quantities which are functions of the observations is called a *sufficient statistic*. The sample itself is of course a sufficient, while uninteresting statistic.

According to R. A. Fisher, a statistic is sufficient for one or several parameters, if by addition of arbitrary other statistics of the same data sample, the parameter estimation cannot be improved. More precise is the following definition [1]: A statistic $\boldsymbol{t}(x_1, \ldots, x_N) \equiv \{t_1(x_1, \ldots, x_N), \ldots, t_M(x_1, \ldots, x_N)\}$ is sufficient for a parameter set $\boldsymbol{\theta}$, if the distribution of a sample $\{x_1, \ldots, x_N\}$, given $\boldsymbol{t}$, does not depend on $\boldsymbol{\theta}$:

$$f(x_1, \ldots, x_N|\boldsymbol{\theta}) = g(t_1, \ldots, t_M|\boldsymbol{\theta})h(x_1, \ldots, x_N). \tag{6.13}$$

The distribution $g(\boldsymbol{t}|\boldsymbol{\theta})$ then contains all the information which is relevant for the parameter estimation. This means that for the estimation process we can replace the sample by the sufficient statistic. In this way we may reduce the amount of data considerably. In the standard situation where all parameter components are constraint by the data, the dimension of $\boldsymbol{t}$ must be larger or equal to the dimension of the parameter vector $\boldsymbol{\theta}$. Every set of uniquely invertible functions of $\boldsymbol{t}$ is also a sufficient statistic.

The relevance of sufficiency is expressed in a different way in the so-called *sufficiency principle*:

If two different sets of observations have the same values of a sufficient statistic, then the inference about the unknown parameter should be the same.

Of special interest is a *minimal sufficient statistic*. It consists of a minimal number of components, ideally only of one element per parameter.

In what follows, we consider the case of a one-dimensional sufficient statistic $t(x_1, \ldots, x_N)$ and a single parameter θ. The likelihood function can according to (6.13) be written in the following way:

$$L = L_1(\theta|t(\boldsymbol{x})) \cdot L_2(\boldsymbol{x}). \tag{6.14}$$

It is easy to realize that the second factor L_2 which is independent of θ,[6] has no bearing on the likelihood ratios of different values of θ. We obtain a data reduction of N to 1. This means that all samples of size N which have the same value of the statistic t lead to the same likelihood function and thus to the same MLE and the same likelihood ratio interval.

[6]Note that also the domain of x has to be independent of θ.

If a minimal sufficient statistic of one element per parameter exists, then the MLE itself is a minimal sufficient statistic and the MLE together with the sample size N fix the likelihood function up to an irrelevant factor. (For the Cauchy distribution the full sample is a minimal sufficient statistic. No further reduction in size is possible. Thus its MLE is not sufficient.)

If in the general situation with P parameters a minimal sufficient statistic $\boldsymbol{t}$ of P components exists, the data reduction is N to P and the MLE for the P parameters will be a unique function of $\boldsymbol{t}$ and is therefore itself a sufficient statistic.

Example 6.15. Sufficient statistic and expected value of a normal distribution

Let $x_1, \ldots, x_N$ be N normally distributed observations with width σ. The parameter of interest be the expected value μ of the distribution. The likelihood function is

$$L(\mu|x_1, \ldots, x_N) = c \prod_{i=1}^{N} \exp[-(x_i - \mu)^2/(2\sigma^2]$$
$$= c \exp[-\sum_{i=1}^{N}(x_i - \mu)^2/(2\sigma^2)],$$

with $c = (\sqrt{2\pi}\sigma)^{-N}$. The exponent can be expressed in the following way:

$$\sum_{i=1}^{N}(x_i - \mu)^2/(2\sigma^2) = N(\overline{x^2} - 2\overline{x}\mu + \mu^2)/(2\sigma^2).$$

Now the likelihood L factorizes:

$$L(\mu, \sigma|x_1, \ldots, x_N) = \exp[-N(-2\overline{x}\mu + \mu^2)/(2\sigma^2)] \cdot c \exp[-N\overline{x^2}/(2\sigma^2)]. \tag{6.15}$$

Only the first factor depends on μ. Consequently the experimental quantity $\overline{x}$ contains the full information on μ and thus is a one-dimensional sufficient statistic. Setting the derivative of the first factor equal to zero, we obtain the MLE $\hat{\mu} = \overline{x}$.

In the following example we show that a sufficient two-dimensional statistic can be found when besides the expectation value also the width σ is to be estimated.

Example 6.16. Sufficient statistic for the mean value and the width of a normal distribution

Let $x_1, \ldots, x_N$ be N normally distributed observations. The mean value μ and the width σ be the parameters of interest. From (6.15) we deduce that $\overline{x}$ and $\overline{x^2}$ together form a sufficient statistic $\{\overline{x}, \overline{x^2}\}$. Alternatively, also $\overline{x}$ and $v^2 = \sum(x_i - \overline{x})^2$ form a sufficient statistic. The MLE in the two-dimensional parameter space μ, σ^2 is

$$\hat{\mu} = \overline{x}, \ \hat{\sigma}^2 = \frac{1}{N}(\overline{x^2} - \overline{x}^2).$$

There is no one-dimensional sufficient statistic for σ.

Remark: In the examples which we have discussed, the likelihood function is fixed up to an irrelevant multiplicative factor if we consider the sample size N as a constant. In case N is also a random variable, then N is part of the sufficient statistic, e.g. in the last example it is $\{\overline{x}, \overline{x^2}, N\}$. Usually N is given and is then an *ancillary statistic.*

Definition: A statistic y is called *ancillary*, if $f(y|\theta) = f(y)$, i.e. the p.d.f. of y is independent of the parameter of interest.[7]

The value of the ancillary statistic has no influence on the MLE but is relevant for the shape of the likelihood function and thus for the precision of the estimation. The sample size is in most cases an ancillary statistic and responsible for the accuracy of the estimation.

6.6.2 *The Conditionality Principle*

Imagine that a measurement is performed either with the precise device A or with the imprecise device B. The device is selected by a stochastic process. After the measurement has been realized, we know the device which had been selected. Let us assume this was device B. The *conditionality principle* tells us that for the parameter inference we are allowed to use this information which means that we may act as if device A had not existed. The analysis is not "blind". Stochastic results influence the way we evaluate the parameters.

[7]Note that the combination of two ancillary statistics is not necessarily ancillary.

More generally, the conditionality principle states:

If an experiment concerning the inference about θ is chosen randomly from a collection of possible experiments, independently of θ, then any experiment not chosen is irrelevant to the inference.

Example 6.17. Conditionality

We measure the position coordinate of the trajectory of an ionizing particle passing a drift chamber. A certain wire responds. Its position provides a rough coordinate. In 90 % of all cases a drift time is registered and we obtain a much more precise value of the coordinate. The conditionality principle tells us that in this case we are allowed to use the drift time information without considering the worse resolution of a possible but not realized failure of the time measurement.

The conditionality principle seems to be trivial. Nevertheless the belief in its validity is not shared by all statisticians because it leads to the *likelihood principle* with its far reaching consequences which are not always intuitively obvious.

6.6.3 *The Likelihood Principle*

We now discuss a principle which concerns the foundations of statistical inference and which plays a central role in Bayesian statistics.

The likelihood principle (LP) states the following:

Given a p.d.f. $f(x|\theta)$ containing an unknown parameter of interest θ and an observation x, all information relevant for the estimation of the parameter θ is contained in the likelihood function $L(\theta|x) = f(x|\theta)$. (Here x stands for many observations and θ for a set of parameters.

Furthermore, two likelihood functions which are proportional to each other, contain the same information about θ. The general form of the p.d.f. is considered as irrelevant. The p.d.f. at variate values which have not been observed has no bearing on the parameter inference.

Correspondingly, for discrete hypotheses H_i the full experimental information relevant for discriminating between them is contained in the likelihoods L_i.

The following examples are intended to make the LP plausible.

Example 6.18. Likelihood principle, dice

We have a bag of two biased dice A and B. Die A produces the numbers 1 to 6 with probabilities 1/12, 1/6, 1/6, 1/6, 1/6, 3/12. The corresponding probabilities for die B are 3/12, 1/6, 1/6, 1/6, 1/6, 1/12. The result of an experiment where one of the dice is selected randomly is "3". We are asked to bet for A or B. We are unable to draw a conclusion from the observed result because both dice produce this number with the same probability, the likelihood ratio is equal to *one*. The LP tells us – what intuitively is clear – that for a decision the additional information, i.e. the probabilities of the two dice to yield values different from "3", are irrelevant.

Example 6.19. Likelihood principle, $V - A$

We come back to an example which we had discussed already in Section 6.3. An experiment investigates $\tau^- \to \mu^- \nu_\tau \bar{\nu}_\mu$, $\mu^- \to e^- \nu_\mu \bar{\nu}_e$ decays and measures the slope $\hat{\alpha}$ of the cosine of the electron direction with respect to the muon direction in the muon center-of-mass. The parameter α depends on the $\tau - \mu$ coupling. Is the τ decay proceeding through $V - A$ or $V + A$ coupling? The LP implies that the probabilities $f_-(\alpha)$, $f_+(\alpha)$ of the two hypotheses to produce values α different from the observed value $\hat{\alpha}$ do not matter. When we now allow that the decay proceeds through a mixture $r = g_V/g_A$ of V and A interaction, the inference of the ratio r is based solely on the observed value $\hat{\alpha}$, i.e. on $L(r|\hat{\alpha})$.

The LP follows inevitably from the sufficiency principle and the conditioning principle. It goes back to Fisher, has been reformulated and derived several times [45–48]. Some of the early promoters (Barnard, Birnbaum) of the LP later came close to rejecting it or to restrict its applicability. The reason for the refusal of the LP has probably its origin in its incompatibility with some concepts of the classical statistics. A frequently expressed argument against the LP is that the confidence intervals of the frequentist statistics cannot be derived from the likelihood function alone and thus contradict the LP. But this fact merely shows that certain statistical methods do not use the full information content of a measurement or use irrelevant information. Another reason lies in problems which sometimes occur if the LP is applied in social sciences, in medicine or biology. There it is often not possible to parametrize the empirical models in a stringent way. But uncertainties in the model prohibit the application of the LP. The exact validity of the model is a basic requirement for the application of the LP.

In the literature examples are presented which are pretended to contradict the LP. These examples are not really convincing and rather strengthen the LP. Anyway, they often contain quite exotic distributions which are irrelevant in physics applications and which lead when treated in a frequentist way to unacceptable results [48].

We abstain from a reproduction of the rather abstract proof of the LP and limit us to present a simple and transparent illustration of it:

The quantity which contains all the information we have on θ after the measurement is the p.d.f. of θ,

$$g(\theta) = \frac{L(\theta|x)\pi(\theta)}{\int L(\theta|x)\pi(\theta)\mathrm{d}\theta}.$$

It is depends on the prior density and the likelihood function. The prior does not depend on the data, thus the complete information that can be extracted *from the data*, and which is relevant for $g(\theta)$, must be contained in the likelihood function.

Another plausibility argument in favor of the LP is the following: With increasing number of observations the Likelihood function shrinks and therefore the prior density becomes less and less important and can be replaced by a constant. The likelihood function approaches up to a normalization factor the p.d.f. of the parameters and thus exhausts the information contained in the data (see A.3). Since the likelihood function is the product of the likelihood functions of many small subsamples, also these likelihood functions cannot have neglected information.

A direct consequence of the LP is that in the absence of prior information, optimal parameter inference has to be based solely upon the likelihood function. It is then logical to select for the estimate the value of the parameter which maximizes the likelihood function and to choose the error interval such that the likelihood is constant at the border, i.e. is smaller everywhere outside than inside. (see Chapter 8). All approaches which are not based on the likelihood function are inferior to the likelihood method or at best equivalent to it.

6.6.4 *Stopping Rules*

An experiment searches for a rare reaction. Just after the first successful observation at time t the experiment is stopped. Is the resulting rate $1/t$ systematically biased? Do we have to consider the stopping rule in the inference process? The answer is "no" but many people have a different opinion. This is the reason why we find the expression *stopping rule paradox* in the literature.

The possibility to stop an experiment without compromising the data analysis, for instance because a detector failed, no money was left or because the desired precision has been reached, means a considerable simplification of the data analysis.

The validity of the independence of the parameter estimation on a large class of stopping conditions is a direct consequence of the LP because the likelihood function of a parameter determined from sequentially obtained data is independent of the reason why data taking has been stopped.

In this context we examine a simple example.

Example 6.20. Stopping rule: four decays in a fixed time interval

In two similar experiments the lifetime of the same unstable particle is measured. In experiment A the time interval t is fixed and 4 decays are observed. In experiment B the time t is measured which is required to observe 4 decays. Accidentally the two times coincide. Thus in both experiments 4 decays are registered in the time interval t but in experiment A the number n of decays is the random variable while in experiment B it is the time t. Do both experiments find the same rate, namely $\tau = t/4$ and the same error interval? We could think "no" because in the first experiment the fourth decay has happened earlier than in the second. The likelihood functions for the two situations are deduced for experiment A from the Poisson distribution and for experiment B from the exponential time distribution:

$$\begin{aligned} L_A(\theta|n) &= \mathcal{P}_{\theta t}(n) \\ &= \frac{\mathrm{e}^{-\theta t}(\theta t)^4}{4!} \sim \theta^4 \mathrm{e}^{-\theta t}, \\ L_B(\theta|t) &= \theta^4 \mathrm{e}^{-\theta t} \sim L_A(\theta|n). \end{aligned}$$

The likelihood functions are equal up to an irrelevant factor and consequently also the results are the same. The stopping rule does not influence the analysis. The only relevant data are the number of decays and the time interval.

The fact that an arbitrary sequential stopping rule does not change the expectation value is illustrated with the example presented in Fig. 6.9. A rate is determined. The measurement is stopped if a sequence of 3 decays occurs within a short time interval of only one second. One might argue that the rate is biased towards a high rate, but this is not true. If we perform many such experiments with i.i.d. observations, one after the other, their combination is equivalent to a single very long experiment where the stopping rule cannot influence the result and from which we can estimate the mean value of the rate with high precision. Since the log-likelihood of the long experiment is equal to the sum of the log-likelihoods of the short experiments, the log-likelihoods of the short experiments obviously have to be unbiased. The stopping condition does not enter into the likelihood function and therefore must be irrelevant.

Why does the fact that neglecting the stopping rule is justified, contradict our intuition? Well, for example, playing roulette and stopping after having won, biases the result but the reason is that the mean gain is known to be zero. The LP does not apply.

6.7 The Moments Method of Parameter Estimation

The moments of a distribution which depends on a parameter θ, usually also depend on θ:

$$\mu_n(\theta) = \int x^n f(x|\theta)\,\mathrm{d}x. \tag{6.16}$$

The empirical moments

$$\hat{\mu}_n = \frac{1}{N}\sum_i x_i^n ,$$

e.g. the sample mean or the mean of squares, which we can extract trivially for a sample, are estimators of the moments of the distribution. From the inverse function μ^{-1} we obtain a consistent estimate of the parameter,

$$\hat{\theta} = \mu^{-1}(\hat{\mu}),$$

because according to the law of large numbers we have (see Appendix A.1)

$$\lim_{N\to\infty} P\{|\hat{\mu} - \mu| > \varepsilon\} = 0.$$

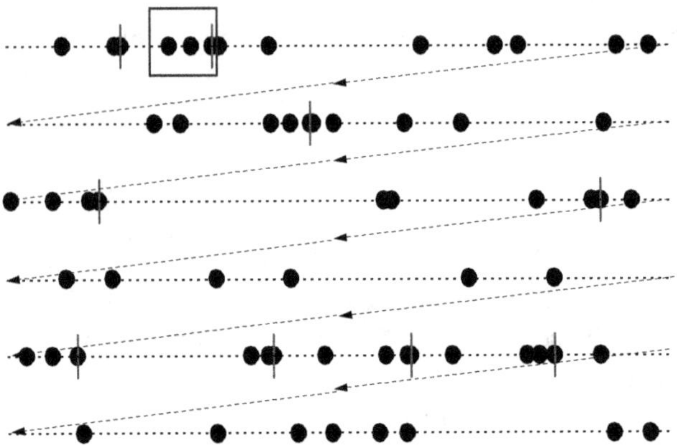

Fig. 6.9 An experiment is stopped when 3 observations are registered within a short time interval (indicated by a box or a bar) A arbitrarily long experiment can be subdivided into many such sub-experiments following the sopping rule.

It is clear that any function $u(x)$, for which expected value and variance exist, and where $\langle u \rangle$ is an invertible function of θ, can be used instead of x^n. Therefore the method is somewhat more general than suggested by its name.

If the distribution has several parameters to be estimated, we must use several moments or expected values, approximate them by empirical averages, and solve the resulting system of – in general non-linear – equations for the unknown parameters.

The estimators derived from the lower moments are usually more precise than those computed from the higher ones. Parameter estimation from the moments is usually inferior to that of the ML method. Only if the moments used, form a sufficient statistic, the two approaches produce the same result.

The uncertainties of the fitted parameters have to be estimated from the covariance matrix of the corresponding moments and subsequently by error propagation or alternatively by a Monte Carlo simulation, generating the measurement several times. Also the *bootstrap method* which will be introduced in Chapter 12, can be employed. Sometimes the error calculation is a bit annoying and reproduces the ML error intervals only in the large sample limit.

Example 6.21. Moments method: Mean and variance of the normal distribution

We come back to the example from Section 6.5.2. For a sample $\{x_1, \ldots, x_N\}$, following the distribution

$$f(x|\mu, \sigma) = \frac{1}{\sigma\sqrt{2\pi}} \exp\left[-(x-\mu)^2/(2\sigma^2)\right],$$

we determine independently the parameters μ and σ. With the abbreviations $\overline{x}$ for the sample mean and $\overline{x^2}$ for the mean of the squares and $v^2 = \overline{(x-\overline{x})^2} = \overline{x^2} - \overline{x}^2$ for the empirical variance. The relation between the moment μ_1 and the parameter of the distribution μ is simply $\mu_1 = \mu$, therefore

$$\hat{\mu} = \hat{\mu_1} = \overline{x}.$$

In Chapter 3, we have derived the relation (3.15) $\langle v^2 \rangle = \sigma^2(N-1)/N$ between the expectation of the empirical variance and the variance of the distribution; inverting it, we get

$$\hat{\sigma} = v\sqrt{\frac{N}{N-1}}.$$

The two estimates are uncorrelated. The error of $\hat{\mu}$ is derived from the estimated variance

$$\delta_\mu = \frac{\hat{\sigma}}{\sqrt{N}},$$

and the error of $\hat{\sigma}$ is determined from the expected variance of v. We omit the calculation, the result is:

$$\delta_\sigma = \frac{\hat{\sigma}}{\sqrt{2(N-1)}}.$$

In the special case of the normal distribution, the independent point estimates of μ and σ of the moments method are identical to those of the maximum likelihood method. The errors differ for small samples but coincide in the limit $N \to \infty$.

The moments method has the advantage that it is very simple, especially in the case of distributions which depend linearly on the parameters – see next example below:

Example 6.22. Moments method: Asymmetry of an angular distribution

Suppose we have to determine the asymmetry parameter α of a distribution $f(x) = (1 + \alpha x)/2$ linear in $x = \cos\beta$ from a sample of N measurements. The first moment of the distribution is $\mu_1 = \alpha/3$. Thus we can compute the parameter from the sample mean $\overline{x} = \sum x_i/N$:

$$\hat{\alpha} = 3\,\overline{x}.$$

The mean square error from an individual measurement x is proportional to the variance of the distribution:

$$\mathrm{var}(\hat{\alpha}) = 9\,\mathrm{var}(x) = 3 - \alpha^2. \tag{6.17}$$

Using instead of α its estimate, we get

$$\delta_{\hat{\alpha}} = 3\,\delta_{\overline{x}} = \sqrt{\frac{3 - 9\overline{x}^2}{N}}.$$

A likelihood fit, according to the likelihood principle, is more accurate and reflects much better the result of the experiment which, because of the kinematic constraint $|\alpha| < 1$, cannot be described very well by symmetric errors; especially when the sample size is small and the estimate happens to lie near the boundary. In this case the maximum likelihood method should be applied. In the asymptotic limit $N \to \infty$ the variance of the moments estimate $\hat{\alpha}$ does not approach the limit that is provided by the Cramer–Rao inequality, see (A.6) in Appendix A.2, which is achieved by the MLE:

$$\mathrm{var}(\hat{\alpha}_{ML}) \approx \frac{\alpha^2}{N}\left[\frac{1}{2\alpha}\ln(\frac{1+\alpha}{1-\alpha}) - 1\right]^{-1}.$$

A comparison with (6.17) shows that the asymptotic efficiency of the moments method, defined as

$$\varepsilon = \frac{\mathrm{var}(\hat{\alpha}_{ML})}{\mathrm{var}(\hat{\alpha})},$$

is unity only for $\alpha = 0$. It is 0.92 for $\alpha = 0.5$ and drops to 0.73 for $\alpha = 0.8$. (At the boundary, $|\alpha| = 1$ the Cramer–Rao relation cannot be applied.) Note that the p.d.f. of our example is a special case of the usual expansion of an angular distribution into Legendre polynomials $P_l(\cos\beta)$:

$$f(x|\boldsymbol{\theta}) = (1 + \sum_{l=1}^{L} \theta_l P_l(x))/2.$$

From the orthogonality of the P_l with the usual normalization

$$\int_{-1}^{1} P_l(x)P_m(x)\mathrm{d}x = \frac{2}{2l+1}\delta_{l,m}$$

it is easy to see that $\theta_l = (2l+1)\langle P_l\rangle$. In the case $l = 1$, $P_1 = x$, this is the first moment of the distribution and we reproduce $\mu_1 = \alpha/3$.

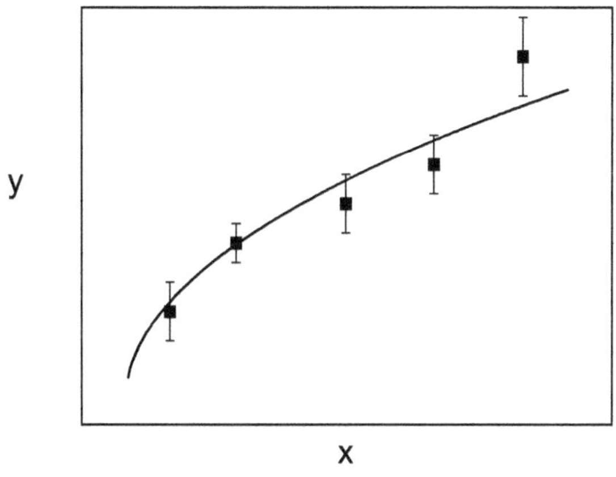

Fig. 6.10 Fit of a curve to measurements.

6.8 The Least Square Method

A frequently occurring problem is that a curve has to be fitted to given measured points with error margins as shown in Fig. 6.10. The standard solution of this regression problem is provided by the least square method which fixes the parameters of a given function by minimizing the sum of the normalized square deviations of the function from the measured points.

Given N measured points $x_i, y_i \pm \delta_i$, and a function $t(x, \boldsymbol{\theta})$, known up to some free parameters $\boldsymbol{\theta}$, the latter are determined such that

$$\chi^2 = \sum_{i=1}^{N} \frac{(y_i - t(x_i, \boldsymbol{\theta}))^2}{\delta_i^2} \tag{6.18}$$

takes its minimal value.

Remark: Frequently the errors δ_i depends on the true values t_i and therefore on the unknown parameter $\boldsymbol{\theta}$. Then also the denominators $\delta_i^2(\boldsymbol{\theta})$ are adjusted in the fitting process.

The least square method goes back to Gauß. Historically it has successfully been applied to astronomical problems and is still the best method we have to adjust parameters of a curve to measured points if only the variance of the error distribution is known. It is closely related to the likelihood method if the errors are normally distributed. Then we can write the p.d.f.

of the measurements in the following way:

$$f(y_1, \dots, y_N|\boldsymbol{\theta}) \propto \exp\left[-\sum_{i=1}^{N} \frac{(y_i - t(x_i, \boldsymbol{\theta}))^2}{2\delta_i^2}\right],$$

and the log-likelihood is

$$\begin{aligned}\ln L(\boldsymbol{\theta}|\boldsymbol{y}) &= -\frac{1}{2}\sum_{i=1}^{N} \frac{(y_i - t(x_i, \boldsymbol{\theta}))^2}{\delta_i^2}, \\ &= -\frac{1}{2}\chi^2. \end{aligned} \tag{6.19}$$

Thus minimizing χ^2 is equivalent to maximizing the likelihood if the errors are normally distributed and independent of the free parameter, a condition which frequently is approximately satisfied. From (6.19) we conclude that the standard deviation errors of the parameters in a least square fit correspond to one unit, $\Delta\chi^2 = 1$, twice the value $1/2$ of the maximum likelihood method. In Section 3.6.7 we have seen that χ^2 follows approximately a χ^2-distribution of $f = N - Z$ (Z is the number of free parameters) degrees of freedom, provided the normality of the errors is satisfied, see Section 3.6.7. Thus we expect χ^2 to be of the order of f, large values indicate possible problems with the data or their description. We will investigate this in Chapter 10.

The standard deviation of the χ^2-distribution for f degrees of freedom is $\sigma = \sqrt{2f}$ which, for example, is equal to 10 for 50 degrees of freedom. With such large fluctuations of the value of χ^2 from one sample to the other, it appears paradoxical at first sight that a parameter error of one standard deviation corresponds to such a small change of χ^2 as one unit, while a variation of χ^2 by 10 is compatible with the prediction. The obvious reason for the good resolution is that the large fluctuations from sample to sample are unrelated to the value of the parameter. In case we would compare the prediction after each parameter change in a minimum searching routine to a new measurement sample, we would not be able to obtain a precise result for the estimate of the parameter $\boldsymbol{\theta}$.

That the least square method can lead to false results if the condition of Gaussian uncertainties is not fulfilled, is illustrated in the following example.

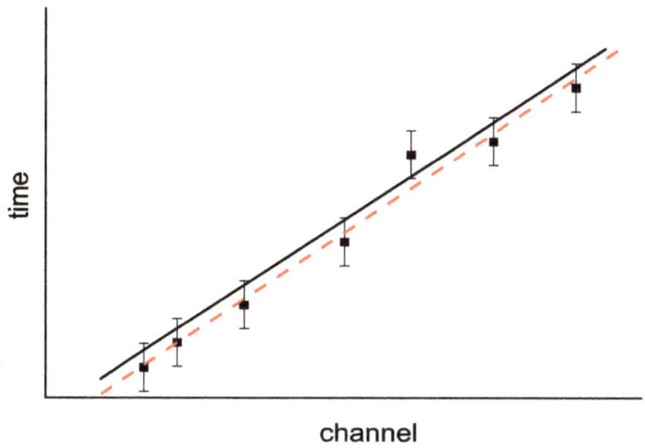

Fig. 6.11 χ^2–Fit (dashed) of a straigt line to digital measurements.

Example 6.23. Counter example to the least square method

A digital clock has to be gauged. Fig. 6.11 shows the time channel as a function of the true time and a least square fit by a straight line (red in the figure). The error bars in the figure are not error bars in the usual sense but indicate the channel width . The fit fails to meet the allowed range of the fifth point and therefore is not compatible with the data. All straight lines which meet all "error bars" have the same likelihood. A correct solution, the black line, is indicated in the figure.

We can easily generalize the expression (6.18) to the case of correlated errors. Then we have with $t_i = t(x_i, \boldsymbol{\theta})$

$$\chi^2 = \sum_{i,j=1}^{N} (y_i - t_i) C_{ij} (y_j - t_j)$$

where $\mathsf{C} = \mathsf{V}^{-1}$, the weight matrix. is the inverse of the covariance matrix. The quantity χ^2 is up to a factor *two* equal to the negative log-likelihood of a multivariate normal distribution,

$$f(y_1, \dots, y_N | \boldsymbol{\theta}) \propto \exp\left[-\frac{1}{2} \sum_{i,j=1}^{N} (y_i - t_i) C_{ij} (y_j - t_j)\right],$$

see Section 7.1.1. Maximizing the likelihood is again equivalent to minimizing χ^2 if the errors are normally distributed.

The sum χ^2 is not invariant against a non-linear variable transformation $y'(y)$.

Example 6.24. Least square method: Fit of a straight line

We fit the parameters a, b of the straight line

$$y(x) = ax + b \tag{6.20}$$

to a sample of points (x_i, y_i) with uncertainties δ_i of the ordinates. We minimize χ^2:

$$\chi^2 = \sum_i \frac{(y_i - a\,x_i - b)^2}{\delta_i^2},$$

$$\frac{\partial\chi^2}{\partial a} = \sum_i \frac{(-y_i + a\,x_i + b)2x_i}{\delta_i^2},$$

$$\frac{\partial\chi^2}{\partial b} = \sum_i \frac{(-y_i + a\,x_i + b)2}{\delta_i^2}.$$

We set the derivatives to zero and introduce the following abbreviations. (In parentheses we put the expressions for the special case where all uncertainties are equal, $\delta_i = \delta$):

$$\overline{x} = \sum_i \frac{x_i}{\delta_i^2} \Big/ \sum_i \frac{1}{\delta_i^2} \quad \left(\sum_i x_i/N\right),$$

$$\overline{y} = \sum_i \frac{y_i}{\delta_i^2} \Big/ \sum_i \frac{1}{\delta_i^2} \quad \left(\sum_i y_i/N\right),$$

$$\overline{x^2} = \sum_i \frac{x_i^2}{\delta_i^2} \Big/ \sum_i \frac{1}{\delta_i^2} \quad \left(\sum_i x_i^2/N\right),$$

$$\overline{xy} = \sum_i \frac{x_i y_i}{\delta_i^2} \Big/ \sum_i \frac{1}{\delta_i^2} \quad \left(\sum_i x_i y_i/N\right).$$

We obtain

$$\hat{b} = \overline{y} - \hat{a}\,\overline{x},$$

$$\overline{xy} - \hat{a}\,\overline{x^2} - \hat{b}\,\overline{x} = 0,$$

and

$$\hat{a} = \frac{\overline{xy} - \overline{x}\,\overline{y}}{\overline{x^2} - \overline{x}^2},$$

$$\hat{b} = \frac{\overline{x^2}\,\overline{y} - \overline{x}\,\overline{xy}}{\overline{x^2} - \overline{x}^2}.$$

The problem is simplified when we put the origin of the abscissa at the center of gravity $\overline{x}$:

$$x' = x - \overline{x},$$

$$\hat{a}' = \frac{\overline{x'y}}{\overline{x'^2}},$$

$$\hat{b}' = \overline{y}.$$

Now the equation of the straight line reads

$$y = \hat{a}'(x - \overline{x}) + \hat{b}'. \tag{6.21}$$

We gain an additional advantage, the errors of the estimated parameters are no longer correlated.

$$\delta^2(\hat{a}') = 1/\sum_i \frac{x_i^2}{\delta_i^2},$$

$$\delta^2(\hat{b}') = 1/\sum_i \frac{1}{\delta_i^2}.$$

We recommend to use always the form (6.21) instead of (6.20).

6.8.1 *Linear Regression*

If the prediction depends only linearly on the parameters, we can compute the parameters which minimize χ^2 analytically. We put

$$\boldsymbol{y}(\boldsymbol{\theta}) = \mathsf{A}\boldsymbol{\theta} + \boldsymbol{e}. \tag{6.22}$$

Here $\boldsymbol{\theta}$ is the P-dimensional parameter vector, $\boldsymbol{e}$ is a N-dimensional error vector with expectation zero, $\boldsymbol{y} = \boldsymbol{t}$ is the N-dimensional vector of predictions. To simplify the formulas, usually $\boldsymbol{y} - \boldsymbol{e}$ is used as random vector. It is then sufficient to consider the case with no additive constant. It has now the expected value

$$\langle \boldsymbol{y} \rangle = \mathsf{A}\boldsymbol{\theta}. \tag{6.23}$$

The given rectangular matrix A with P columns and N rows, also called the design matrix, defines the above mentioned *linear* mapping from the P-dimensional parameter space into the N-dimensional sample space.

The straight line fit discussed in Example 6.24 is a special case of (6.22) with $\mathrm{E}(y_i) = \sum_{j=1}^{P=2} A_{ij}\theta_j = \theta_1 x_i + \theta_2,\ i = 1, \ldots, N$, and

$$\mathsf{A} = \begin{pmatrix} x_1 & \cdots & x_N \\ 1 & \cdots & 1 \end{pmatrix}^T .$$

We have to find the minimum in $\boldsymbol{\theta}$ of

$$\chi^2 = (\boldsymbol{y} - \mathsf{A}\boldsymbol{\theta})^T \mathsf{C} (\boldsymbol{y} - \mathsf{A}\boldsymbol{\theta})$$

where, as usual, C is the weight matrix. In the example above it is a diagonal $N \times N$ matrix with elements $1/\delta_i^2$ where δ_i is the standard deviation of the observed value y_i [8]. We differentiate χ^2 with respect to the parameters $\boldsymbol{\theta}$ and set the derivatives equal to zero:

$$\frac{1}{2}\frac{\partial \chi^2}{\partial \boldsymbol{\theta}}\Big|_{\hat{\boldsymbol{\theta}}} = \mathbf{0} = -\mathsf{A}^T \mathsf{C}(\boldsymbol{y} - \mathsf{A}\hat{\boldsymbol{\theta}}). \tag{6.24}$$

From these so-called normal equations we get the estimate for the P parameters $\hat{\boldsymbol{\theta}}$:

$$\hat{\boldsymbol{\theta}} = (\mathsf{A}^T \mathsf{C} \mathsf{A})^{-1} \mathsf{A}^T \mathsf{C} \boldsymbol{y}. \tag{6.25}$$

Note that $\mathsf{A}^T\mathsf{CA}$ is a symmetric $P \times P$ matrix which turns out to be the inverse of the the error- or covariance matrix V_P of $\hat{\boldsymbol{\theta}}$. This matrix is (see Section 4.2 relation (4.5)) $\mathsf{V}_P = \mathsf{DVD}^T$ with D the derivative matrix

$$\mathsf{D} = (\mathsf{A}^T \mathsf{C} \mathsf{A})^{-1} \mathsf{A}^T \mathsf{C}$$

derived from (6.25). After some simplifications we obtain:

$$\mathsf{V}_P = (\mathsf{A}^T \mathsf{C} \mathsf{A})^{-1}.$$

A feature of the linear model is that the result (6.25) for $\hat{\boldsymbol{\theta}}$ turns out to be linear in the measurements $\boldsymbol{y}$. Using it together with (6.23) one easily finds $\left\langle \hat{\boldsymbol{\theta}} \right\rangle = \boldsymbol{\theta}$, i.e. the estimate is unbiased.[9] The Gauss–Markov–theorem states that every estimate obeying these two assumptions will have an error matrix with larger or equal diagonal elements than the above estimate (also called BLUE: best linear unbiased estimate).

[8]We keep here the notation χ^2, which is strictly justified only in case of Gaussian error distributions or asymptotically for large N. Only then it obeys a χ^2-distribution with $N-P$ degrees of freedom. The index of quadratic matrices indicates its dimension.

[9]This is true for any N, not only asymptotically.

Linear regression provides an optimal solution only for normally distributed, known errors. Often, however, the latter depend on the parameters.

Strictly linear problems are therefore rare. When the prediction is a non-linear function of the parameters, the problem can be linearized by a Taylor expansion as a first rough approximation. By iteration the precision can be improved.

The importance of non-linear parameter inference by iterative linear regression has decreased considerably. The minimum searching routines,which we find in all computer libraries, are more efficient and easier to apply. Some basic minimum searching approaches are presented in Appendix A.10.

6.9 Properties of Estimators

The content of this section is resumed in the Appendices A.2 and A.2.2.

6.9.1 *Consistency*

An estimator is consistent if in the large number limit the estimator approaches the true parameter value θ, $\lim_{N\to\infty}(\hat{\theta}-\theta)=0$. Consistency is a necessary condition for a useful estimator. The MLE is consistent (see Appendix A.2.2).

6.9.2 *Transformation Invariance*

We require that the estimate $\hat{\theta}$ and the estimate of a function of θ, $\widehat{f(\theta)}$, satisfy the relation $\widehat{f(\theta)} = f(\hat{\theta})$. For example the mean lifetime τ and the decay rate γ of a particle are related by $\gamma = 1/\tau$. Therefore their estimates from a sample of observations have to be related by $\hat{\gamma} = 1/\hat{\tau}$. If they were different, the prediction for the number of decays in a given time interval would depend on the choice of $\hat{\tau}$ or $\hat{\gamma}$ used to evaluate the number. Similarly in the computation of a cross section which depends on different powers of a coupling constant g, we would get inconsistent results unless $\hat{g}^n = \widehat{g^n}$. Estimators applied to constants of nature have to be transformation invariant. The MLE and the likelihood ratio error intervals satisfy this condition.

Remark that the transformation invariance is not important in most statistical applications outside the natural sciences. This is why it is not always considered as necessary.

6.9.3 *Accuracy and Bias of Estimators*

The bias b of an estimate $\hat{\theta}$ is the deviation of its expectation value from the true value θ of the parameter:

$$b = \mathrm{E}(\hat{\theta}) - \theta .$$

Example 6.25. Bias of the estimate of a decay parameter

We estimate the decay parameter γ from 5 observed decays of an unstable particle. We have seen in a previous example that the MLE $\hat{\gamma}$ is the inverse of the average of the individual decay times, $\hat{\gamma} = 1/\bar{t}$. The mean value $\bar{t}$ follows a gamma distribution (see Section 3.6.8).

$$f(\bar{t}|\gamma) = \frac{(5\gamma)^5}{4!}\bar{t}^4 \exp(-5\gamma\bar{t}),$$

and thus the expectation value $E(\hat{\gamma})$ of $\hat{\gamma}$ is

$$\begin{aligned} E(\hat{\gamma}) &= \int_0^\infty \hat{\gamma} f(\bar{t}|\gamma)\,\mathrm{d}\bar{t} \\ &= \int_0^\infty \frac{(5\gamma)^5 4! \bar{t}^3}{\exp}(-5\gamma\bar{t})\,\mathrm{d}\bar{t} = \frac{5}{4}\gamma . \end{aligned}$$

When in a large number of similar experiments with 5 observed events the MLE of the decay time is determined, then the arithmetic mean differs from the true value by 25 %, the bias of the MLE is $b = E(\hat{\gamma}) - \gamma = \gamma/4$. For a single decay the bias is infinite.

The MLE of the decay constant of an exponential decay distribution is biased while the MLE of the mean lifetime is unbiased.

Similarly, we may define as a measure of accuracy a as the expected value of the mean squared deviation of the estimate from the true value.

$$a = \mathrm{E}[(\hat{\theta} - \theta)^2] . \tag{6.26}$$

In both cases the estimate is considered as a random variable. An estimate with the property that a in the large sample limit, $N \to \infty$, is smaller than the result of any other estimator is called *efficient* (see Appendix A.2). Efficient estimators have to be unbiased. In an exponential decay the MLE $\hat{\tau}$ of the lifetime is both unbiased and efficient while the MLE of the decay rate $\gamma = 1/\tau$ is neither unbiased nor efficient.

Biases occur quite frequently at small samples. With increasing number of observations the bias decreases (see Appendix A.2.2).

The word *bias* somehow suggests that something is wrong and thus it appears quite disturbing at first sight that estimates may be systematically biased. In fact in most of the conventional statistical literature it is recommended to correct for the bias. One reason given for the correction is the expectation that averaging many biased results the error would decrease, but the bias would remain. However, there is no obvious reason for a correction and a closer study reveals that bias corrections lead rather to difficulties when we combine different measurements $\hat{\theta}_i$ in the usual way, weighting the results by the inverse covariance matrix, or in the one dimensional case according to (4.7) simply by the inverse error squared.

$$\overline{\theta} = \frac{\sum \hat{\theta}_i/\delta_i^2}{\sum 1/\delta_i^2}.$$

Since usually the estimated errors depend on the value of the MLE, the weighting introduces a bias which may partially compensate a bias of the MLE or it may increase it.

Let us resume our last example and assume that many experiments measure the decay rate from samples of size $N = 5$. The estimates $\hat{\gamma}_i$ will vary from experiment to experiment. Each experiment will, apart from the estimate, evaluate the error δ_i which will turn out to be proportional to $\hat{\gamma}_i$, namely $\delta_i = \hat{\gamma}_i/\sqrt{5}$. Averaging without bias correction according to our prescription, we will obtain $E(\overline{\gamma}) = 5/6\,\gamma$, thus the bias is reduced, while averaging the bias corrected estimates would lead to the expectation $E(\overline{\gamma}) = 2/3\,\gamma$, a result which is considerably worse. Table 6.1 summarizes the expected mean values from 10 observed decay times from particles with true lifetime 1. (The table includes results from averaging procedures that will be discussed in Section 8)

We have to conclude that bias corrections should not be applied to MLEs.

The accuracy as defined by (6.26) and the bias are important quantities in frequentist statistics, but are less relevant in Bayesian and likelihood based statistics. Why is this so?

The frequentist statistics uses properties like a and b of the estimate *given the true parameter value*, while we are interested in the properties of the unknown true value *given the measurement*. The inversion of probabilities can lead to contradictions as becomes obvious in our lifetime example. As the decay rate is biased towards high values, one might conclude that the

Table 6.1 Expected weighted mean of 10 decay time measurements.

method	τ	γ
mean	1.00	1.11
weighted mean	0.80	0.91
weighted mean, bias corrected	0.80	0.82
weighted mean PDG	0.88	0.95
weighted mean extended	0.93	0.97

true value is likely to be located below the estimate. As a consequence the true value of $\tau = 1/\gamma$, should be located above its estimate $\hat{\tau}$, the estimate should be negatively biased, however it is unbiased.

The requirements of unbiasedness and maximal efficiency violate transformation invariance and for this reason are not relevant for the estimates of constants of nature. Bias corrections should not be applied to MLEs. It is not conceivable to insert, for instance, in a power expansion of the strong coupling constant α in each power term a different value of α because the bias correction depends on the power. Biases occur if the number of events are small. In this situation the uncertainty of measurements should be represented by asymmetric errors, or even better, the full likelihood function should recorded.

In the following we discuss some examples where the likelihood function is very asymmetric.

Example 6.26. Bias of the estimate of a Poisson rate with observation zero

We search for a rare decay but we do not observe any. The likelihood for the mean rate λ is according to the Poisson statistic

$$L(\lambda) = \frac{e^{-\lambda}\lambda^0}{0!} = e^{-\lambda} .$$

When we normalize the likelihood function to obtain the Bayesian p.d.f. with a uniform prior, we obtain the expectation value $\langle\lambda\rangle = 1$ while the value $\hat{\lambda} = 0$ corresponds to the maximum of the likelihood function. (It may seem astonishing, that an expectation value *one* follows from a null-measurement. This result is a consequence of the assumption of a uniform distribution of the prior which is not unreasonable because had we not

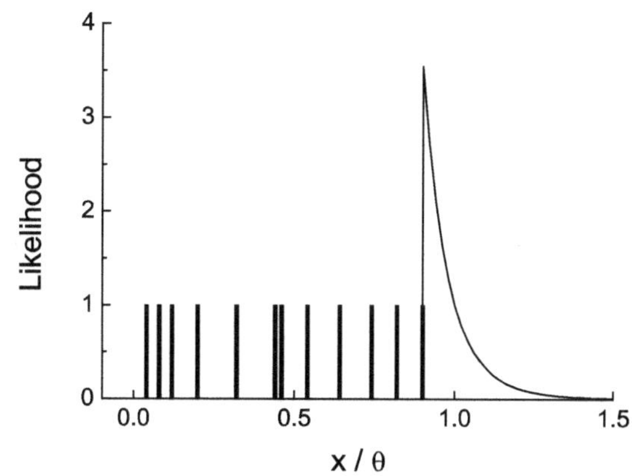

Fig. 6.12 Likelihood funktion of the width of a uniform distribution for 12 observations.

anticipated the possibility of a decay, we would not have performed the measurement. Since also mean rates different from zero may lead to the observation *zero* it is natural that the expectation value of λ is different from *zero*.) Now if none of 10 similar experiments would observe a decay, a naive averaging of the expected values alone would again result in a mean of *one*, a crazy value. Strictly speaking, the likelihoods of the individual experiments should be multiplied, or, equivalently the null rate would have to be normalized to ten times the original time with the Bayesian result 1/10.

Example 6.27. Bias of the measurement of the width of a uniform distribution

Let $x_1, \ldots, x_N$ be N observations of a sample following a uniform distribution $f(x) = 1/\theta$ with $0 < x < \theta$. We search for the value of the parameter θ. Figure 6.12 shows the observations and the likelihood function for $N = 12$. The likelihood function is

$$\begin{aligned} L &= 0 \ \text{ for } \theta < x_{\max}, \\ &= \frac{1}{\theta^N} \ \text{ for } \theta \geq x_{\max}. \end{aligned}$$

Obviously, the likelihood has its maximum when θ coincides with the largest observation x_{max} of the sample: $\hat{\theta} = x_{max}$. (Here x_{max} is a sufficient statistic.) At smaller values of x, the likelihood is zero. The estimate is biased towards small values. Given a sample size of N, we obtain $N + 1$ gaps between the observations and the borders $[0, \theta]$. The average distance of the largest observation from θ thus is $\theta/(N + 1)$. The bias is $-\hat{\theta}/N$. There is no reason to correct for the bias. We rather prefer to present the biased result with a one-sided error

$$\theta = x_{max}\,{}^{+x_{max}/N}_{-0}$$

or, alternatively, the full likelihood function.

6.10 Comparison of Estimation Methods

The following table contains an evaluation of the virtues and properties of the estimation approaches which we have been discussing.

Table 6.2 Virtues and caveats of different methods of parameter estimation.

	moments	χ^2	max. likelihood
simplicity	++	+	−
precision	−	+	++
individual observations	+	−	+
measured points	−	+	−
histograms	+	+	+
upper and lower limits	−	−	+
external constraints	−	+	+
background included	+	+	−
error assignment	error propag.	$\chi^2_{min} + 1$	$\ln L_{max} - 0.5$
requirement	full p.d.f.	variance	full p.d.f.

Whenever possible, the likelihood method should be applied. It requires a sample of observations and a p.d.f. in analytic or well defined numerical form and is very sensitive to wrongly assigned observations in the sample. When the theoretical description of the data is given in form of a simulated histogram, the Poisson likelihood adjustment of the simulation to the bin content should be chosen, see following chapter. When we have to fit a function to measured data points, we use the least square method. If computing time is a limitation like in some on-line applications, the moments method lends itself. In many situations all three methods are equivalent.

All methods are sensitive to spurious background. Especially robust methods have been invented to solve this problem. An introduction and references are given in Appendix A.16. For completeness we present in Appendix A.3.1 some frequentist criteria of point and interval estimation which are relevant when parameters of many objects of the same type, for instance particle tracks, are measured. Algorithms for minimum search are sketched in Appendix A.10.

Chapter 7

Estimation II

7.1 Likelihood of Histograms

For large samples it is more efficient to analyze the data in form of histograms than to compute the likelihood for many single observations. The individual observations are classified and collected into bins where all events of a bin have approximately the same likelihood. We then compare the number of entries in a bin with the parameter dependent prediction. Often the prediction is available only as a Monte Carlo simulation in form of a histogram. We will discuss the comparison of data to a Monte Carlo simulation in some detail in the following section.

We denote the total number of events by N, the number of events in bin i by d_i and the number of bins by B. In the following all sums run over all bins, $i = 1, ..., B$.

We have to distinguish different situations:

i) We have an absolute prediction $t_i(\boldsymbol{\theta})$ for the number of events d_i in bin i. The numbers d_i are described by Poisson distributions with mean t_i.

ii) The absolute particle flux is not known. The prediction $ct_i(\boldsymbol{\theta})$ of the number of events in bin i contains an unknown normalization factor c. The numbers d_i are described by Poisson distributions with mean ct_i. The parameter c is a free parameter in the fit.

The second case is much more frequent than the first. Think for instance of the measurement of a particle lifetime from a sample of events where the flux is not predicted.

Remark: The case with unknown normalization can also be formulated in the following way: The relative probabilities $p_i(\boldsymbol{\theta}) = t_i(\boldsymbol{\theta})/\Sigma_i t_i(\boldsymbol{\theta})$ for the number of events of the bins are predicted. Then the observed data follow a multinomial distribution where N events are distributed into the

B bins with probabilities p_i and the normalization parameter is eliminated. As a consequence, in case the normalization c is kept as a free parameter of the fit, it is not correlated with $\boldsymbol{\theta}$. The multinomial treatment and the Poissonian treatment with c as a free parameter are equivalent, see Appendix A.9.1. The latter is preferable because the error treatment is much simpler than in the multinomial case where the constraint $\Sigma_i p_i = 1$ has to be satisfied.

We start with case i):

The likelihood for t_i expected and d_i observed entries according to the Poisson distribution is given by

$$L_i(\boldsymbol{\theta}) = \frac{\mathrm{e}^{-t_i} t_i^{d_i}}{d_i!},$$
$$\ln L_i(\boldsymbol{\theta}) = -t_i + d_i \ln t_i - \ln(d_i!).$$

Since factors not depending on $\boldsymbol{\theta}$ are irrelevant for the likelihood inference (see Section 6.5.1), we are allowed to omit the term with the factorial. The log-likelihood of the complete histogram with B bins is then

$$\ln L(\boldsymbol{\theta}) = \sum_{i=1}^{B} (-t_i + d_i \ln t_i) \,. \tag{7.1}$$

The parameter dependence is hidden in the quantities t_i. The maximum of this function is determined by numerical methods.

In the second case with unknown normalization we have to replace t_i by ct_i:

$$\ln L(\boldsymbol{\theta}) = \sum_{i=1}^{B} (-ct_i + d_i \ln(ct_i)) \,. \tag{7.2}$$

We differentiate the log-likelihood with respect to c, set the derivative equal to zero, and obtain the proper estimate for the normalization: $\hat{c} = \Sigma d_i / \Sigma t_i$. Te parameter c is not correlated with the parameters of interest $\boldsymbol{\theta}$. Therefore the error estimates of $\boldsymbol{\theta}$ are independent of c.

Example 7.1. Adjustment of a linear distribution to a histogram

The cosine $u = \cos\alpha$ of an angle α be linearly distributed according to

$$f(u|\lambda) = \frac{1}{2}(1 + \lambda u), \ -1 \le u \le 1, \ |\lambda| < 1.$$

We want to determine the parameter λ which best describes the observed distribution of 500 entries d_i into 20 bins (Fig. 7.1). In the Poisson approximation we expect t_i entries for the bin i corresponding to the value $u_i = -1 + (i - 0.5)/10$ of the cosine in the center of the bin:

$$t_i = \frac{500}{20}(1 + \lambda u_i).$$

We obtain the likelihood function by inserting this expression into (7.1). The likelihood function and the MLE are indicated in Figure 7.1.

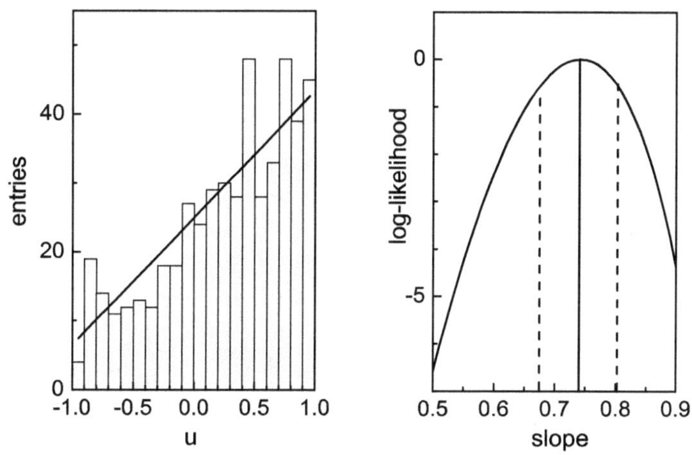

Fig. 7.1 Linear distribution with adjusted straight line (left) and likelihood function (right). The error of the adjusted slope is indicated by the dashed lines.

7.1.1 *The χ^2 Approximation*

We have seen in Section 3.6.3 that with increasing mean value t, the Poisson distribution asymptotically approaches a normal distribution with variance t_i. Thus for high statistics histograms the number of events d in a bin with prediction $t(\theta)$ is described by

$$f(d) = \frac{1}{\sqrt{2\pi t}} \exp\left[-\frac{(d-t)^2}{2t}\right].$$

Contrary to the case of relation (6.18) the denominator of the exponent and the normalization now depend on the parameters.

The corresponding log-likelihood is

$$\ln L = -\frac{(d-t)^2}{2t} - \frac{1}{2}\ln(2\pi) - \ln t .$$

For large t, the logarithmic term is an extremely slowly varying function of t. In situations where the Poisson distribution can be approximated by a normal distribution, it can safely be neglected. Omitting it and the constant term, we find for the whole histogram

$$\begin{aligned} \ln L &= -\frac{1}{2}\sum_{i=1}^{B}\frac{(d_i - t_i)^2}{t_i} \\ &= -\frac{1}{2}\chi^2 \end{aligned}$$

with

$$\chi^2 = \sum_{i=1}^{B}\frac{(d_i - t_i)^2}{t_i} . \tag{7.3}$$

If the approximation of the Poisson distribution by a normal distribution is justified, the likelihood estimation of the parameters is equivalent to a least square fit and the standard errors are given by an increase of χ^2 by one unit.

Often histograms contain some bins with few entries. Then a binned likelihood fit is to be preferred to a χ^2 fit, since the above condition of large t_i is violated. It is recommended to perform always a likelihood adjustment.

7.2 Extended Likelihood

When we record N independent multi-dimensional observations, $\{\boldsymbol{x}_i\}$, $i = 1, \dots, N$, of a distribution depending on a set of parameters $\boldsymbol{\theta}$, then it may happen that these parameters also determine the rate, i.e. the expected rate $\lambda(\boldsymbol{\theta})$ is a function of $\boldsymbol{\theta}$. In this situation N is no longer a fixed parameter but a random variable like the $\boldsymbol{x}_i$.[1] This means that we have to multiply two probabilities, the probability to find N observations which follow the Poisson statistics $\mathcal{P}_\lambda(N)$ and the probability to observe a certain distribution of the variates x_i. Given a p.d.f. $f(\boldsymbol{x}|\boldsymbol{\theta})$ for a single observation, we obtain the extended likelihood function [34, 35]

$$L(\boldsymbol{\theta}) = \frac{\mathrm{e}^{-\lambda(\boldsymbol{\theta})}\lambda(\boldsymbol{\theta})^N}{N!}\prod_{i=1}^{N} f(\boldsymbol{x}_i|\boldsymbol{\theta})$$

[1] In the statistical literature this is called a *compound distribution*, see Section 3.58.

and its logarithm

$$\ln L(\boldsymbol{\theta}) = -\lambda(\boldsymbol{\theta}) + N\ln(\lambda(\boldsymbol{\theta})) + \sum_{i=1}^{N} \ln f(\boldsymbol{x}_i|\boldsymbol{\theta}) - \ln N! . \tag{7.4}$$

Again we can omit the last term in the likelihood analysis, because it does not depend on $\boldsymbol{\theta}$.

Example 7.2. Fit of the particle composition of an event sample [36]

We consider the distribution $f(x)$ of a mixture of K different types of particles. The p.d.f. of the identification variable x (This could be for example the energy loss) for particles of type k be $f_k(x)$. The task is to determine the numbers λ_k of the different particle species in the sample from the observed values x_i of N detected particles. The p.d.f. of x is

$$f(x) = \sum_{k=1}^{K} \lambda_k f_k(x) / \sum_{k=1}^{K} \lambda_k$$

and the probability to observe N events is

$$\exp\left(-\sum_{k=1}^{K}\lambda_k\right)\frac{\left(\sum_{k=1}^{K}\lambda_k\right)^N}{N!} .$$

The extended log-likelihood is

$$\begin{aligned}\ln L &= -\sum_{k=1}^{K}\lambda_k + N\ln\sum_{k=1}^{K}\lambda_k + \sum_{i=1}^{N}\ln\sum_{k=1}^{K}\lambda_k f_k(x_i) - N\ln\sum_{k=1}^{K}\lambda_k \\ &= -\sum_{k=1}^{K}\lambda_k + \sum_{i=1}^{N}\ln\sum_{k=1}^{K}\lambda_k f_k(x_i) . \end{aligned} \tag{7.5}$$

To find the MLE, we differentiate $\ln L$:

$$\frac{\partial \ln L}{\partial \lambda_m} = -1 + \sum_{i=1}^{N}\frac{f_m(x_i)}{\sum_{k=1}^{K}\lambda_k f_k(x_i)} = 0,$$

$$1 = \sum_{i=1}^{N}\frac{f_m(x_i)}{\sum_{k=1}^{K}\lambda_k f_k(x_i)} . \tag{7.6}$$

The solution of (7.6) can be obtained iteratively [36]

$$\lambda_m^{(n)} = \sum_{i=1}^{N} \frac{\lambda_m^{(n-1)} f_m(x_i)}{\sum_{k=1}^{K} \lambda_k^{(n-1)} f_k(x_i)}$$

or with a standard maximum searching program applied to (7.5). Alternatively, we can base the fit on (7.5) and constrain the parameters, i.e. require $\Sigma\lambda_k = N$. This solution will be explained in Section 7.5.

As a special case, let us assume that the cross section for a certain reaction is equal to $g(\boldsymbol{x}|\boldsymbol{\theta})$. Then we get the p.d.f. by normalization of g:

$$f(\boldsymbol{x}|\boldsymbol{\theta}) = \frac{g(\boldsymbol{x}|\boldsymbol{\theta})}{\int g(\boldsymbol{x}|\boldsymbol{\theta})\mathrm{d}\boldsymbol{x}}. \tag{7.7}$$

The production rate λ is equal to the normalization factor multiplied with the luminosity S which is a constant:

$$\lambda(\boldsymbol{\theta}) = S \int g(\boldsymbol{x}|\boldsymbol{\theta})\mathrm{d}\boldsymbol{x}. \tag{7.8}$$

The relations (7.7) and (7.8) have to be inserted into (7.4).

7.3 Comparison of Measurements to a Monte Carlo Simulation

7.3.1 *Motivation*

Measurements usually suffer from event losses due to a limited acceptance and limited efficiency of the detectors and from distortions due to the limited resolution of the detectors. Modern research in natural sciences requires more and more complex experimental setups with the consequence that these effects cannot be corrected for analytically. Therefore we simulate the data taking. The correction for the acceptance losses is straight forward, but the correction of smearing effects is more complex. The general problem of unfolding is treated in Chapter 9. In this Section we concentrate on the problem of parameter inference from distorted data. We follow Ref. [37].

7.3.2 *The Likelihood Function*

The theoretical models are represented by Monte Carlo samples where each event is represented by the true value $\boldsymbol{x}$ and its measured value $\boldsymbol{x}'$. The parameter inference is carried out by a comparison of experimental and simulated histograms of the observed variable $\boldsymbol{x}'$. For d_i observed and m_i Monte Carlo events in bin i and a normalization parameter c_m we get for the likelihood instead of (7.1).

$$\ln L = \sum_{i=1}^{B} \left(-c_m m_i + d_i \ln(c_m m_i)\right) \tag{7.9}$$

assuming that the statistical error of the simulation can be neglected, i.e. $m_i \gg d_i$ applies. In some rare cases the normalization c_m is known, if not, it is a free parameter in the likelihood fit. The parameters of interest are hidden in the Monte Carlo predictions $m_i(\boldsymbol{\theta})$.

7.3.3 *The χ^2 Approximation*

If the number of the entries in all bins is large enough to approximate the Poisson distribution by the normal distribution, we can as well minimize the corresponding χ^2 expression (7.3)

$$\chi^2 = \sum_{i=1}^{B} \frac{(d_i - c_m m_i)^2}{c_m m_i}. \tag{7.10}$$

The simulation programs usually consist of two different parts. The first part describes the physical process which depends on the parameters of interest. The second models the detection process. Both parts often require large program packages, the so-called event generators and the detector simulators. The latter usually consume considerable computing power. Limitations in the available computing time then may result in non-negligible statistical fluctuations of the simulation.

7.3.4 *Weighting the Monte Carlo Observations*

When we fit parameters, every parameter change obviously entails a modification of the Monte Carlo prediction. We do not want to repeat the full simulation with every fitting step. Apart from the fact that we want to avoid the computational effort there is another more important reason: With the χ^2 fit, we find the standard error interval by letting vary χ^2 by one unit. On the other hand when we compare experimental data with an optimal simulation, we expect a contribution to χ^2 from the simulation of the order of $\sqrt{2BN/M}$ for B histogram bins, N data events and M Monte Carlo events. Even with a simulation sample which is a hundred times larger than the data sample this value is of the order of one. This means that a repetition of the simulation causes considerable fluctuations of the χ^2 value which have nothing to do with parameter changes. These fluctuations can only be reduced if the same Monte Carlo sample is used for all parameter values. We have to adjust the simulation to the modified parameters by weighting its events.

The weights are computed in the following way: For each generated Monte Carlo event $\boldsymbol{x}, \boldsymbol{x}'$ we store also the true values $\boldsymbol{x}$ of the variates and the corresponding p.d.f. $f(\boldsymbol{x}|\boldsymbol{\theta}_0)$ for the parameter $\boldsymbol{\theta}_0$, which had been used at the generation. When we modify the parameter, we weight each observation by $w(\boldsymbol{\theta}) = f(\boldsymbol{x}|\boldsymbol{\theta})/f(\boldsymbol{x}|\boldsymbol{\theta}_0)$. The sum of the weights then describes the modified prediction. For the histogram bin i with K_i Monte Carlo events it is

$$\begin{aligned} m_i &= \sum_{k=1}^{K_i} w_{ik} \\ &= \sum_{k=1}^{K_i} f(\boldsymbol{x}_{ik}|\boldsymbol{\theta})/f(\boldsymbol{x}_{ik}|\boldsymbol{\theta}_0)\,. \end{aligned}$$

Also re-weighting produces additional fluctuations, which are larger than $\sqrt{K_i}$ These, however, should be tolerable if the weights do not vary too much inside a bin and if the Monte Carlo sample is much larger than the data sample. If we are not sure that this assumption is justified, we can verify it: We reduce the number of Monte Carlo observations and check whether the result remains stable. We know that the contribution of the simulation to the parameter errors scales with the inverse square root of the number of simulated events.

7.3.5 *Inclusion of the Monte Carlo Uncertainty*

In rare cases it is necessary to include the statistical error of the Monte Carlo simulation. The formulas are derived in Appendix A.8 and the problem is discussed in detail in Ref. [26]. We summarize here the relevant relations. The Monte Carlo prediction for a histogram bin is up to a normalization constant $m = \sum w_k$, where the sum runs over all weights w_k of the following quantities $\tilde{m}$, s and $\tilde{c}_m$ have to be evaluated for each bin. We define a scaled number $\tilde{m}$,

$$\tilde{m} = sm$$

with

$$s = \frac{\left[\sum w_k\right]}{\sum w_k^2},$$

and a normalization constant $\tilde{c}$ specific for the bin

$$\tilde{c}_m = c_m / s.$$

The χ^2 expression to be minimized with respect to $\boldsymbol{\theta}$ and c_m is then

$$\chi^2 = \sum_{i=1}^{B} \left[\frac{1}{\tilde{c}_m} \frac{(n - \tilde{c}_m \tilde{m})^2}{(n + \tilde{m})} \right]_i .$$

If resolution effects are absent or small and only acceptance losses have to be taken care of, all weights in bin i can be approximated by

$$w_i = f(\overline{x}_i|\theta)/f(\overline{x}_i|\theta_0)$$

. Here $\overline{x}_i$ is the mean value of all K_i entries of the bin i. The previous relation simplifies with K_i Monte Carlo entries in bin i to

$$\chi^2 = \sum_{i=1}^{B} \frac{1}{c_m w_i} \frac{(n_i - c_m m_i)^2}{(n_i + K_i)} . \tag{7.11}$$

7.3.6 *Fitting a Superposition of Monte Carlo Distributions*

Statistical problems decrease with increasing event numbers, but computational requirements may increase. The numerical minimum search that is required to estimate the wanted parameters can become quite slow. It may happen that we have of the order of 10^6 or more simulated events. This means that, for say 10^3 changes of a parameter value during the extremum search, that 10^9 weights have to be computed. This is feasible, but we may want to speed up the fitting procedure. This can be achieved in situations where the Monte Carlo uncertainties can be neglected. We represent the prediction by a superposition of Monte Carlo histograms with factors that depend on the parameters. To this end it is useful to expand the p.d.f. $f(x|\theta)$ in a Taylor expansion with respect to the parameter at some preliminary estimate θ_0:

$$f(x|\theta) = f(x|\theta_0) + \Delta\theta \frac{df(x|\theta)}{d\theta}|_{\theta_0} + \frac{(\Delta\theta)^2}{2!}\frac{d^2 f(x|\theta)}{d\theta^2}|_{\theta_0} + \cdots \tag{7.12}$$

$$= f(x|\theta_0)\left\{1 + \Delta\theta \frac{1}{f_0}\frac{df(x|\theta)}{d\theta}|_{\theta_0} + \frac{(\Delta\theta)^2}{2!}\frac{1}{f_0}\frac{d^2 f(x|\theta)}{d\theta^2}|_{\theta_0} + \cdots\right\}. \tag{7.13}$$

We generate events according to $f_0(x) = f(x|\theta_0)$ and obtain simulated events with the observed kinematic variable x'. We histogram x' and obtain the histogram m_{0i}. Weighting each event by $\omega_1(x)$, we obtain the histogram m_{1i} and weighting by $\omega_2(x)$ the histogram m_{2i} with the weights

$$\omega_1(x) = \frac{1}{f_0}\frac{df}{d\theta}(x|\theta_0), \tag{7.14}$$

$$\omega_2(x) = \frac{1}{2f_0}\frac{d^2 f}{d\theta^2}(x|\theta_0). \tag{7.15}$$

The parameter inference of $\Delta\theta$ is performed by comparing $m_i = (m_{0i} + \Delta\theta m_{1i} + (\Delta\theta)^2 m_{2i})$ with the experimental histogram d_i as explained in Section 2:

$$\chi^2 = \sum_{i=1}^{B} \frac{(d_i - cm_i)^2}{cm_i}. \tag{7.16}$$

In many cases the quadratic term can be neglected. In other situations it might be necessary to iterate the procedure.

We illustrate the method with two examples.

Example 7.3. Fit of the slope of a linear distribution with Monte Carlo correction

The p.d.f. be

$$f(x|\theta) = \frac{1+\theta x}{1+\theta/2}, \quad 0 \le x \le 1.$$

We generate observations x uniformly distributed in the interval $0 \le x \le 1$, simulate the experimental resolution and the acceptance, and histogram the distorted variable x' into bins i and obtain contents m_{0i}. The same observations are weighted by x and summed up to the histogram m_{1i}. These two distributions are shown in Fig. 7.2 a, b. The dotted histograms correspond

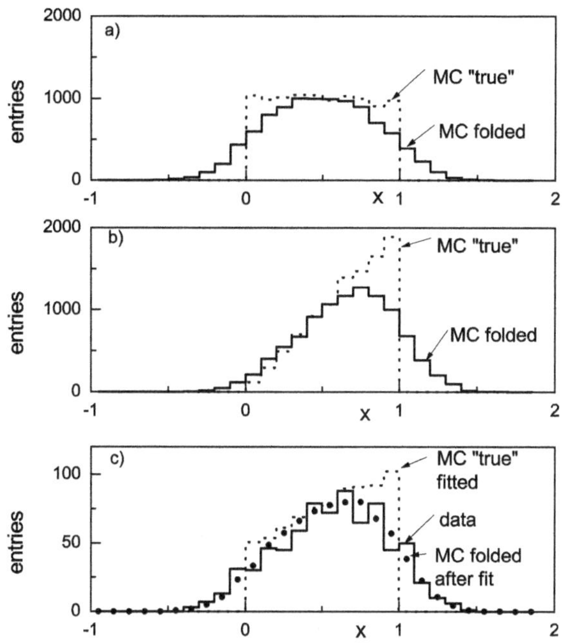

Fig. 7.2 The superposition of two Monte Carlo distributions, a) flat and b) triangular is adjusted to the experimental data.

to the distributions before the distortion by the measurement process. In Fig. 7.2 c is also depicted the experimental distribution. It should be possible to describe it by a superposition m_i of the two Monte Carlo distributions:

$$d_i \sim m_i = m_{0i} + \theta m_{1i}. \tag{7.17}$$

We optimize the parameter θ such that the histogram d_i is described up to a normalization constant as well as possible by a superposition of the two Monte Carlo histograms. We have to insert m_i from (7.17) into (7.9) and set $c_m = N / \sum_i m_i$.

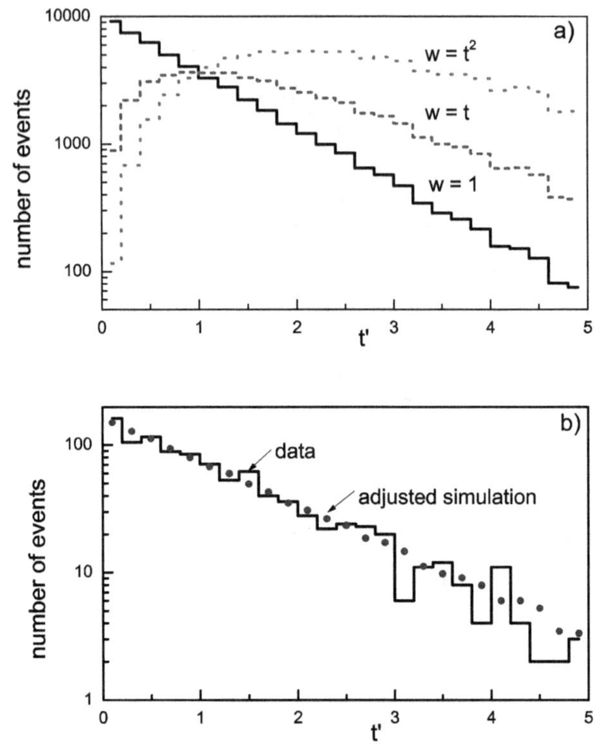

Fig. 7.3 Lifetime fit. The dotted histogram in b) is the superposition of the three histograms of a) with weights depending on $\Delta\lambda$.

Example 7.4. Lifetime fit with Monte Carlo correction

We expand the p.d.f.

$$f(t|\gamma) = \gamma e^{-\gamma t}$$

into a Taylor expansion at γ_0 which is a first guess of the decay rate γ:

$$f(t|\gamma) = \gamma_0 e^{-\gamma_0 t}\left\{1 + \frac{\Delta\gamma}{\gamma_0}(1-\gamma_0 t) + (\frac{\Delta\gamma}{\gamma_0})^2(-\gamma_0 t + \frac{\gamma_0^2 t^2}{2}) + \cdots\right\}.$$

The Monte Carlo simulation follows the distribution $f_0 = \gamma_0 e^{-\gamma_0 t}$. Weighting the events by $(1/\gamma_0 - t)$ and $(-t/\gamma_0 + t^2/2)$, we obtain the distributions $f_1 = (1-\gamma_0 t)e^{-\gamma_0 t}$, $f_2 = (-t + \gamma_0 t^2/2)e^{-\gamma_0 t}$ and

$$f(t|\gamma) = f_0(t) + \Delta\gamma f_1(t) + (\Delta\gamma)^2 f_2(t) + \cdots .$$

If it is justified to neglect the higher powers of $\Delta\gamma/\gamma_0$, we can again describe our experimental distribution this time by a superposition of three distributions $f'_0(t')$, $f'_1(t')$, $f'_2(t')$ which are the distorted versions of f_0, f_1, f_2. The parameter $\Delta\gamma$ is determined by a χ^2 or likelihood fit. In our special case it is even simpler to weight f_0 by t, and t^2, respectively, and to superpose the corresponding distributions f_0, $g_1 = tf_0$, $g_2 = t^2 f_0$ with the factors given in the following expression:

$$f(t|\gamma) = f_0(t)\left(1 + \frac{\Delta\gamma}{\gamma_0}\right) - \gamma_0 g_1(t)\left(\frac{\Delta\gamma}{\gamma_0} + (\frac{\Delta\gamma}{\gamma_0})^2\right) + \frac{1}{2} g_2(t)\gamma_0^2\left(\frac{\Delta\gamma}{\gamma_0}\right)^2 .$$

The parameter $\Delta\gamma$ is then modified until the correspondingly weighted sum of the distorted histograms agrees optimally with the data. Figure 7.3 shows an example. In case the quadratic term can be neglected, two histograms are sufficient.

7.4 Parameter Estimation of a Signal Contaminated by Background

7.4.1 *Introduction*

Frequently, an interesting signal is located above a continuum produced by an uninteresting or unknown physics source. If this background follows Poisson statistics with known mean b_i in bin i of a histogram with B bins, we simply have to modify the expression (7.1) for the log-likelihood to

$$\ln L = \sum_{i=1}^{B} [-(t_i(\theta) + b_i) + d_i \ln(t_i(\theta) + b_i)].$$

The corresponding formula in the LS formalism is:

$$\chi^2 = \sum_{i=1}^{B} \frac{[t_i(\theta) + b_i - d_i]^2}{t_i(\theta) + b_i}.$$

A simple subtraction of the average background from the data d_i would underestimate the uncertainties.

If we are lucky, we have independent experimental information about the background from a separate experiment, if not, we have to parametrize the background distribution.

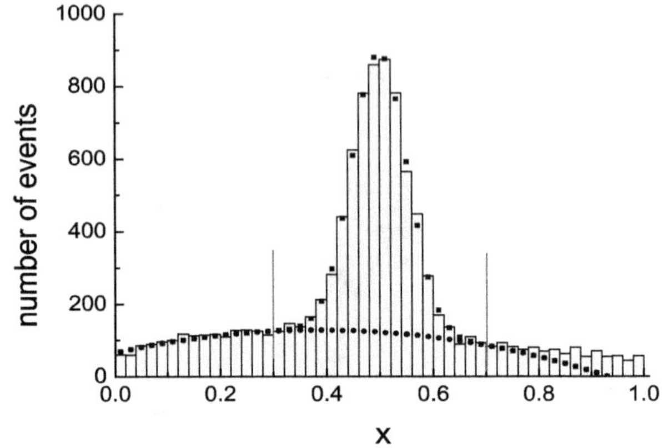

Fig. 7.4 Normally distributed signal contaminated by background.

7.4.2 *Parametrization of the Background*

We have to estimate the background from the shape of the distribution. For a signal consisting of a narrow peak, a possibility is to interpolate the background from the two sides of the peak and to subtract it. A common procedure is to use side bands. This method is not very professional. Instead we should fit the peak together with a linear background distribution. The result is more precise and the error is automatically provided by the fit. Depending on the shape of the distribution, it may be necessary to adjust quadratic or higher order polynomials. There is no absolute save way to estimate the background. We have to accept that there are systematic uncertainties.

Example 7.5. Fit of the parameters of a peak above background

Fig. 7.4 shows a normally distributed peak superposed to a smooth background. The parameters of interest are the number of events α in the peak, its location μ and the corresponding standard deviation σ. We parametrize the background distribution by a quadratic polynomial and fit the parameters of the function

$$f(x) = \frac{\alpha}{\sqrt{2\pi}\sigma} \exp\left[-\frac{(x-\mu)^2}{2\sigma^2}\right] + \beta_0 + \beta_1(x-0.5) + \beta_2(x-0.5)^2$$

to the observed histogram. The following table summarizes the results for linear and quadratic background subtractions and different ranges of x.

backg.	range	$\hat{\alpha}$	$\hat{\mu}$	$\hat{\sigma}$	χ^2/ndf
quad.	[0.2, 0.8]	5033(122)	0.4994(10)	0.0515(11)	0.82
quad.	[0.3, 0.7]	4975(242)	0.4996(10)	0.0512(16)	1.10
linear	[0.3, 0.7]	5131(104)	0.4996(10)	0.0521(10)	1.14
linear	[0.34, 0.66]	5165(133)	0.5006(11)	0.0524(12)	0.88

For the linear background subtraction the fitted amount of background and the width of the peak are larger than for the quadratic background interpolation. The quadratic background function leaves more freedom to the fit than the linear one and consequently the errors become larger. We choose the conservative solution with quadratic background shape and narrow range, $\hat{\alpha} = 4975 \pm 242$, $\hat{\mu} = 0.9996 \pm 0.0010$, $\hat{\sigma} = 0.512 \pm 0.0016$. The error margins cover also the results of the linear background subtraction. Part of the errors are of systematic type caused by the uncertainties in the background parametrization. The purely statistical errors can be estimated by fixing the parameters of the background function. They are $\delta_\alpha = 73$, $\delta_\mu = 0.0008$, $\delta_\sigma = 0.0008$. As expected the precision of the number of events suffers primarily from the uncertain shape of the background. As the statistical and the systematic errors squared add up to the total error squared, we can calculate the systematic contributions $\delta_\alpha^{(sys)} = 231$, $\delta_\mu^{(sys)} = 0.0006$, $\delta_\sigma^{(sys)} = 0.0014$. Except for the location parameter μ, the systematic errors dominate. The values of χ^2 are acceptable in all cases. (The χ^2 goodness-of-fit test will be discussed in Chapter 10.) In our Monte Carlo experiment we know the true parameter values $\alpha = 5000$, $\mu = 0.5$, $\sigma = 0.05$. The linear fits underestimate the background contribution and therefore lead to too large values of σ.

7.4.3 *Histogram Fits with Separate Background Measurement*

In rare cases we have the chance to record independently from a signal sample also a reference sample containing pure background. The measuring times or fluxes, i.e. the relative normalization of the two experiments are either known or to be determined from the data distributions. In this lucky situation, we do not need to parametrize the background distribution and thus are independent of assumptions about its shape.

We introduce B additional parameters β_i for the unknown background prediction. The relative flux normalization c can either be known, or be an unknown parameter in the fit. Our model predicts $t_i(\boldsymbol{\theta}) + \beta_i$ for bin i of the signal histogram and β_i/c for the background histogram. Our LS statistic is

$$\chi^2 = \sum_{i=1}^{B} \frac{[t_i(\boldsymbol{\theta}) + \beta_i - d_i]^2}{t_i(\theta) + \beta_i} + \frac{(\beta_i/c - b_i)^2}{\beta_i/c}$$

for Poisson distributed numbers d_i and b_i.

Especially in low statistics experiments, it is better to avoid the normal approximation and to switch to the Poisson likelihood formalism.

The likelihood is $L \sim \Pi_i \mathcal{P}(d_i|t_i + \beta_i)\Pi_i \mathcal{P}(b_i|\beta_i/c)$ and its logarithm is up to constants

$$\ln L = \sum_{i=1}^{B} [d_i \ln (t_i(\boldsymbol{\theta}) + \beta_i) - (t_i(\boldsymbol{\theta}) + \beta_i) + b_i \ln(\beta_i/c) - \beta_i/c].$$

7.4.4 *A Binning-Free Likelihood Approach*

If the number of events is very small, we may apply a binning-free likelihood fit following a suggestion found in the Russian translation of the book by Eadie *et al.* [8] and which has been introduced probably by the Russian editors [38].

The idea behind the method is simple: The log-likelihood of the wanted signal parameter as derived for the full signal sample is a superposition of the log-likelihood of the genuine signal events and the log-likelihood of the background events. The latter can be estimated from the reference sample and subtracted from the full log-likelihood.

To illustrate the procedure, imagine we want to measure the signal response of a radiation detector by recording a sample of signal heights $x_1, \ldots, x_N$ from a mono-energetic source. For a pure signal, the x_i would follow a normal distribution with resolution σ:

$$f(x|\mu) = \frac{1}{\sqrt{2\pi}\sigma} \mathrm{e}^{-(x-\mu)^2/(2\sigma^2)}.$$

The unknown parameter μ is to be estimated. After removing the source, we can – under identical conditions – take a reference sample $x'_1, \ldots, x'_M$ of background events. They follow a distribution which is of no interest to us.

If we knew, which observations x_i in our signal sample were signal ($x_i^{(S)}$), respectively background ($x_i^{(B)}$) events, we could take only the S

signal events and calculate the correct log-likelihood function which is up to constants

$$\ln L = \sum_{i=1}^{S} \ln f(x_i^{(S)}|\mu) = -\sum_{i=1}^{S} \frac{(x_i^{(S)} - \mu)^2}{2\sigma^2}$$
$$= \sum_{i=1}^{N} \frac{(x_i - \mu)^2}{2\sigma^2} - \sum_{i=1}^{B} \frac{(x_i^{(B)} - \mu)^2}{2\sigma^2},$$

with $S + B = N$. The second unknown term can be estimated from the control sample

$$\sum_{i=1}^{B} \frac{(x_i^{(B)} - \mu)^2}{2\sigma^2} \approx \sum_{i=1}^{M} \frac{(x'_i - \mu)^2}{2\sigma^2}.$$

The logarithm of our corrected log-likelihood becomes:

$$\ln \tilde{L} = \sum_{i=1}^{N} \frac{(x_i - \mu)^2}{2\sigma^2} - \sum_{i=1}^{M} \frac{(x'_i - \mu)^2}{2\sigma^2}.$$

We call it pseudo log-likelihood, $\ln \tilde{L}$, to distinguish it from a genuine log-likelihood. To obtain the estimate $\hat{\mu}$ of our parameter, we look for the parameter $\hat{\mu}$ which maximizes $\ln \tilde{L}$ and find the expected simple function of the mean values $\overline{x}$, $\overline{x'}$:

$$\hat{\mu} = \frac{\sum_{i=1}^{N} x_i - \sum_{i=1}^{M} x'}{N - M}$$
$$= \frac{N\overline{x} - M\overline{x'}}{N - M}. \tag{7.18}$$

The general problem where the sample and parameter spaces could be multidimensional and with different fluxes of the signal and the reference sample, is solved in complete analogy to our example: Given a contaminated signal distribution of size N and a reference distribution of size M and flux c times larger than that of the signal sample, we put

$$\ln \tilde{L} = \sum_{i=1}^{N} \ln f(\boldsymbol{x}_i|\boldsymbol{\theta}) - \frac{1}{c}\sum_{i=1}^{M} \ln f(\boldsymbol{x}'_i|\boldsymbol{\theta}). \tag{7.19}$$

The expected value of $\ln \tilde{L}$ and $\ln L$ agree. The formula (7.19) is completely general and avoids histogramming which is problematic for low event counts. The LS method with subtraction of a background histogram from the signal histogram often fails in such a situations.

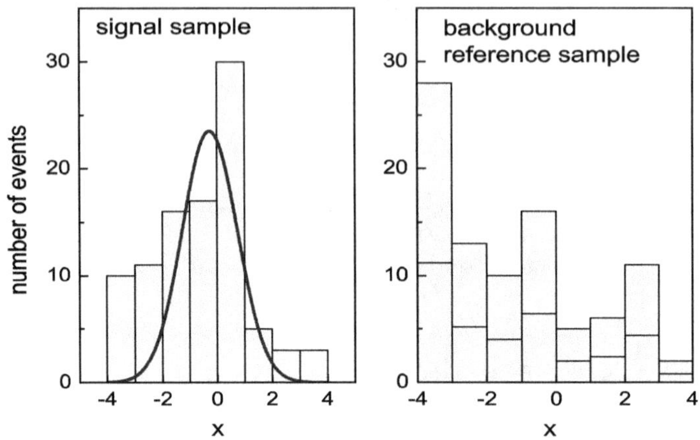

Fig. 7.5 Experimental distribution of a normally distributed signal over background (left) and background reference sample (right). The lower histogram is scaled to the signal flux.

The shape itself of the pseudo likelihood cannot be used directly to estimate the parameter errors. It has to be determined by error propagation or alternatively with the bootstrap method, where we take a large number of bootstrap samples from the experimental distributions of both the signal- and the control experiment and calculate the background corrected parameter estimate for each pair of samples, see Section 12.2.

Example 7.6. Fit of the parameters of a peak with a background reference sample

Fig. 7.5 shows an experimental histogram of a normally distributed signal of width $\sigma = 1$ contaminated by background, together 95 events with mean $\overline{x} = 0.61$ and empirical variance $v^2 = 3.00$. The right hand side is the distribution of a background reference sample with $1/r = 2.5$ times the flux of the signal sample, containing 91 events with mean $\overline{x'} = -1.17$ and variance $v'^2 = 4.79$. The mean of the signal is obtained from the flux corrected version of (7.18):

$$\hat{\mu} = \frac{N\overline{x} - rM\overline{x'}}{N - rM}$$
$$= \frac{95 \cdot 0.61 - 0.4 \cdot 91 \cdot 1.17}{95 - 0.4 \cdot 91} = -0.26 \pm 0.33.$$

The error of the result can be estimated with the bootstrap method. One would have to generate about hundred pairs of bootstrap samples corresponding to the signal and the background and repeat each time the analysis. From the distributions of the estimated parameters the errors could be extracted, see Section 12.2.

The distributions of the example were generated with nominally 60 pure signal plus 40 background events and 100 background reference events. The signal corresponds to a normal distribution, $\mathcal{N}(x|0,1)$, and the background to an exponential, $\sim \exp(-0.2x)$.

A different method, where the shape of the background distribution is approximated by probability density estimation (PDE) will be presented in Section 12.1.1.

7.5 Inclusion of Constraints

7.5.1 *Introduction*

The interesting parameters are not always independent of each other but are often constrained by physical or geometrical laws.

As an example let us look at the decay of a Λ particle into a proton and a pion, $\Lambda \to p + \pi$, where the direction of flight of the Λ hyperon and the momentum vectors of the decay products are measured. The momentum vectors of the three particles which participate in the reaction are related through the conservation laws of energy and momentum. Taking into account the conservation laws, we add information and can improve the precision of the momentum determination.

In the following we assume that we have N direct observations x_i which are predicted by functions $t_i(\boldsymbol{\theta})$ of a parameter vector $\boldsymbol{\theta}$ with P components and K constraints of the form $h_k(\boldsymbol{\theta}) = 0$. Let us assume further that the uncertainties Δ_i of the observations are normally distributed and that the constraints are fulfilled with the precision δ_k,

$$\begin{aligned} \left\langle (t_i(\boldsymbol{\theta}) - x_i)^2 \right\rangle &= \Delta_i^2 , \\ \left\langle h_k^2(\boldsymbol{\theta}) \right\rangle &= \delta_k^2 . \end{aligned} \tag{7.20}$$

Then χ^2 is:

$$\chi^2 = \sum_{i=1}^{N} \frac{[x_i - t_i(\boldsymbol{\theta})]^2}{\Delta_i^2} + \sum_{k=1}^{K} \frac{h_k^2(\boldsymbol{\theta})}{\delta_k^2} . \tag{7.21}$$

We minimize χ^2 by varying the parameters and obtain their best estimates at the minimum of χ^2. The procedure works also when the constraints contain more than N parameters, as long as the number of parameters P does not exceed the the number of terms $N+K$. We assume that there is a single minimum.

A corresponding likelihood fit would maximize

$$\ln L = \sum_{i=1}^{N} \ln f(x_i|\boldsymbol{\theta}) - \frac{1}{2}\sum_{k=1}^{K} \frac{h_k^2(\boldsymbol{\theta})}{\delta_k^2} .$$

In most cases the constraints are obeyed exactly, $\delta_k = 0$, and the second term in (7.21) diverges. This difficulty is avoided in the following three procedures:

(1) The constraints are used to reduce the number of parameters and consider the limit $\delta_k \to 0$.
(2) We approximate the constraints by narrow Gaussians.
(3) Lagrange multipliers are adjusted to satisfy the constraints.

We will discuss the problem in terms of a χ^2 minimization. The solutions can also be applied to likelihood fits.

7.5.2 *Eliminating Redundant Parameters*

Sometimes it is possible to eliminate parameters by expressing them by an unconstrained subset.

Example 7.7. Fit with constraint: Two pieces of a rope

A rope of exactly 1 m length is cut into two pieces. A measurement of both pieces yields $l_1 = 35.3\,\mathrm{cm}$ and $l_2 = 64.3\,\mathrm{cm}$, both with the same Gaussian uncertainty of $\delta = 0.3$. We have to find the estimates $\hat{\lambda}_1, \hat{\lambda}_2$ of the lengths. We minimize

$$\chi^2 = \frac{(l_1 - \lambda_1)^2}{\delta^2} + \frac{(l_2 - \lambda_2)^2}{\delta^2}$$

including the constraint $\lambda_1 + \lambda_2 = l = 100\,\mathrm{cm}$. We simply replace λ_2 by $l - \lambda_1$ and adjust λ_1, minimizing

$$\chi^2 = \frac{(l_1 - \lambda_1)^2}{\delta^2} + \frac{(l - l_2 - \lambda_1)^2}{\delta^2} .$$

The minimization relative to λ_1 leads to the result:

$$\hat{\lambda}_1 = \frac{l}{2} + \frac{l_1 - l_2}{2} = 35.5 \pm 0.2\,\text{cm}$$

and the corresponding estimate of λ_2 is just the complement of $\hat{\lambda}_1$ with respect to the full length. Note that due to the constraint the error of λ_i is reduced by a factor $\sqrt{2}$, as can easily be obtained by error propagation. The constraint has the same effect as a double measurement, but with the modification that now the results are (maximally) anti-correlated: one finds $\text{cov}(\lambda_1\lambda_2) = -\text{var}(\lambda_i)$.

Example 7.8. Fit of the particle composition of an event sample, reconsidered

A particle identification variable x has different distributions $f_m(x)$ for different particles. The p.d.f. given the relative particle abundance λ_m for particle species m out of M different particles is

$$f(x|\lambda_1, \ldots, \lambda_M) = \sum_{m=1}^{M} \lambda_m f_m(x),$$

$$\sum_{m=1}^{M} \lambda_m = 1. \tag{7.22}$$

As the constraint relation is linear, we can easily eliminate the parameter λ_M to get rid of the constraint $\sum \lambda_m = 1$:

$$f'(x|\lambda_1, \ldots, \lambda_{M-1}) = \sum_{m=1}^{M-1} \lambda_m f_m(x) + (1 - \sum_{m=1}^{M-1} \lambda_m) f_M(x).$$

The log-likelihood for N particles is

$$\ln L = \sum_{i=1}^{N} \ln \left[\sum_{m=1}^{M-1} \lambda_m f_m(x_i) + (1 - \sum_{m=1}^{M-1} \lambda_m) f_M(x_i) \right].$$

From the MLE we obtain in the usual way the first $M - 1$ parameters and their error matrix E. The remaining parameter λ_M and the related error matrix elements E_{Mj} are derived from the constraint (7.22) and the corresponding relation $\sum \Delta\lambda_m = 0$. The diagonal error is the expected value of $(\Delta\lambda_M)^2$:

$$\Delta\lambda_M = -\sum_{m=1}^{M-1} \Delta\lambda_m,$$

$$(\Delta\lambda_M)^2 = \left[\sum_{m=1}^{M-1} \Delta\lambda_m\right]^2,$$

$$E_{MM} = \sum_{m=1}^{M-1} E_{mm} + \sum_{m}^{M-1}\sum_{l\neq m}^{M-1} E_{ml}.$$

The remaining elements are computed analogously:

$$E_{Mj} = E_{jM} = -E_{jj} - \sum_{m\neq j}^{M-1} E_{mj}.$$

An iterative method, called *channel likelihood*, to find the particle contributions, is given in [34].

These trivial examples are not really representative for the typical problems we have to solve in particle or astrophysics. Indeed, it is often complicated or even impossible to reduce the parameter set analytically to an unconstrained subset, but we can introduce a new unconstrained parameter set which then predicts the measured quantities. To find such a set is straight forward in the majority of problems: We just have to think how we would simulate the corresponding experimental process. A simulation is always based on a minimum set of parameters. The constraints are satisfied automatically.

Example 7.9. Kinematic fit with constraints: eliminating parameters

A neutral particle c is decaying into two charged particles a and b, for instance $\Lambda \to p + \pi^-$. The masses m_c, m_a, m_b are known. Measured are the decay vertex $\boldsymbol{\rho}$ and the momenta $\boldsymbol{\pi}_a, \boldsymbol{\pi}_b$ of the decay products. The measurements of the components of the momentum vectors are correlated. The inverse error matrices be V_a and V_b. The origin of the decaying particle be at the origin of the coordinate system. Thus we have 9 measurements $(\boldsymbol{r}, \boldsymbol{p}_a, \boldsymbol{p}_b)$, 10 parameters, namely the 3 momentum vectors and the distance $(\boldsymbol{\pi}_c, \boldsymbol{\pi}_a, \boldsymbol{\pi}_b, \rho)$, and 4 constraints from momentum and energy conservation:

$$\boldsymbol{\pi}(\boldsymbol{\pi}_c, \boldsymbol{\pi}_a, \boldsymbol{\pi}_b) \equiv \boldsymbol{\pi}_c - \boldsymbol{\pi}_a - \boldsymbol{\pi}_b = 0,$$

$$\varepsilon(\boldsymbol{\pi}_c, \boldsymbol{\pi}_a, \boldsymbol{\pi}_b) \equiv \sqrt{\pi_c^2 + m_c^2} - \sqrt{\pi_a^2 + m_a^2} - \sqrt{\pi_b^2 - m_b^2} = 0.$$

The corresponding χ^2 expression is

$$\chi^2 = \sum_{i=1}^{3} \left(\frac{r_i - \rho_i}{\delta r_i} \right)^2 + \sum_{i,j=1}^{3} (p_{ai} - \pi_{ai}) V_{aij} (p_{aj} - \pi_{aj}) + \sum_{i,j=1}^{3} (p_{bi} - \pi_{bi}) V_{bij} (p_{bj} - \pi_{bj}).$$

A correlation of the Cartesian components of the momenta of particles a and b are taken into account by the weight matrices V_a and V_b. The vertex parameters ρ_i are fixed by the vector relation $\boldsymbol{\rho} = \rho \boldsymbol{\pi}_c / |\boldsymbol{\pi}_c|$. We would like to remove 4 out of the 10 parameters using the 4 constraints. A Monte Carlo simulation of the Λ decay would proceed as follows: First we would select the Λ momentum vector (3 parameters). Next the decay length would be generated (1 parameter). The decay of the Λ hyperon into proton and pion is fully determined when we choose the proton direction in the lambda center of mass system (2 parameters). All measured laboratory quantities and thus also χ^2 can then be expressed analytically by these 6 unconstrained quantities (we omit here the corresponding relations) which are varied in the fitting procedure until χ^2 is minimal. Of course in the fit we would not select random starting values for the parameters but the values which we compute from the experimental decay length and the measured momentum vectors. Once the reduced parameter set has been adjusted, it is straight forward to compute also the remaining laboratory momenta and their errors which, obviously, are strongly correlated.

Often the reduced parameter set is more relevant than the set corresponding to the measurement, because a simulation usually is based on parameters which are of scientific interest. For example, the investigation of the Λ decay might have the goal to determine the Λ decay parameter which depends on the center of mass direction of the proton relative to the Λ polarization, i.e. on one of the directly fitted quantities.

7.5.3 *Gaussian Approximation of Constraints*

The direct inclusion of the constraint through a penalty term in the fit is technically very simple and efficient.

We have to minimize:

$$\chi^2 = \lim_{\delta k \to 0, k=1,K} \sum_{i=1}^{N} \frac{[x_i - t_i(\boldsymbol{\theta})]^2}{\Delta_i^2} + \sum_{k=1}^{K} \frac{h_k^2(\boldsymbol{\theta})}{\delta_k^2}. \tag{7.23}$$

The parameters δ_k have to be small compared to the experimental resolution of the constraint. Parameter estimation is performed by numerical approximation in computer programs following methods like Simplex. The require precision is steered by a parameter provided by the user. The widths of the Gaussian approximation should not affect the anticipated precision of the fit.

To estimate the resolution, the constraint is evaluated from the observed data, $\tilde{h}(x_i, ..., x_N)$ and we require $\delta_k^2 << \overline{\tilde{h}^2}$. The precise choice of the constraint precision δ_k is not at all critical, but extremely small values of δ_k could lead to numerical problems. In case the minimum search is slow, or does not converge, one should start initially with loose constraints which subsequently could be tightened.

The value of χ^2 in the major part of the parameter space is dominated by the contributions from the constraint terms. In the minimum searching programs the parameter point will therefore initially move quickly from its starting value towards the subspace defined by the constraint equations and then proceed towards the minimum of χ^2.

Remark that the minimum of χ^2 is found at parameter values that satisfy the constraints much better than naively expected from the set of constraint tolerances. The reason is the following: Once the parameters are close to their estimates, small changes which reduce the χ^2 contribution of the penalty terms, will hardly affect the remaining terms. Thus the minimum will be observed very close to $h_k = 0$. As a consequence, the contribution of the K constraint terms in (7.21) to the minimum value of χ^2 is negligible.

Example 7.10. Example 7.7 continued

We approximate the constraint by a narrow Gaussian. The value $\delta_k^2 = 10^{-10}\delta^2$ is chosen small compared to the measurement error δ.

$$\chi^2 = \frac{(l_1 - \lambda_1)^2}{\delta^2} + \frac{(l_2 - \lambda_2)^2}{\delta^2} + \frac{(\lambda_1 + \lambda_2 - l)^2}{(10^{-5}\delta)^2}.$$

The results of the χ^2 fit reproduces the results of the fit where one parameter is eliminated.

For a numerical test we consider the decay of a Λ hyperon into a proton and a pion and simplify example 7.9.

Example 7.11. Example 7.9 continued

To simplify the equations, we can fix the decay length z of the lambda hyperon and the absolute value of its momentum because these two quantities are not related to the constraint equations. In the simulation the Λ particle moves along the z axis. The measured coordinates x, y are small compared to the decay length $r \approx z$ which is fixed. The χ^2 expression is

$$\chi^2 = \frac{(x-\xi)^2}{\delta_x^2} + \frac{(y-\theta)^2}{\delta_y^2} +$$
$$\sum_{i,j=1}^{3} (p_{ai} - \pi_{ai}) V_{aij} (p_{aj} - \pi_{aj}) + \sum_{i,j=1}^{3} (p_{bi} - \pi_{bi}) V_{bij} (p_{bj} - \pi_{bj}) +$$
$$\frac{(\xi/z - \pi_{cx}/\pi_{cz})^2}{\delta_\alpha^2} + \frac{(\theta/z - \pi_{cy}/\pi_{cz})^2}{\delta_\alpha^2} + \frac{(m_{p\pi} - m_\Lambda)^2}{\delta_m^2}. \tag{7.24}$$

The first two terms of (7.24) compare the x and y components of the Λ path vector with the corresponding parameters ξ and θ. The next two terms measure the difference between the observed and the fitted momentum components of the proton and the pion. The following two terms constrain the direction of the Λ hyperon flight path to the direction of the momentum vector $\boldsymbol{\pi} = \boldsymbol{\pi}_a + \boldsymbol{\pi}_b$ and the last term constrains the invariant mass $m_{p\pi}(\boldsymbol{\pi}_a, \boldsymbol{\pi}_b)$ of the decay products to the Λ mass. We generate 10^4 events, all with the same nominal parameter values but different normally distributed measurement errors. The velocity of the Λ particle is parallel to the z axis with a Lorentz factor $\gamma = 9$. The decay length is fixed to $1\,m$. The direction of the proton in the Λ center of mass is defined by the polar and azimuthal angles $\theta = 1.5$, $\phi = 0.1$. The measurement errors of the x and y coordinates are $\delta_x = \delta_y = 1\,cm$. The momentum error is assumed to be the sum of a term proportional to the momentum p squared, $\delta_{pr} = 2p^2/(GeV)^2$ and a constant term $\delta_{p0} = 0.02\,GeV$ added to each momentum component. The tolerances for the constraints are $\delta_\alpha = 0.001$

and $\delta_m = 0.1\,MeV$, i.e. about 10^{-3} times the experimental uncertainty. The minimum search is performed with a combination of a simplex and a parabolic minimum searching routine. The starting values for the parameters are the measured values. The fit starts with a typical value of χ_0^2 of 2×10^8 and converges for all events with a mean value of χ^2 of 2.986 and a mean value of the standard deviation of 2.446 to be compared to the nominal values 3 and $\sqrt{6} = 2.450$. The contribution from each of the three constraint terms to χ^2 is 10^{-4}. Thus the deviation from the constraints is about 10^{-7} times the experimental resolution.

7.5.4 *The Method of Lagrange Multipliers*

This time we choose the likelihood presentation of the problem. The likelihood function is extended to

$$\ln L = \sum_{i=1}^{N} \ln f(x_i|\boldsymbol{\theta}) + \sum_{k} \alpha_k h_k(\boldsymbol{\theta}). \tag{7.25}$$

We have appended an expressions that in the end should be equal to zero, the constraint functions multiplied by the so-called *Lagrange multipliers*. The MLE as obtained by setting $\partial \ln L/\partial\theta_j = 0$ yields parameters that depend on the Lagrange multipliers α. We can now use the free parameters α_k to fulfill the constraints, or in other words, we use the constraints to eliminate the Lagrange multiplier dependence of the MLE.

Example 7.12. Example 7.7 continued

Our full likelihood function is now

$$\ln L = -\frac{(l_1 - \lambda_1)^2}{2\delta^2} - \frac{(l_2 - \lambda_2)^2}{2\delta^2} + \alpha(\lambda_1 + \lambda_2 - l)$$

with the MLE $\hat{\lambda}_{1,2} = l_{1,2} - \delta^2\alpha$. Using $\hat{\lambda}_1 + \hat{\lambda}_2 = l$ we find $\delta^2\alpha = (l_1 + l_2 - l)/2$ and, as before, $\hat{\lambda}_1 = (l + l_1 - l_2)/2$, $\hat{\lambda}_2 = (l + l_2 - l_1)/2$.

Of course, in general the situation is much more complicated than that of our trivial example. An analytic solution will hardly be possible. Instead we can set the derivative of the log-likelihood not only with respect to the parameters $\boldsymbol{\theta}$ but also with respect to the multipliers α_k equal to zero, $\partial \ln L/\partial \alpha_k = 0$, which automatically implies, see (7.25), that the constraints are satisfied. Unfortunately, the zero of the derivative corresponds to a saddle point and cannot be found by a maximum searching routine. More subtle numerical methods have to be applied.

Most methods avoid this complication and limit themselves to linear regression models which require a linear dependence of the observations on the parameters and linear constraint relations. Non-linear problems are then solved iteratively. The solution then is obtained by a simple matrix calculus.

Linear regression has be sketched in Section 6.8.1 and the inclusion of constraints in Appendix A.11. For a detailed discussion see Ref. [39].

7.5.5 *Conclusion*

By far the simplest method is the one where the constraint is directly included and approximated by a narrow Gaussian. With conventional minimizing programs the full error matrix is produced automatically.

The approach using a reduced parameter set is especially interesting when we are primarily interested in the parameters of the reduced set. This is the case in most kinematic fits. Due to the reduced dimension of the parameter space, it is faster than the other methods. The determination of the errors of the original parameters through error propagation is sometimes tedious, but in most applications only the reduced set is of interest.

It is recommended to either eliminate redundant parameters or to use the simple method where we represent constraints by narrow Gaussians. The application of Lagrange multipliers is unnecessarily complicated and the linear approximation requires additional assumptions and iterations.

7.6 Reduction of the Number of Variates

7.6.1 *The Problem*

A statistical analysis of a uni-variate sample is obviously much simpler than that of a multi-dimensional one. This is not only true for the qualitative comparison of a sample with a parameter dependent p.d.f. but also for the quantitative parameter inference. Especially when the p.d.f. is distorted by the measurement process and a Monte Carlo simulation is required, the direct ML method cannot be applied as we have seen above. The parameter inference then happens by comparing histograms with the problem that in multidimensional spaces the number of entries can be quite small in some bins. Therefore, we have an interest to reduce the dimensionality of the variable space by appropriate transformations, of course, if possible, without loss of information. However, it is not always easy to find out which variable or which variable combination is especially important for the parameter estimation.

7.6.2 *Two Variables and a Single Linear Parameter*

A p.d.f. $f(x,y|\theta)$ of two variates with a linear parameter dependence can always be written in the form

$$f(x,y|\theta) = v(x,y)[1 + u(x,y)\theta]. \tag{7.26}$$

The parameter θ depends only on the variable u and can be estimated by minimizing the log-likelihood function of the observations x_i, y_i. It is not necessary, to perform the analytic variable transformation $f(x,y|\theta) \to g(u,v|\theta)$ that is not always possible.

$$\ln L(\theta) = \sum_i \ln(1 + u_i(x_i, y_i)\theta).$$

Equivalent and not more complicated is a maximum likelihood fit in the x, y space, but often we have to correct for acceptance losses or resolution effects. Then we are forced to apply a LSF of the data to a Monte-Carlo simulation which is simpler in one dimension. We then compare a histogram of the observed values u_i to a superposition of two simulated histograms generated with two different values of θ.

To estimate θ the following recipe can be applied:

(1) Compute for each observation x_i, y_i the variable $u_i = u(x_1, y_i)$ and build the histogram d of u_i.
(2) Select two values θ_1, θ_2 of the parameter, generate events, compute u and construct histograms $\boldsymbol{t_1}, \boldsymbol{t_2}$.
(3) Perform a LS fit of the superposition of the two simulated histograms to the observed histogram,

$$\chi^2 = \sum \frac{[d_j - (\alpha t_{1j} + \beta t_{2j})]^2}{\delta_j^2},$$

with δ_j the error of the bracket in the numerator. Estimate $\hat{\alpha}, \hat{\beta}$ and their errors.
(4) compute

$$\hat{\theta} = \frac{\hat{\alpha}\theta_1 + \hat{\beta}\theta_2}{\hat{\alpha} + \hat{\beta}}.$$

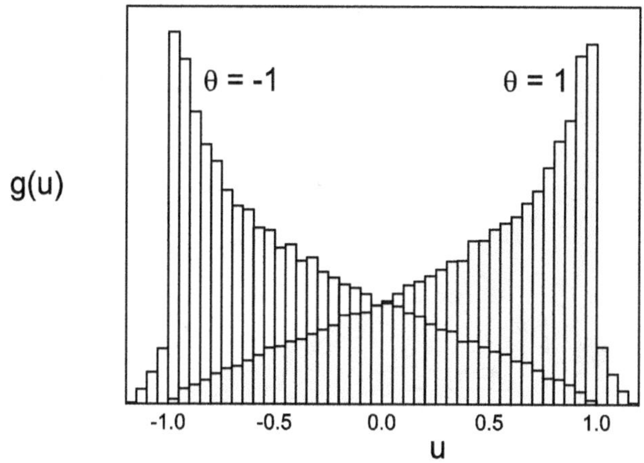

Fig. 7.6 Simulated p.d.f.s of the reduced variable u for the values $\pm\theta$ of the parameter.

The steps 3 and 4 can be combined. Then the parameter θ is adjusted directly. Alternatively we can perform a Poisson likelihood fit, as in an example below and apply the methods discussed in Section 7.1.

7.6.3 *Generalization to Several Variables and Parameters*

The generalization to N variates which we combine to a vector $\boldsymbol{x}$ is trivial:

$$f(\boldsymbol{x}|\theta) = v(\boldsymbol{x})\,[1 + u(\boldsymbol{x})\theta] .$$

Again we can reduce the variate space to a single significant variate u without loosing relevant information. If simultaneously P parameters have to be determined, we usually will need also P new variates $u_1, \ldots, u_P$:

$$f(\boldsymbol{x}|\boldsymbol{\theta}) = v(\boldsymbol{x}) \left[1 + \sum_p u_p(\boldsymbol{x})\theta_p\right] .$$

Thus our procedure makes sense only if the number of parameters is smaller than the dimension of the variate space.

Example 7.13. Reduction of the variate space

We consider the p.d.f.

$$f(x, y, z|\theta) = \frac{1}{\pi}\left[(x^2 + y^2 + z^2)^{1/2} + (x + y^3)\theta\right], \; x^2 + y^2 + z^2 \leq 1, \quad (7.27)$$

which depends on three variates and one parameter. For a given sample of observations in the three dimensional Cartesian space we want to determine the parameter θ. The substitutions

$$u = \frac{x + y^3}{(x^2 + y^2 + z^2)^{1/2}}, \ |u| \le \sqrt{2},$$
$$v = (x^2 + y^2 + z^2)^{1/2}, \ 0 \le v \le 1,$$
$$z = z$$

lead to the new p.d.f. $g'(u, v, z)$

$$g'(u, v, z|\theta) = \frac{v}{\pi}\,[1 + u\,\theta]\,\frac{\partial(x, y, z)}{\partial(u, v, z)},$$

which after integrating out v and z yields the p.d.f. $g(u|\theta)$:

$$g(u|\theta) = \int \mathrm{d}z\,\mathrm{d}v\,g'(u, v, z|\theta).$$

This operation is not possible analytically but we do not need to compute g explicitly. We are able to determine the MLE and its error from the simple log-likelihood function of θ

$$\ln L(\theta) = \sum_i \ln(1 + u_i\theta).$$

In case we have to account for acceptance effects, we have to simulate the distribution of u and to apply the method explained in Section 7.3, where an example for a linear parameter fit is presented.

7.6.4 *Non-Linear Parameters*

The example which we just investigated is especially simple because the p.d.f. depends linearly on a single parameter. Linear dependencies are quite frequent because distributions often consist of a superposition of several processes, and the interesting parameters are the relative weights of those.

For the general, non-linear case we restrict ourselves to a single parameter to simplify the notation. We expand the p.d.f. into a Taylor series at a first rough estimate θ_0:

$$\begin{aligned} f(\boldsymbol{x}|\theta) &= f(\boldsymbol{x}|\theta_0) + \frac{\partial f}{\partial \theta}|_{\theta_0}\,\Delta\theta + \frac{1}{2}\frac{\partial^2 f}{\partial \theta^2}|_{\theta_0}\,\Delta\theta^2 + \cdots \\ &= V\left[1 + u_1\Delta\theta + u_2\Delta\theta^2 + \cdots\right]. \end{aligned} \tag{7.28}$$

As before, we choose the coefficients u_i as new variates. Neglecting quadratic and higher terms, the estimate $\hat{\theta} = \theta_0 + \widehat{\Delta\theta}$ depends only on the new variate u_1,

$$u_1(\boldsymbol{x}) = \frac{\partial f(\boldsymbol{x}|\theta)/\partial\theta\,|_{\theta_0}}{f(\boldsymbol{x}|\theta_0)}$$

which is a simple function of $\boldsymbol{x}$.

If the linear approximation is not sufficient, a second variate u_2 corresponding to the second term of the expansion should be added. Alternatively, the solution can be iterated. The generalization to several parameters is straight forward.

A more detailed description of the method with application to a physics process can be found in Refs. [40, 41]. The corresponding choice of the variate is also known under the name *optimal variable method* [42].

7.7 Approximated Likelihood Estimators

As in Section 7.3 we investigate the situation where we have to estimate parameters in presence of acceptance and resolution effects. The idea of the method is the following: We fit the parameters ignoring the distortion and obtain a biased estimate θ'. The bias is then corrected based on a Monte Carlo simulation which provides the relation $\theta(\theta')$ between the parameter of interest θ and the observed quantity θ'. The method should become clear in the following example.

Example 7.14. Approximated likelihood estimator: Lifetime fit from a distorted distribution

The sample mean $\bar{t}$ of a sample of N undistorted exponentially distributed lifetimes t_i is a sufficient estimator: It contains the full information related to the parameter τ, the mean lifetime (see Section 6.6.1). In case the distribution is distorted by resolution and acceptance effects (Fig. 7.7), the mean value

$$\overline{t'} = \sum t'_i/N$$

of the distorted sample t'_i will usually still contain almost the full information relative to the mean life τ. The relation $\tau(\overline{t'})$ between τ and its approximation $\overline{t'}$ (see insert of Fig. 7.7) is generated by a Monte Carlo

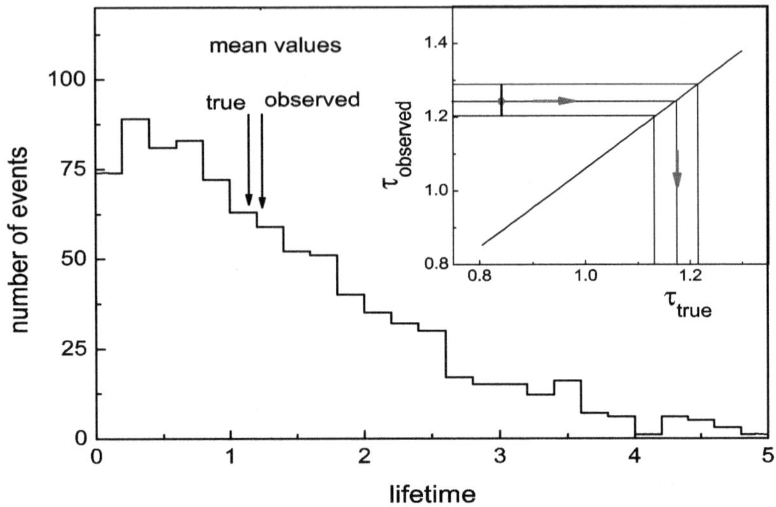

Fig. 7.7 Measured lifetime distribution. The insert indicates the transformation of the measured lifetime to the corrected one.

simulation. The uncertainty $\delta\tau$ is obtained by error propagation from the uncertainty $\delta\overline{t'}$ of $\overline{t'}$,

$$(\delta\overline{t'})^2 = \frac{(\overline{t'^2} - \overline{t'}^2)}{N-1},$$

$$\text{with} \quad \overline{t'^2} = \frac{1}{N}\sum t_i'^2$$

using the Monte Carlo relation $\tau(\overline{t'})$.

This approach has several advantages:

- We do not need to histogram the observations.
- Problems due to small event numbers for bins in a multivariate space are avoided.
- It is robust, simple and requires little computing time.

For these reasons the method is especially suited for online applications, provided that we find an efficient estimator.

If the distortions are not too large, we can use the likelihood estimator extracted from the observed sample $\{x'_1, \ldots, x'_N\}$ and the undistorted distribution $f(x|\lambda)$:

$$\begin{aligned} L(\lambda) &= \prod f(x'_i|\lambda), \\ \frac{\mathrm{d}L}{\mathrm{d}\lambda}|_{\hat{\lambda}'} &= 0. \end{aligned} \tag{7.29}$$

This means concretely that we perform the usual likelihood analysis where we ignore the distortion. We obtain $\hat{\lambda}'$. Then we correct the bias by a Monte Carlo simulation which provides the relation $\hat{\lambda}(\hat{\lambda}')$.

In rare cases where the experimental resolution is very bad, it can happen that $f(x|\lambda)$ is undefined for some extremely distorted observations. This problem can be cured by scaling $\hat{\lambda}'$ or by eliminating particular observations.

Acceptance losses $\alpha(x)$ alone without resolution effects do not necessarily entail a reduction in the precision of our approach. For example, as has been shown in Section 6.5.2, cutting an exponential distribution at some maximum value of the variate, the mean value of the observations is still a sufficient statistic. But there are cases where sizable acceptance losses have the consequence that our method deteriorates. In these cases we have to take the losses directly into account. We only sketch a suitable method.

The observed p.d.f. $f'(x|\lambda)$ for the variate x is

$$f'(x|\lambda) = \frac{\alpha(x)f(x|\lambda)}{\int \alpha(x)f(x|\lambda)\mathrm{d}x},$$

where $\alpha(x)$ is the local acceptance. The denominator is the global acceptance and provides the correct normalization. We abbreviate it by $A(\lambda)$. The log-likelihood of N observations is

$$\ln L(\lambda) = \sum \ln \alpha(x_i) + \sum \ln f(x_i|\lambda) - NA(\lambda).$$

The first term can be omitted. The acceptance $A(\lambda)$ can be determined by a Monte Carlo simulation. Again a rough estimation is sufficient, at most it reduces the precision but does not introduce a bias, since all approximations are automatically corrected with the transformation $\lambda(\lambda')$.

Frequently, the relation (7.29) can only be solved numerically, i.e. we find the maximum of the likelihood function in the usual manner. We are also allowed to approximate this relation such that an analytic solution is possible. The resulting error is compensated in the simulation.

Example 7.15. Approximated likelihood estimator: Linear and quadratic distributions

A sample of events x_i is distributed linearly inside the interval $[-1, 1]$, i.e. the p.d.f. is $f(x|b) = 0.5 + bx$. The slope $b, |b| < 1/2$, is to be fitted. It is located in the vicinity of b_0. We expand the likelihood function

$$\ln L = \sum \ln(0.5 + bx_i)$$

at b_0 with

$$b = b_0 + \beta$$

and differentiate it with respect to β to find the value $\hat{\beta}$ at the maximum:

$$\sum \frac{x_i}{0.5 + (b_0 + \hat{\beta})x_i} = 0.$$

Neglecting quadratic and higher order terms in $\hat{\beta}$, we can solve this equation for $\hat{\beta}$ and obtain

$$\hat{\beta} \approx \frac{\sum x_i/f_{0i}}{\sum x_i^2/f_{0i}^2} \tag{7.30}$$

where we have set $f_{0i} = f(x_i|b_0)$. If we allow also for a quadratic term

$$f(x|a,b) = a + bx + (1.5 - 3a)x^2,$$

we write, in obvious notation,

$$f(x|a,b) = f_0 + \alpha(1 - 3x^2) + \beta x$$

and get, after differentiation of $\ln L$ with respect to α, two linear equations for $\hat{\alpha}$ and $\hat{\beta}$:

$$\begin{aligned} \hat{\alpha} \sum A_i^2 + \hat{\beta} \sum A_i B_i &= \sum A_i, \\ \hat{\alpha} \sum A_i B_i + \hat{\beta} \sum B_i^2 &= \sum B_i, \end{aligned} \tag{7.31}$$

with the abbreviations $A_i = (1 - 3x_i^2)/f_{0i}$, $B_i = x_i/f_{0i}$. From the observed data using (7.31), we get $\hat{\beta}'(x')$, $\hat{\alpha}'(x')$, and the simulation provides the parameter estimates $\hat{b}(\hat{\beta}'), \hat{a}(\hat{\alpha}')$ and their uncertainties.

The calculation is much faster than a numerical minimum search and almost as precise. If $\hat{\alpha}, \hat{\beta}$ are large, we have to iterate.

7.8 Nuisance Parameters

Frequently a p.d.f. $f(\boldsymbol{x}|\boldsymbol{\theta}, \boldsymbol{\nu})$ contains several parameters, from which only some, namely $\boldsymbol{\theta}$, are of interest, while the other parameters $\boldsymbol{\nu}$ are not, but influence the estimate of the former. Those are called *nuisance parameters.* A typical example is the following.

Example 7.16. Nuisance parameter: Decay distribution with background

We want to infer the decay rate γ of a certain particle from the decay times t_i of a sample of N events. Unfortunately, the sample contains an unknown amount of background. The decay rate γ_b of the background particles be known. The nuisance parameter is the number of background events η. For a fraction of background events of η/N, the p.d.f. for a single event with lifetime t is

$$f(t|\gamma, \eta) = \left(1 - \frac{\eta}{N}\right) \gamma e^{-\gamma t} + \frac{\eta}{N} \gamma_b e^{-\gamma_b t}, \ \eta \leq N,$$

from which we derive the likelihood for the sample:

$$L(\gamma, \eta) = \prod_{i=1}^{N} \left[\left(1 - \frac{\eta}{N}\right) \gamma e^{-\gamma t_i} + \frac{\eta}{N} \gamma_b e^{-\gamma_b t_i}\right].$$

A contour plot of the log-likelihood of a specific data sample of 20 events and $\gamma_b = 0.2$ is depicted in Fig. 7.8. The two parameters γ and η are correlated. The question is then: What do we learn about γ, what is a sensible point estimate of γ and how should we determine its uncertainty?

We will re-discuss this example in the next subsection and present in the following some approaches which permit to eliminate the nuisance parameters. First we will investigate exact methods and then we will turn to the more problematic part where we have to apply approximations.

7.8.1 *Nuisance Parameters with Given Prior*

If we know the p.d.f. $\pi(\boldsymbol{\nu})$ of a nuisance parameter vector $\boldsymbol{\nu}$, the prior of $\boldsymbol{\nu}$, then we can eliminate $\boldsymbol{\nu}$ simply by integrating it out, weighting $\boldsymbol{\nu}$ with its probability $\pi(\boldsymbol{\nu})$ to occur.

$$f_\theta(\boldsymbol{x}|\boldsymbol{\theta}) = \int f(\boldsymbol{x}|\boldsymbol{\theta}, \boldsymbol{\nu}) \pi(\boldsymbol{\nu}) d\boldsymbol{\nu}.$$

In this way we obtain a p.d.f. depending solely on the parameters of interest $\boldsymbol{\theta}$. The corresponding likelihood function of $\boldsymbol{\theta}$ is

$$L_{\boldsymbol{\theta}}(\boldsymbol{\theta}|\boldsymbol{x}) = \int L(\boldsymbol{\theta},\boldsymbol{\nu}|\boldsymbol{x})\pi(\boldsymbol{\nu})\mathrm{d}\boldsymbol{\nu} = \int f(\boldsymbol{x}|\boldsymbol{\theta},\boldsymbol{\nu})\pi(\boldsymbol{\nu})\mathrm{d}\boldsymbol{\nu}. \tag{7.32}$$

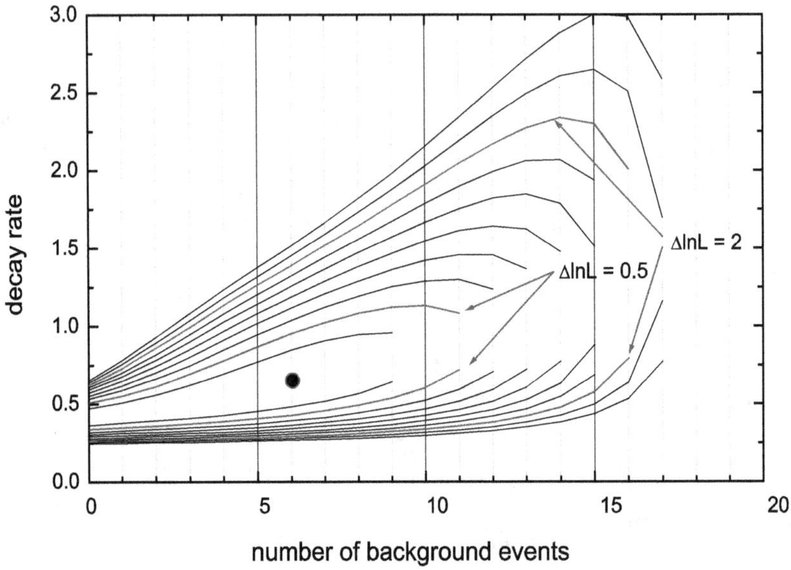

Fig. 7.8 Log-likelihood contour as a function of the decay rate and the number of background events. For better visualization the discrete values of the event numbers are connected.

Example 7.17. Nuisance parameter: Measurement of a Poisson rate with a digital clock

An automatic monitoring device measures a Poisson rate θ with a digital clock counting in units of Δ. For n observed reactions within a time interval ν the p.d.f. is given by the Poisson distribution $\mathcal{P}_{\theta\nu}(n)$. If we consider both, the rate parameter θ and the length of the time interval ν as unknown parameters, the corresponding likelihood function is

$$L(\theta,\nu) = \frac{\mathrm{e}^{-\theta\nu}\,[\theta\nu]^n}{n!}.$$

For a clock reading t_0, the true measurement time is contained in the time interval $t_0 \pm \Delta/2$. We can assume that all times ν within that interval are equally probable and thus the prior of ν is $\pi(\nu) = 1/\Delta$ for ν in the interval $[t_0 - \Delta/2,\, t_0 + \Delta/2]$ and equal to zero elsewhere. We eliminate constant factors, and, integrating over ν,

$$L_\theta(\theta) = \int_{t_0-\Delta/2}^{t_0+\Delta/2} \mathrm{e}^{-\theta\nu} \left[\theta\nu\right]^n \mathrm{d}\nu,$$

we get rid of the nuisance parameter.

Example 7.18. Nuisance parameter: Decay distribution with background sample

Let us resume the problem discussed in the introduction. We now assume that we have prior information on the amount of background: The background expectation had been determined in an independent experiment to be 10 with sufficient precision to neglect its uncertainty. The actual number of background events follows a binomial distribution. The likelihood function is

$$L(\gamma) = \sum_{\eta=0}^{20} \mathcal{B}_{0.5}^{20}(\eta) \prod_{i=1}^{20} \left[\left(1 - \frac{\eta}{20}\right) \gamma \mathrm{e}^{-\gamma t_i} + \frac{\eta}{20} 0.2 \mathrm{e}^{-0.2 t_i} \right].$$

Since our nuisance parameter η is discrete, we have replaced the integration in (7.32) by a sum.

7.8.2 *Factorizing the Likelihood Function*

Very easy is the elimination of the nuisance parameter if the p.d.f. is of the form

$$f(\boldsymbol{x}|\theta, \nu) = f_\theta(\boldsymbol{x}|\theta) f_\nu(\boldsymbol{x}|\nu), \tag{7.33}$$

i.e. only the first factor f_θ depends on θ. Then we can write the likelihood as a product

$$L(\theta, \nu) = L_\theta(\theta) L_\nu(\nu)$$

with

$$L_\theta = \prod f_\theta(\boldsymbol{x}_i|\theta),$$

independent of the nuisance parameter ν.

Example 7.19. Elimination of a nuisance parameter by factorization of a two-dimensional normal distribution

A sample of space points (x_i, y_i), $i = 1, \dots, N$ follows a normal distribution

$$f(x,y|\theta,\nu) = \frac{ab}{2\pi} \exp\left(-\frac{1}{2}\left[a^2(x-\theta)^2 + b^2(y-\nu)^2\right]\right)$$
$$= \frac{ab}{2\pi} \exp\left(-\frac{a^2}{2}(x-\theta)^2\right) \exp\left(-\frac{b^2}{2}(y-\nu)^2\right).$$

with θ the parameter which we are interested in. The normalized x distribution depends only on θ. Whatever value ν takes, the shape of this distribution remains always the same. Therefore we can estimate θ independently of ν. The likelihood function is proportional to a normal distribution of θ,

$$L(\theta) \sim \exp\left(-\frac{a^2}{2}(\theta - \hat{\theta})^2\right),$$

with the estimate $\hat{\theta} = \overline{x} = \sum x_i/N$.

7.8.3 *Parameter Transformation, Restructuring*

Sometimes we manage by means of a parameter transformation $\nu' = \nu'(\theta, \nu)$ to bring the p.d.f. into the desired form (7.33) where the p.d.f. factorizes into two parts which depend separately on the parameters θ and ν' [19]. We have already sketched an example in Section 4.2.4: When we are interested in the slope θ and not in the intersection ν with the y-axis of a straight line $y = \theta x + \nu$ which should pass through measured points, then we are able to eliminate the correlation between the two parameters. To this end we express the equation of the straight line by the slope and the ordinate at the center of gravity, see Example 6.24.

A simple transformation $\nu' = c_1\nu + c_2\theta$ also helps to disentangle correlated parameters of a Gaussian likelihood

$$L(\theta,\nu) \sim \exp\left(-\frac{a^2(\theta-\hat{\theta})^2 - 2ab\rho(\theta-\hat{\theta})(\nu-\hat{\nu}) + b^2(\nu-\hat{\nu})^2}{2(1-\rho^2)}\right),$$

With suitable chosen constants c_1, c_2 it produces a likelihood function that factorizes in the new parameter pair θ, ν'. In the notation where the

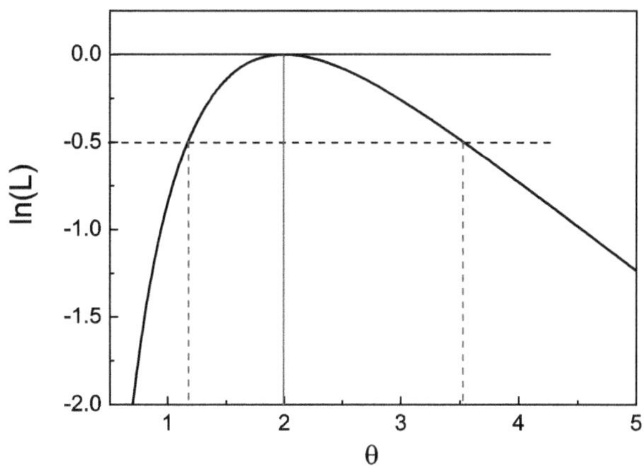

Fig. 7.9 Log-likelihood function of an absorption factor.

quantities $\hat{\theta}, \hat{\nu}$ maximize the likelihood function, the transformation produces the result

$$L_\theta(\theta) \sim \exp\left(-\frac{a^2}{2}(\theta - \hat{\theta})^2\right) .$$

We omit the proof of this assertion.

It turns out that this procedure yields the same result as simply integrating out the nuisance parameter and as the profile likelihood method which we will discuss below. This is an important observation in the following respect: In many situations the likelihood function is nearly of Gaussian shape. As is shown in Appendix A.3, the likelihood function approaches a Gaussian with increasing number of observations. Therefore, integrating out the nuisance parameter, or better to apply the profile likelihood method, is a sensible approach in many practical situations. Thus nuisance parameters are a problem only if the sample size is small.

The following example is frequently discussed in the literature [19].

Example 7.20. Elimination of a nuisance parameter by restructuring: absorption measurement

The absorption factor θ for radioactive radiation by a plate is determined from the numbers of events r_1 and r_2, which are observed with and without the absorber within the same time intervals. The numbers r_1, r_2 follow Poisson distributions with mean values ρ_1 and ρ_2:

$$f_1(r_1|\rho_1) = \frac{\mathrm{e}^{-\rho_1}\rho_1^{r_1}}{r_1!},$$
$$f_2(r_2|\rho_2) = \frac{\mathrm{e}^{-\rho_2}\rho_2^{r_2}}{r_2!}.$$

The interesting parameter is the expected absorption $\theta = \rho_2/\rho_1$. In first approximation we can use the estimates r_1, r_2 of the two independent parameters ρ_1 and ρ_2 and their errors to calculate in the usual way through error propagation θ and its uncertainty:

$$\hat{\theta} = \frac{r_2}{r_1},$$
$$\frac{(\delta\hat{\theta})^2}{\hat{\theta}^2} = \frac{1}{r_1} + \frac{1}{r_2}.$$

For large numbers r_1, r_2 this method is justified but the correct way is to transform the parameters ρ_1, ρ_2 of the combined distribution

$$f(r_1, r_2|\rho_1, \rho_2) = \frac{\mathrm{e}^{-(\rho_1+\rho_2)}\rho_1^{r_1}\rho_2^{r_2}}{r_1!r_2!}$$

into the independent parameters $\theta = \rho_2/\rho_1$ and $\nu = \rho_1 + \rho_2$. The transformation yields:

$$\tilde{f}(r_1, r_2|\theta, \nu) = \mathrm{e}^{-\nu}\nu^{r_1+r_2}\frac{(1+1/\theta)^{-r_2}(1+\theta)^{-r_1}}{r_1!r_2!},$$
$$L(\theta, \nu|r_1, r_2) = L_\nu(\nu|r_1, r_2)L_\theta(\theta|r_1, r_2).$$

Thus the log-likelihood function of θ is

$$\ln L_\theta(\theta|r_1, r_2) = -r_2\ln(1+1/\theta) - r_1\ln(1+\theta).$$

It is presented in Fig. 7.9 for the specific values $r_1 = 10$, $r_2 = 20$. The maximum is located at $\hat{\theta} = r_2/r_1$, as obtained with the simple estimation above. However the errors are asymmetric.

Example 7.21. Eliminating a nuisance parameter by restructuring: Slope of a straight line with the y-axis intercept as nuisance parameter

We come back to one of our standard examples which can, as we have indicated, be solved by a parameter transformation. Now we solve it in a simpler way. Points (x_i, y_i) are distributed along a straight line. The x coordinates are exactly known, the y coordinates are the variates. The p.d.f. $f(y_1, \ldots, y_n|\theta, \nu)$ contains the slope parameter θ and the uninteresting intercept ν of the line with the y axis. It is easy to recognize that the statistic $\{\tilde{y}_1 = y_1 - y_n, \tilde{y}_2 = y_2 - y_n, \ldots, \tilde{y}_{n-1} = y_{n-1} - y_n\}$ is independent of ν. In this specific case, the new statistic is also sufficient relative to the slope θ which clearly depends only on the differences of the ordinates. We leave the details of the solution to the reader.

Further examples for the elimination of a nuisance parameter by restructuring have been given already in Section 6.5.2, Examples 6.10 and 6.11.

7.8.4 *Conditional Likelihood*

This method is closely related to the restructuring method.

In case we can find a sufficient statistic S of the nuisance parameter ν, we may condition $f(x|\theta, \nu)$ on S. If S does not depend on θ, we can fix ν to the value required to satisfy S

Example 7.22. Fitting the width of a normal distribution with the mean as nuisance parameter

The sample mean $\bar{x}$ of measurements is a sufficient statistic for μ of the normal distribution $\mathcal{N}(x|\mu, \sigma)$. We can replace μ by $\bar{x}$ in the Gaussian and are left with the wanted parameter only, see also Example 6.11.

If $S(\nu, \theta)$ depends on theta, we can again condition on S, but in this situation, we prefer to switch to the profile likelihood which is described in the following subsection.

7.8.5 *Profile Likelihood*

We now turn to approximate solutions.

Some scientists propose to replace the nuisance parameter by its estimate. This corresponds to a delta function for the prior of the nuisance parameter and is for that reason quite exotic and dangerous. It leads to an illegitimate reduction of the error limits whenever the nuisance parameter and the interesting parameter are correlated. Remark, that a correlation always exists unless a factorization is possible. In the extreme case of full correlation the error would shrink to zero.

A much more sensible approach to eliminate the nuisance parameter uses the so-called profile likelihood [43]. To explain it, we choose an example with a single nuisance parameter.

The likelihood function is maximized with respect to the nuisance parameter ν as a function of the wanted parameter θ. The function $\hat{\nu}(\theta)$ which maximizes L then satisfies the relation

$$\frac{\partial L(\theta,\nu|x)}{\partial \nu}|_{\hat{\nu}} = 0 \rightarrow \hat{\nu}(\theta).$$

It is inserted into the likelihood function and provides the profile likelihood L_p,

$$L_p = L\left(\theta, \hat{\nu}(\theta)|x\right),$$

which depends solely on θ.

This method has the great advantage that only the likelihood function enters and no assumptions about priors have to be made. It also takes correlations into account. Graphically we can visualize the error interval of the profile likelihood $\Delta \ln L_p(\theta,\nu) = 1/2$ by drawing the tangents of the curve $\Delta \ln L = 1/2$ parallel to the ν axis. These tangents include the standard error interval.

Example 7.23. Profile likelihood, absorption measurement

We reformulate the absorption example 7.20 with the nuisance parameter ρ_1 and the parameter of interest $\theta = \rho_2/\rho_1$. The log-likelihood, up to constants, is:

$$\ln L(\rho_1,\theta) = -\rho_1(1+\theta) + (r_1+r_2)\ln\rho_1 + r_2\ln\theta.$$

The maximum of ρ_1 as a function of θ is $\hat{\rho}_1 = (r_1+r_2)/(1+\theta)$ and the profile likelihood becomes

$$\ln L_p(\theta) = -(r_1+r_2)\ln(1+\theta) + r_2\ln\theta.$$

The function $\hat{\rho}_1$, the profile likelihood and the one standard deviation error contour are shown in Fig. 7.8.5. The result coincides with that of the exact factorization. (In the figure the nuisance parameter is denoted by ν.)

In the literature we find methods which orthogonalize the parameters at the maximum of the likelihood function [44] which means to diagonalize a multidimensional Gaussian. The result is similar to that of the profile likelihood approach. Errors derived from the profile likelihood are computed in the program MINUIT [49].

In the limit of a large number of observations where the likelihood function approaches the shape of a normal distribution, the profile likelihood method is identical to restructuring and factorizing the likelihood. The profile likelihood is not a genuine likelihood. For instance, it does not always have the property that the product of the likelihoods of subsamples is equal to the likelihood of the full sample.

7.8.6 *Integrating Out the Nuisance Parameter*

If the methods fail which we have discussed so far, we are left with only two possibilities: Either we give up the elimination of the nuisance parameter or we integrate it out. The result of a simple integration

$$L_\theta(\theta|x) = \int_{-\infty}^{\infty} L(\theta, \nu|x) d\nu$$

implicitly contains the assumption of a uniform prior of ν and therefore depends to some extend on the validity of this condition. Problematic are asymmetric likelihood functions. It is interesting to notice that in many cases integrating out the nuisance parameter assuming a uniform prior leads to the same result as restructuring the problem.

7.8.7 *Explicit Declaration of the Parameter Dependence*

It is not always possible to eliminate the nuisance parameter in such a way that the influence of the method on the result can be neglected. When the likelihood function has a complex structure, we are obliged to document the full likelihood function. In many cases it is possible to indicate the dependence of the estimate θ and its error limits θ_1, θ_2 on the nuisance parameter ν explicitly by a simple linear function

$$\hat{\theta} = \hat{\theta}_0 + c(\nu - \hat{\nu}),$$

$$\theta_{1,2} = \theta_{1,2} + c(\nu - \hat{\nu}).$$

Usually the error limits will show the same dependence as the MLE which means that the width of the interval is independent of ν.

However, publishing a dependence of the parameter of interest on the nuisance parameter is useful only if ν corresponds to a physical constant and not to an internal parameter of an experiment like *efficiency* or *background*.

7.8.8 *Recommendation*

If it is impossible to eliminate the nuisance parameter explicitly and if the shape of the likelihood function does not differ dramatically from that of a Gaussian, the profile likelihood approach should be used for the parameter and interval estimation. In case the deviation from a Gaussian is considerable, the dependence of the ML estimate of the parameter of interest and its error on the nuisance parameter should be given.

Chapter 8

Interval Estimation and Combination of Results

We have discussed interval estimation in previous sections. In this chapter we repeat, summarize and extend the results. We treat p.d.f.s of continuous variables and if not stated differently, parameters with given likelihood functions.

As mentioned before, a measurement or a point estimate of a parameter in order to be useful has to be accompanied by an interval that indicates the accuracy of the measurement and that ideally also contains the true parameter value with given probability. However, both requirements cannot be fulfilled simultaneously without the introduction of partially subjective assumptions or approximations. We distinguish between *error intervals* as defined in Chapter 4, *likelihood ratio intervals*, *confidence intervals* and *credibility intervals*.

Error intervals should fulfill the following essential conditions: They should be compatible with the measured value or point estimate, indicate the precision of the measurement and permit the combination of results. There are further requirements that have to be considered: They are related to invariance under variable transformations, to the asymptotic behavior of errors with increasing number of elements of a sample and to asymmetries.

In the wast majority of applications and consequently in the statistical literature the point estimate and error are usually attributed to estimates of the mean value and of the standard deviation of the supposed distribution. When a distance, the time, a magnetic field etc is measured, the estimate of the standard deviation is best choice to indicate the precision of the result as long as variable transformations are no issue.

The aim of confidence intervals is to provide a region that contains true parameter value with a predefined probability. The Bayesian school introduces prior probabilities, constructs a p.d.f. of the parameter and credibility intervals that, like confidence intervals, allow to make or test predictions.

8.1 Error Intervals of Adjusted Parameters

In basic research the focus is on the measurement of constants of nature, particle masses, lifetimes, cross sections, coupling constants etc. and inference proceeds mostly through the likelihood method where, what we called *statistical errors*, are evaluated. According to the likelihood principle all information about the parameter of interest is contained in the likelihood function. Then it is logical to choose as point estimate the MLE and to base the error interval on the likelihood ratio as done in Section 6.5.1. In Appendix A.3, it is shown that with increasing sample size the MLE and the $-1/2$ interval approach the true parameter value and the standard deviation of the probability density, respectively. Experiments with low statistics lead to large and asymmetric errors. Then the likelihood ratio limits do not conform to the standard deviation error interval. Likelihood ratio errors and the MLE are invariant under variable transformation and are internally consistent. Results from different experiments can be combined by adding the log-likelihood functions. For a monotonic function $\lambda(\theta)$ and error limits of θ and λ equal to θ_1, θ_2 and λ_1, λ_2, we require that $\lambda_{1,2} = \lambda(\theta_{1,2})$. (This condition that is violated by the standard deviation, unless the errors are so small that a linear transformation of the errors is adequate.)

8.1.1 *Parabolic Approximation of the Log-Likelihood Function*

With increasing sample size the width of the likelihood function decreases, and we can neglect possible variations of the prior density in the region where the likelihood is significant. Under this condition we obtain normally distributed p.d.f.s of the parameters. The standard deviation error includes the parameter with probability 68.3% (see Section 4.6).

In the parabolic approximation with a constant prior, the MLE and the expectation value coincide, as well as the likelihood ratio error squared and the variance. Thus we can also derive the standard deviation $\delta\theta$ from the curvature of the likelihood function at its maximum. For a single parameter we can approximate the log-likelihood function by the expression

$$-\ln L_{par} = \frac{1}{2}C(\theta - \hat{\theta})^2 + const.. \tag{8.1}$$

Consequently, a change of $\ln L_{par}$ by $1/2$ corresponds to the second derivative of $\ln L$ at $\hat{\theta}$:

$$(\delta\theta)^2 = C^{-1} = -\left(\frac{\mathrm{d}^2 \ln L}{\mathrm{d}\theta^2}|_{\hat{\theta}}\right)^{-1} .$$

For several parameters the parabolic approximation can be expressed by

$$-\ln L_{par} = \frac{1}{2}\sum_{i,j}(\theta_i - \hat{\theta}_i)C_{ij}(\theta_j - \hat{\theta}_j) + const..$$

We obtain the symmetric weight matrix[1] C from the derivatives

$$C_{ij} = -\frac{\partial^2 \ln L}{\partial\theta_i \partial\theta_j}|_{\hat{\theta}} \tag{8.2}$$

and the covariance or error matrix as its inverse $\mathsf{V} = \mathsf{C}^{-1}$.

Uninteresting nuisance parameters can simply be erased im the covariance matrix. This is a consequence of the Gaussian shape of the error distribution and the considerations from Section 7.8.

In most cases the likelihood function is not known analytically. In one dimension it is easy to determine the error limits numerically from the decrease of the log-likelihood by 0.5 from its maximum. In multi-dimensional parameter spaces it is easier to estimate the second derivative and the weight matrix C computing the likelihood function at parameter points close to the MLE and subsequently the curvature. To ensure that the parabolic approximation is valid, the distance of the points from the MLE should be modified and checked whether the result remains consistent.

In publications we find sometimes statements like "The measurement excludes the theoretical prediction by four standard deviations." Assertions of this kind are often misleading. Their validity relies on the assumption that the log-likelihood is parabolic over a very wide parameter range. Neglecting tails can lead to completely wrong conclusions.

The sum of the parabolic log-likelihoods of several independent measurements $\hat{\theta}_i$ is again a quadratic parabola.

$$\frac{1}{2}\sum C_i(\theta - \hat{\theta}_i)^2 = \frac{1}{2}\tilde{C}(\theta - \hat{\theta})^2 + const..$$

The comparison of the coefficients of θ^2 and θ yields

$$\tilde{C} = \sum C_i ,$$
$$\tilde{C}\hat{\theta} = \sum C_i \hat{\theta}_i ,$$

[1] It is also called Fisher information.

the weighted mean (4.7) and its error (4.8). Consequently, we should aim at approximately parabolic log-likelihood functions when we present experimental results. Sometimes this is possible by a suitable choice of the parameter. For example, we are free to quote either the estimate of the mass or of the mass squared, of the momentum or of its inverse.

In the following section we address more problematic situations which usually occur with small data samples where the asymptotic solutions are not appropriate. Fortunately, they are rather the exception. We keep in mind that a relatively rough estimate of the error often is sufficient and that therefore approximate methods in most cases are justified.

8.1.2 *General Situation*

Many measurements have asymmetric errors. An extreme example is the following.

Example 8.1. Asymmetric errors: Error of small Poisson numbers

We want to measure the decay rate of a radioactive compound. After one hour we have recorded one decay. Given such small rates, it is not correct to compute the error from a Poisson distribution (see Section 3.6.3) in which we replace the mean value by the observed measurement. The declaration $R = 1 \pm 1$ does not reflect the result correctly because $R = 0$ is excluded by the observation while $R = 2.5$ on the other hand is consistent with it.

As above, we again use the likelihood ratio to define the error limits. In the one-dimensional case the two errors δ_- and δ_+ satisfy

$$\ln L(\hat{\theta}) - \ln L(\hat{\theta} - \delta_-) = \ln L(\hat{\theta}) - \ln L(\hat{\theta} + \delta_+) = 1/2. \tag{8.3}$$

If the log-likelihood function deviates considerably from a parabola, the advise is to publish the complete likelihood function.

Example 8.2. Error of a lifetime measurement

We return to one of our standard examples. The likelihood function for the mean lifetime τ of an unstable particle derived from a sample of observed decay times is

$$L_\tau = \prod_{i=1}^{N} \frac{1}{\tau} e^{-t_i/\tau} = \frac{1}{\tau^N} e^{-N\bar{t}/\tau}. \tag{8.4}$$

The corresponding likelihood for the decay rate is

$$L_\lambda = \prod_{i=1}^{N} \lambda e^{-\lambda t_i} = \lambda^N e^{-N\bar{t}\lambda}.$$

The values of the functions are equal at equivalent values of the two parameters τ and λ, i.e. for $\lambda = 1/\tau$:

$$L_\lambda(\lambda) = L_\tau(\tau).$$

Figure 8.1 shows the two log-likelihoods for a small sample of ten events with mean value $\bar{t} = 0.5$. The lower curves for the parameter τ are strongly asymmetric. This is also visible in the limits for changes of the log-likelihood by 0.5 or 2 units which are indicated on the right-hand cut-outs. The likelihood with the decay rate as parameter (upper figures) is less asymmetric than that of the mean life. This means that the decay rate is the more appropriate parameter to document the shape of the likelihood function and to average different measurements.

Generally, it does not matter whether we use one or the other parameter to present the result but for further applications it is always simpler and more precise to work with approximately symmetric limits. For this reason usually $1/p$ (p is the absolute value of a particle momentum) instead of p is used to fit charged particle trajectories to the measured hits in a magnetic spectrometer.

8.2 Error Propagation

8.2.1 *Average of Measurements*

In case the errors of a parameter is correlated with the parameter value which occurs frequently with small event numbers, the standard procedure to compute the weighted mean according to relation (4.7) introduces a bias (see Example 4.15 in Chapter 4). Either the raw data should be combined, or what is equivalent, the log-likelihood functions should be added.

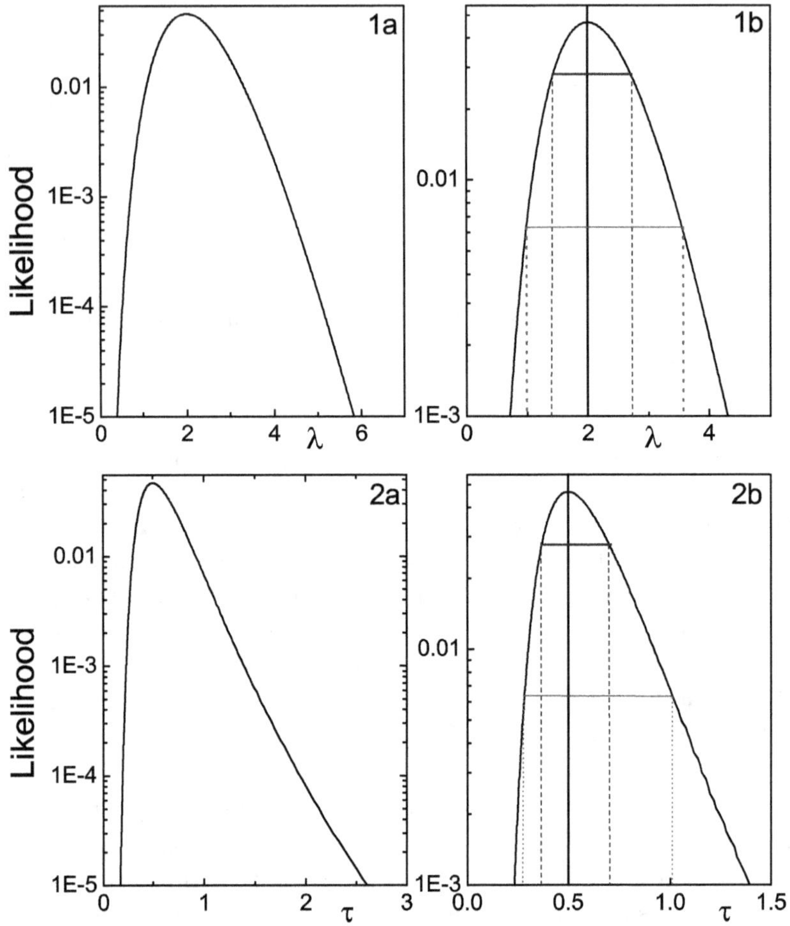

Fig. 8.1 Likelihood functions for the parameters *decay rate* (top) and emphlifetime (below). The error limits are shown in the cut-outs on the right hand side.

Example 8.3. Averaging lifetime measurements

N experiments quote lifetimes $\hat{\tau}_i \pm \delta_i$ of the same unstable particle. The estimates and their errors are computed from the individual measurements t_{ij} of the i-th experiment according to $\hat{\tau}_i = \sum_{j=1}^{n_i} t_{ij}/n_i$, respectively $\delta_i = \hat{\tau}_i/\sqrt{n_i}$ where n_i is the number of observed decays. It is more precise to reconstruct the individual log-likelihood functions and their sum $\ln L$, with n, $n = \sum_{i=1}^{N} n_i$, the overall event number:

$$\begin{aligned}\ln L(\tau) &= \sum_{i=1}^{N} -n_i(\ln \tau + \hat{\tau}_i/\tau) \\ &= -n \ln \tau - \sum \frac{n_i \hat{\tau}_i}{\tau}\end{aligned}$$

with the maximum at

$$\hat{\tau} = \frac{\sum n_i \hat{\tau}_i}{n}$$

and its error

$$\delta = \frac{\hat{\tau}}{\sqrt{n}}.$$

The individual measurements are weighted by their event numbers, instead of weights proportional to $1/\delta_i^2$. As the errors are correlated with the measurements, the standard weighted mean (4.7) with weights proportional to $1/\delta_i^2$ would be biased. In our specific example the correlation of the errors and the parameter values is known and we could use weights proportional to $(\tau_i/\delta_i)^2$.

Example 8.4. Averaging ratios of Poisson distributed numbers

In absorption measurements and many other situations we are interested in a parameter which is the ratio of two numbers which follow the Poisson distribution. Averaging naively these ratios $\hat{\theta}_i = m_i/n_i$ using the weighted mean (4.7) can lead to strongly biased results. Instead we add the log-likelihood functions which we have derived in Section 6

$$\begin{aligned}\ln L &= \sum [n_i \ln(1 + 1/\theta) - m_i \ln(1 + \theta)] \\ &= n \ln(1 + 1/\theta) - m \ln(1 + \theta)\end{aligned}$$

with $m = \sum m_i$ and $n = \sum n_i$. The MLE is $\hat{\theta} = m/n$ and the error limits have to be computed numerically in the usual way or for not too small n, m by linear error propagation, $\delta_\theta^2/\theta^2 = 1/n + 1/m$.

In the common situation where we do not know the full likelihood function but only the MLE and the error limits, we have to be content with an approximate procedure.

8.2.2 *Approximation of the Likelihood Function*

To combine results we add the corresponding likelihood functions. If only the point estimate and asymmetric errors are given, we can try to reconstruct the likelihood function from the three parameters with the additional constraint that the derivative at the MLE is zero.

The simplest parametrization uses two different parabola branches

$$-\ln L(\theta) = \frac{1}{2}(\theta - \hat{\theta})^2/\delta_{\pm}^2 \qquad (8.5)$$

with

$$\delta_{\pm} = \frac{1}{2}\delta_+[1 + \mathrm{sgn}(\theta - \hat{\theta})] + \frac{1}{2}\delta_-[1 - \mathrm{sgn}(\theta - \hat{\theta})],$$

i.e. the parabolas meet at the maximum and form a concave differentiable function.

Parametrizations varying the width σ of a parabola linearly or quadratically with the parameter [50] are usually superior to the simple two branch approximation. Contrary to the simple solution the constructed function also the second derivative exists. We set

$$-\ln L(\theta) = \frac{1}{2}\left[(\theta - \hat{\theta})/\sigma(\theta)\right]^2$$

with

$$\sigma(\theta) = \frac{2\delta_+\delta_-}{\delta_+ + \delta_-} + \frac{\delta_+ - \delta_-}{\delta_+ + \delta_-}(\theta - \hat{\theta}) \qquad (8.6)$$

or

$$(\sigma(\theta))^2 = \delta_+\delta_- + (\delta_+ - \delta_-)(\theta - \hat{\theta}), \qquad (8.7)$$

respectively. The log-likelihood function has poles at locations of θ where the width becomes zero, $\sigma(\theta) = 0$. Thus our approximations are justified only in the range of θ which excludes the corresponding parameter values.

In Fig. 8.2 we present four typical examples of asymmetric likelihood functions. The log-likelihood function of the mean lifetime derived from four randomly chosen exponentially distributed times is shown in Fig. 8.2a. Figure 8.2b is the corresponding log-likelihood function of the decay rate. The asymetric functions that are approximated in Figs. 8.2c and 8.2d have been derived by a parameter transformation from normally distributed

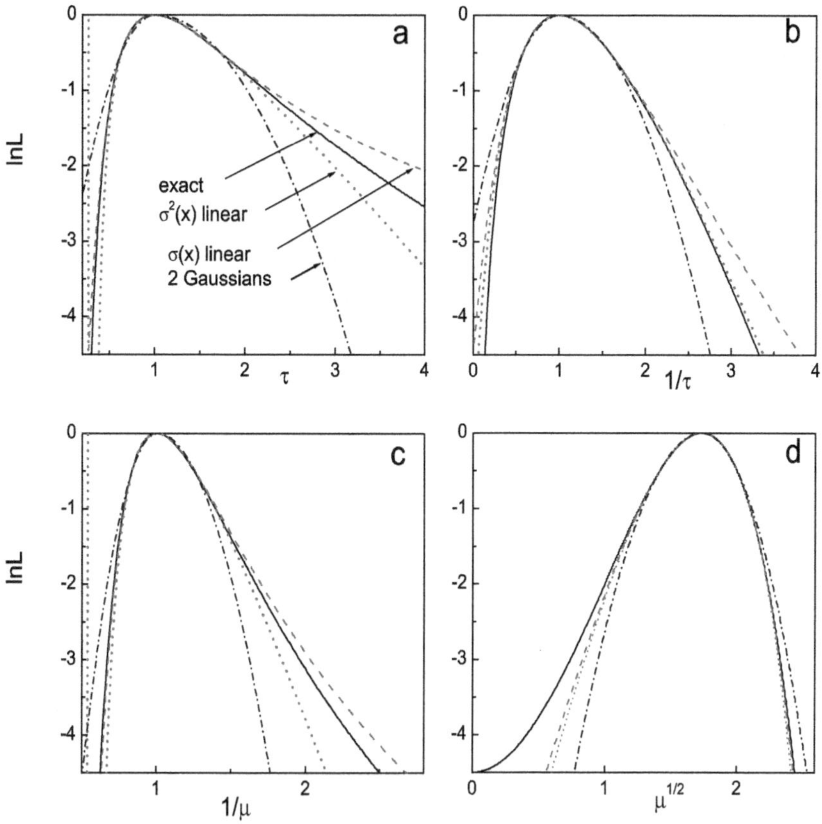

Fig. 8.2 Asymmetric likelihood functions and parametrizations

observations where in one case the new parameter is one over the mean and the square root of the mean[2] in the other case.

A method which is optimum for all cases does not exist. All three approximations fit very well inside the "-1/2 limits".[3] Outside, the two parametrizations (8.6) (8.7) are superior to the two-parabola approximation and reproduce well the "-2 limit".

We propose to use one of the two parametrizations ((8.6) and (8.7)) but to be careful if $\sigma(\theta)$ approaches zero.

[2] An example of such a situation is a fit of a particle mass from normally distributed mass squared observations.

[3] Usually they are called "one standard deviation limits", even though this denotation is not correct.

8.2.3 *Incompatible Measurements*

Before we combine results of different experiments we should make sure that they are statistically compatible. What we mean with *compatible*, is not obvious at this point. It will become clearer in Chapter 10, where we discuss significance tests which lead to the following plausible procedure that has proven to be quite useful in particle physics [31].

We compute the weighted mean value $\tilde{\theta}$ of the N results and form the sum of the quadratic deviations of the individual measurements from their average, normalized to their expected errors squared[4]:

$$\chi^2 = \sum (\theta_i - \tilde{\theta})^2 / \delta_i^2 .$$

The expectation value of this quantity is $N-1$ if the deviations are normally distributed with variances δ_i^2. If χ^2 is (e.g. by 50%) higher than $N-1$, then we can suspect that at least one of the experiments has published a wrong value, or what is more likely, has underestimated the error, for instance when systematic errors have not been detected. Under the premise that none of the experiments can be discarded a priori, we scale-up all declared errors by a common scaling factor $S = \sqrt{\chi^2/(N-1)}$ and publish this factor together with mean value and the scaled error. Large scaling factors indicate problems in one or several experiments. After scaling, χ^2 has the expected value of the χ^2-distribution with $N-1$ degrees of freedom. A similar procedure is applied if the errors are asymmetric even though the condition of normality then obviously is violated. We form

$$\chi^2 = \sum (\theta_i - \tilde{\theta})^2 / \delta_{i\pm}^2 ,$$

where δ_{i+} and δ_{i-}, respectively, are valid for $\theta_i < \tilde{\theta}$ and $\theta_i > \tilde{\theta}$.

8.2.4 *Error Propagation for a Scalar Function of a Single Parameter*

If we have to propagate the MLE and its error limits of a parameter θ to another parameter $\theta' = \theta'(\theta)$, we should apply the direct functional relation which is equivalent to a transformation of the likelihood function:

$$\begin{aligned} \hat{\theta'} &= \theta'(\hat{\theta}) , \\ \hat{\theta'} + \delta'_+ &= \theta'(\hat{\theta} + \delta_+) , \\ \hat{\theta'} - \delta'_- &= \theta'(\hat{\theta} - \delta_-) . \end{aligned}$$

[4]We assume here that the errors are symmetric.

Here we have assumed that $\theta'(\theta)$ is a monotonically increasing function. If it is decreasing, the arguments of θ' have to be interchanged.

The errors of the output quantity are asymmetric either because the input errors are asymmetric or because the functional dependence is non-linear. For instance an angular measurement $\alpha = 87^0 \pm 1^0$ would transform into $\sin\alpha = 0.9986^{+.0008}_{-.0010}$.

8.2.5 *Error Propagation for a Function of Several Parameters*

A difficult problem is the determination of the error of a scalar quantity $\theta(\boldsymbol{\mu})$ which depends on several measured input parameters $\boldsymbol{\mu}$ with asymmetric errors. We have to eliminate nuisance parameters.

If the complete likelihood function $\ln L(\boldsymbol{\mu})$ of the input parameters is available, we derive the error limits from the profile likelihood function of θ as proposed in Section 6.5.1.

The MLE of θ is simply $\hat{\theta} = \theta(\widehat{\boldsymbol{\mu}})$. The profile likelihood of θ has to fulfill the relation $\Delta \ln L(\theta) = \ln L(\hat{\theta}) - \ln L(\theta) = \ln L(\widehat{\boldsymbol{\mu}}) - \ln L(\boldsymbol{\mu})$. To find the two values of θ for the given $\Delta \ln L$, we have to find the maximum and the minimum of θ fulfilling the constraint. The one standard deviation limits are the two extreme values of θ located on the $\Delta \ln L(\theta) = \ln L(\widehat{\boldsymbol{\mu}}) - \ln L(\boldsymbol{\mu}) = 1/2$ surface in the μ space. Since we assumed likelihood functions with a single maximum, this is a closed surface, in two dimensions a closed line. The limits have to be computed numerically using the methods described in Section 7.5.

If the likelihood functions are not known, the only practical way is to resort to a Bayesian treatment, i.e. to make assumptions about the p.d.f.s of the input parameters. In many cases part of the input parameters have systematic uncertainties. Then, anyway, the p.d.f.s of those parameters have to be constructed. Once we have established the complete p.d.f. $f(\boldsymbol{\mu})$, we can also determine the distribution of θ. The analytic parameter transformation and reduction described in Chapter 3 will fail in most cases and we will adopt the simple Monte Carlo solution where we generate a sample of events distributed according to $f(\boldsymbol{\mu})$ and where $\theta(\boldsymbol{\mu})$ provides the θ distribution in form of a histogram and the uncertainty of this parameter. To remain consistent with our previously adopted definitions we would then interpret this p.d.f. of θ as a likelihood function and derive from it the MLE $\hat{\theta}$ and and the likelihood ratio error limits.

8.3 Credibility Intervals, Upper and Lower limits

Credibility- and confidence intervals should contain the value of a parameter with a predefined probability For example, before the discovery of the Higgs particle, its mass could be restricted on the bases of measurements to a certain range with 90 % confidence. Contrary to the measurement error, the emphasis is on *probability* and not on accuracy. As the probability cannot be determined exactly, the terms confidence and credibility are used.

Bayesian statistics provide - given a likelihood functions - p.d.f.s of parameters as indicated in Section 6.4. The results then depend on the prior probability $\pi(\theta)$ of the parameter θ and are of limited use. The subjectivity that is connected with the choice of a priors can be reduced by applying a uniform (uninformative) prior and to avoid parameter transformations. This means that before the measurement is performed all values of the parameter are considered as equally likely. To obtain the p.d.f. $f(\theta)$ of the parameter of interest, we then have to normalize the likelihood function. From $f(\theta)$ a probability interval can be derived, that due to the Bayesian origin is denoted *credibility interval.*

The uniformity of the prior probability is not invariant under transformations of the parameter. As a consequence, the results of such a procedure depend on the choice of the variable. For instance lower limits of a mass u_m and a mass squared u_{m^2}, respectively, would not obey the relation $u_{m^2} = (u_m)^2$. Unfortunately we cannot avoid this annoying defect when we want to present probabilities. Knowing that a uniform prior has been applied, the reader of a publication can interpret the limit as a sensible parametrization of the experimental result and draw his own conclusions. It is recommended to present in addition to the limit also the likelihood function which fully documents the result.

In the following paragraphs upper and lower limits are treated.

If, for instance, we cannot achieve the precision which is necessary to measure a small physical quantity, we usually present an upper limit. A typical example is the measurement of the lifetime of a very short-lived particle which cannot be resolved by the measurement. The result of such a measurement is then quoted by a phrase like "The lifetime of the particle is smaller than ... with 90 % confidence." Upper limits are often quoted for rates of rare reactions if no reaction has been observed or the observation is compatible with background. For masses of hypothetical particles, postulated by theory, but not observed with the limited energy of present accelerators, experiments provide lower limits.

The probability $P\{\theta < \theta_0\}$ computed from $f(\theta)$ is the credibility level α for the upper limit θ_0:

$$\alpha_{up}(\theta_0) = \frac{\int_{-\infty}^{\theta_o} L(\theta)\,\mathrm{d}\theta}{\int_{-\infty}^{\infty} L(\theta)\,\mathrm{d}\theta}. \tag{8.8}$$

Lower limits are computed in an analogous way:

$$\alpha_{low}(\theta_0) = \frac{\int_{\theta_o}^{\infty} L(\theta)\,\mathrm{d}\theta}{\int_{-\infty}^{\infty} L(\theta)\,\mathrm{d}\theta}. \tag{8.9}$$

Given the credibility level α the relations (8.8), (8.9) have to be solved for θ_0.

8.3.1 *Upper Poisson Limits, Simple Case*

Event rates are usually described by the Poisson distribution. As above, the upper limit is obtained by integration of the normalized likelihood function.

For k observed events, we want to determine an upper limit μ_0 with credibility α for the expectation value of the Poisson rate. The normalization integral over the parameter μ of the Poisson distribution $\mathcal{P}(k|\mu) = \mathrm{e}^{-\mu}\,\mu^k/k!$ is equal to one. Thus we obtain:

$$\begin{aligned} \alpha &= \int_0^{\mu_o} \mathcal{P}(k|\mu)\mathrm{d}\mu \\ &= \frac{\int_0^{\mu_o} \mathrm{e}^{-\mu}\,\mu^k\,\mathrm{d}\mu}{k!}. \end{aligned} \tag{8.10}$$

The integral is solved by partial integration,

$$\begin{aligned} \alpha &= 1 - \sum_{i=0}^{k} \frac{\mathrm{e}^{-\mu_0}\,\mu_0^i}{i!} \\ &= 1 - \sum_{i=0}^{k} \mathcal{P}(i|\mu_0). \end{aligned}$$

However, the sum over the Poisson probabilities cannot be computed analytically for μ_0. It has to be computed numerically, or (8.10) is evaluated with the help of tables of the incomplete gamma function.

A special role plays the case $k = 0$, e.g. when no event has been observed. The integral simplifies to:

$$\begin{aligned} \alpha &= 1 - \mathrm{e}^{-\mu_0}, \\ \mu_0 &= -\ln(1-\alpha). \end{aligned}$$

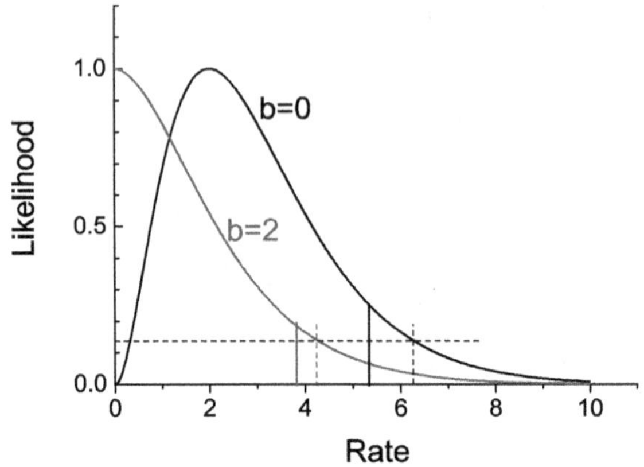

Fig. 8.3 Upper limits for Poisson rates. The dashed lines are likelihood ratio limits (decrease by e^2).

For a 90% credibility limit, $\alpha = 0.90$, this relation is fulfilled for $\mu_0 \approx 2.3$.

Remark that for Poisson limits of rates without background the frequentist statistics and the Bayesian statistics with uniform prior lead to the same result. For the following more general situations, this does not hold anymore.

8.3.2 *Poisson Limit for Data with Background*

When we find in an experiment events which can be explained by a background reaction with a Poisson distributed number of events and expected mean b, we have to modify (8.10) correspondingly. The expectation value of k is then $\mu + b$ and the credibility α is

$$\alpha = \frac{\int_0^{\mu_o} \mathcal{P}(k|\mu + b)\mathrm{d}\mu}{\int_0^{\infty} \mathcal{P}(k|\mu + b)\mathrm{d}\mu}.$$

Again the integrals can be replaced by sums:

$$\alpha = 1 - \frac{\sum_{i=0}^{k} \mathcal{P}(i|\mu_0 + b)}{\sum_{i=0}^{k} \mathcal{P}(i|b)}.$$

Example 8.5. Upper limit for a Poisson rate with background

Expected are two background events and observed are also two events. Thus the mean signal rate μ is certainly small. We obtain an upper limit μ_0 for the signal with 90% credibility by solving numerically the equation

$$0.9 = 1 - \frac{\sum_{i=0}^{2} \mathcal{P}(i|\mu_0 + 2)}{\sum_{i=0}^{2} \mathcal{P}(i|2)} .$$

We find $\mu_0 = 3.88$. The Bayesian probability that the mean rate μ is larger than 3.88 is 10 %. Fig. 8.3 shows the likelihood functions for the two cases $b = 2$ and $b = 0$ together with the limits. For comparison are also given the likelihood ratio limits which correspond to a decrease from the maximum by e^2. (For a normal distribution this would be equivalent to two standard deviations).

We now investigate the more general case that both the acceptance ε and the background are not perfectly known, and that the p.d.f.s of the background and the acceptance f_b, f_ε are given. For a mean Poisson signal μ the probability to observe k events is

$$g(k|\mu) = \int \mathrm{d}b \int \mathrm{d}\varepsilon \, \mathcal{P}(k|\varepsilon\mu + b) f_b(b) f_\varepsilon(\varepsilon) = L(\mu|k) .$$

For k observations this is also the likelihood function of μ. According to our scheme, we obtain the upper limit μ_0 by normalization and integration,

$$\alpha = \frac{\int_0^{\mu_0} L(\mu|k) \mathrm{d}\mu}{\int_0^{\infty} L(\mu|k) \mathrm{d}\mu}$$

which is solved numerically for μ_0.

Example 8.6. Upper limit for a Poisson rate with uncertainty in background and acceptance

Observed are 2 events, expected are background events following a normal distribution $\mathcal{N}(b|2.0, 0.5)$, with mean value $b_0 = 2$ and standard deviation $\sigma_b = 0.5$. The acceptance is assumed to follow also a normal distribution with mean $\varepsilon_0 = 0.5$ and standard deviation $\sigma_\varepsilon = 0.1$. (The normal distributions are assumed to be truncated to eliminate negative values of the random variables.) The likelihood function is

$$L(\mu|2) = \int \mathrm{d}\varepsilon \int \mathrm{d}b \mathcal{P}(2|\varepsilon\mu + b) \mathcal{N}(\varepsilon|0.5, 0.1) \mathcal{N}(b|2.0, 0.5) .$$

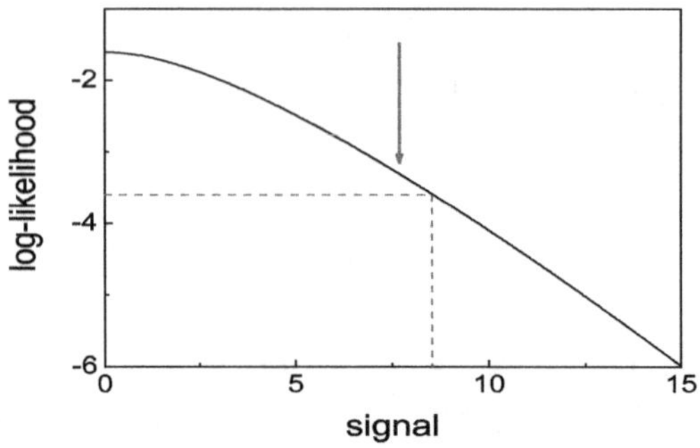

Fig. 8.4 Log-likelihood function for a Poisson signal with uncertainty in background and acceptance. The arrow indicates the upper 90% limit Also shown is the two st. dev likelihood ratio limits (dashed lines).

We solve this integral numerically for values of μ in the range of $\mu_{\rm min} = 0$ to $\mu_{\rm max} = 20$, in which the likelihood function is noticeable different from zero (see Fig. 8.4). Subsequently we determine μ_0 such that the fraction $\alpha = 0.9$ of the normalized likelihood function is located left of μ_0. Since negative values of the normal distributions are unphysical, we cut these distributions and re-normalize them. The computation in our case yields the upper limit $\mu_0 = 7.7$. In the figure we also indicate the e^{-2} likelihood ratio limit.

8.3.3 *Unphysical Parameter Values*

Sometimes the allowed range of a parameter is restricted by physical or mathematical boundaries, for instance it may happen that from the experimental data we infer a negative mass. In these circumstances the parameter range will be cut, and the likelihood function will be normalized to the allowed region. This is illustrated in Fig. 8.5. The integral of the likelihood in the physical region is one. The shaded area is equal to α. The parameter θ is less than $\theta_{\rm max}$ with credibility α.

We have to treat observations which are outside the allowed physical region with caution and check whether the errors have been estimated correctly and no systematic uncertainties have been neglected.

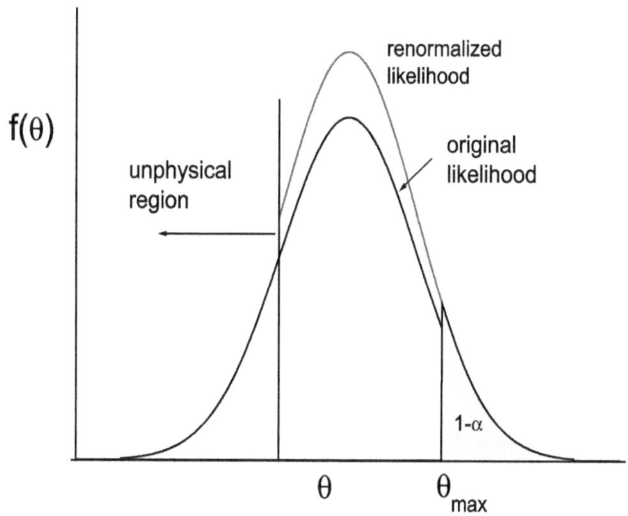

Fig. 8.5 Re-normalized likelihood function and upper limit.

8.4 Confidence Intervals

J. Neyman has proposed a method that circumvents the problematic introduction of prior probabilities and introduced the concept of confidence intervals. Unfortunately, as it is often the case, one problem is traded for another one. The independence of the prior is payed for by a loss of information and in some cases by misleading results.

In most cases we are concerned with a single parameter and well behaved p.d.f.s.[5]

Neyman's confidence intervals have the following defining property: *Confidence intervals of the parameter θ with confidence level α contain θ with probability α.* In other words: Given a true value θ, a measurement $\hat{\theta}$ of θ will include θ in its associated confidence interval $[\theta_1, \theta_2]$ – "cover" it – with probability α.

Typical values chosen for the confidence level are 90%, 95% and 99%. The value 68.3% corresponds to the standard error interval of normally distributed errors.

[5]The variable space is simply connected, the p.d.f. has a single maximum and a concave logarithm.

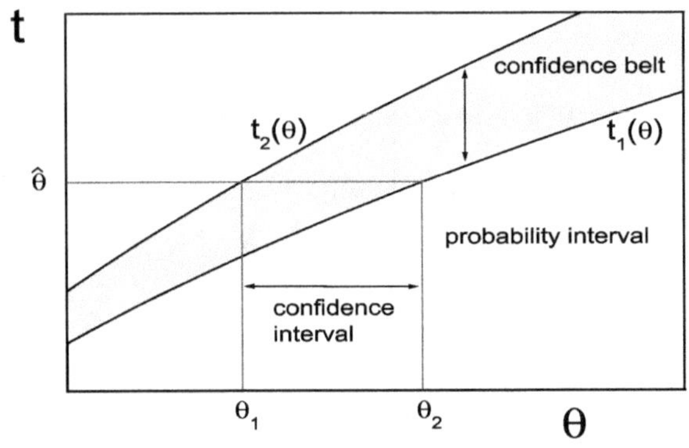

Fig. 8.6 Confidence belt. The area between the functions t_1 and t_2 is the confidence belt. The consisting of the probability intervals $[t_1(\theta), t_2(\theta)]$ for θ. The observation $\hat{\theta}$ leads to the confidence interval $[\theta_1, \theta_2]$. Remark the inversion of the indexes between t and θ.

Given is the p.d.f. $f(t|\theta)$ of the observed value t of the true value θ. A region $[t_1, t_2]$ with probability containment α is selected.

$$P\{t_1(\theta) \leq t \leq t_2(\theta)\} = \int_{t_1}^{t_2} f(t|\theta)dt = \alpha. \tag{8.11}$$

To fix the interval $[t_1, t_2]$ completely, an additional condition is required which is realized by an *ordering rule*. A common procedure is to choose central probability intervals, where both tails correspond to the same probability.

$$P\{t < t_1\} = P\{t > t_2\} = \frac{1-\alpha}{2}.$$

The construction of the confidence interval is explained in Fig. 8.6. The two functions t_1, t_2 are the probability limits as a function of the true parameter θ. They enclose the confidence belt. The confidence limits that correspond to the observed value $\hat{\theta}$ are $t_1(\hat{\theta})$and $t_2(\hat{\theta})$.

Central intervals are invariant against variable transformations. Other ordering conventions are minimum length (promoted by Neyman), equal probability intervals defined by $f(t_1) = f(t_2)$ and likelihood ratio intervals. Probability intervals can also be applied in higher dimensions but are not invariant under variable transformations. More complicated to construct are the likelihood ratio confidence intervals. The probability region that is associated to θ is chosen such that the ratio $f(t|\theta)/f(t|\theta_{best})$ inside the

Table 8.1 Properties of different ordering schemes.

ordering	invariance	multi-dim.	always compact
central	yes	no	yes
equal probability	no	yes	yes
shortest	no	no	yes
likelihood ratio	yes	yes	no

probability interval is larger than outside. The parameter $\theta_{best}(t)$ is the MLE of θ given t. The idea behind this choice is to include preferentially those t-values where $f(t|\theta_{best})$ of competing parameters is relatively low. The likelihood ratio ordering is invariant under parameter transformations but has the problematic feature that the ordering can produce disjoint regions in t and disjoint confidence intervals if the region of the parameter is restricted by physical or mathematical constraints [121].

Table 8.1 summarizes the properties of the different ordering schemes for well behaved likelihood functions

The construction of two-dimensional confidence regions is illustrated in Fig. 8.7. In this example parameters, indicated by boxes, contain

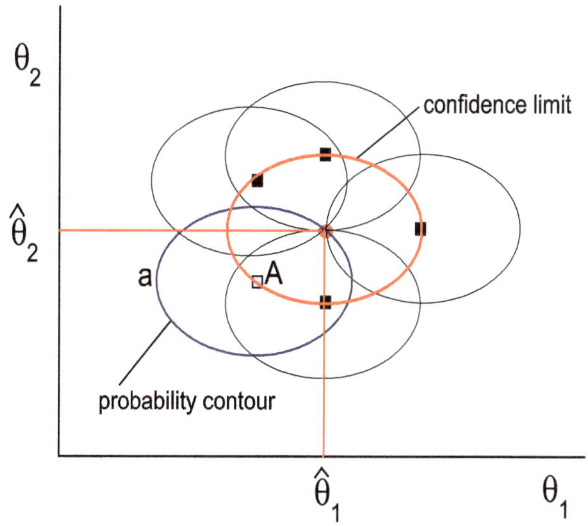

Fig. 8.7 Two-dimensional confidence interval. Given the measurement $\hat{\theta_1}, \hat{\theta_2}$ the red ellipse contains the true parameter values with confidence α.The parameter values indicated by boxes contain measurements inside the associated ellipses with probability α.

measurements inside elliptical regions with probability α. The ellipses touch the actually measured parameter location. In this example the parameter and the measured quantity share the same space.

8.4.1 *Properties*

8.4.1.1 *Confidence Intervals and Error Intervals*

Confidence intervals are not suited as error intervals because they are unrelated to the point estimate, do not allow to apply the usual error calculus and to combine measurements. They cannot easily be extended to higher dimensions.

8.4.1.2 *Coverage*

Coverage is a valuable property of confidence intervals. It guarantees in a series of measurements that a fraction of results are located inside the limits provided by the interval. For example, a 99 % confidence interval for parameters obtained in a blood test may indicate that the parameter is in the save region with at least 99 % probability. Similarly a company that measures parameters of products can guarantee that the true parameter is in the faction α within the specifications. Coverage is important, if a large number of measurements of independent replicates are considered. In natural sciences where typically constants of nature or a unique parameters are of interest, coverage is of minor importance. Knowing, for instance, that out of 3 measurements 2 cover the true value of the mass of a particle is pretty useless. Confidence intervals play an important role in humanities, in medical and commercial applications where quality or performance has to be within given specifications.

8.4.1.3 *Probability*

Confidence intervals suffer from two problems: They ignore relevant information but depend on irrelevant information, namely the prediction of the p.d.f. of data that have not been observed.

The first point is obvious if parts of the interval or the complete interval are located in a region that is forbidden by mathematical or physical laws, but more important is the neglect of the likelihood function. This is illustrated in a discrete analogon:

Two normal distributions $f_1(x) = \mathcal{N}(x|0, 0.5)$ and $f_2(x) = \mathcal{N}(x|5, 4)$ are confronted with a measurement $x = 2.7$, see Fig. 8.8. It is obvious that we

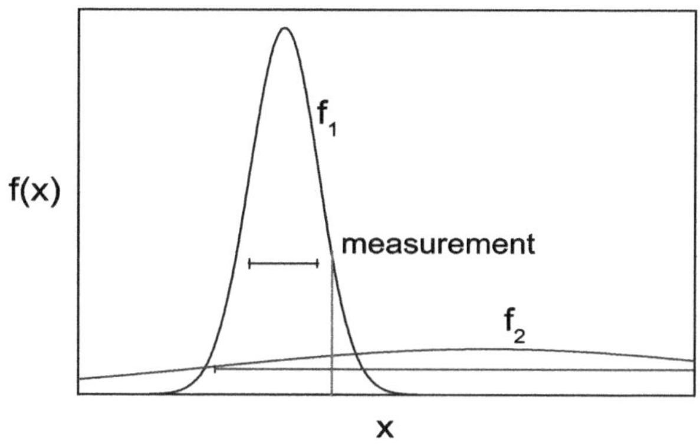

Fig. 8.8 Comparison of two normal distribution with a measurement. The likelihood ratio favors the narrow p.d.f. f1, while a one standard deviation cut would exclude it.

would rather favor the p.d.f. f_1. For a decision between the two hypotheses only the function values at $\hat{x}$ are relevant. However, only f_2 is compatible with the measurement within a one standard deviation confidence interval.

In the following example the parameter distribution is continuous.

Example 8.7. Comparison of a likelihood ratio interval with a confidence interval

The following p.d.f. describes a Gaussian distributions where the width is strongly increasing with the location parameter.

$$f(t) = \frac{25}{\sqrt{2\pi\theta^2}} \exp\left[-\frac{625(x-\theta)^2}{2\theta^4}\right]$$

Fig. 8.9 shows the likelihood function for a measurement $x = 10$ together with the one standard deviation likelihood ratio interval and the confidence interval. The confidence intervals extends to infinity. It includes parameter values that would produce the measurement with a tiny probability that is by a huge factor smaller than parameter values that are excluded.

Confidence intervals favor probability distributions with little predictive power.

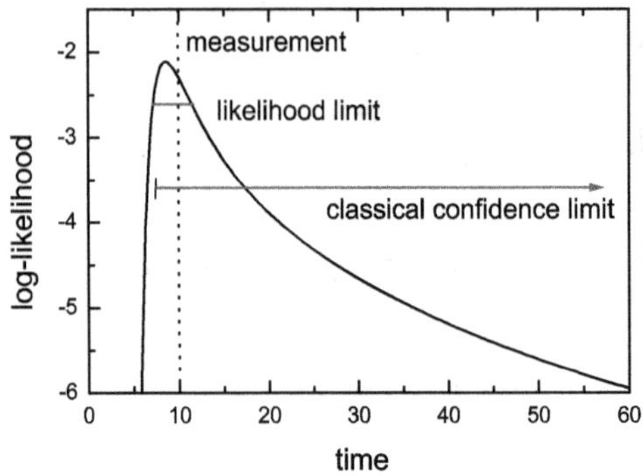

Fig. 8.9 Likelihood function. Comparison of likelihood ratio and confidence intervals.

8.4.2 *One-Sided Limits*

One-sided limits are treated in the same way as two-sided, except that one limit is fixed and the other one is either $-\infty$ (upper limit) or $+\infty$ (lower limit). In Fig. 8.10 left a normal distribution and the 95% upper confidence limit $\theta_{95\%}$ for the mean value corresponding to the observation $\hat{x}$ are shown. The right hand side of the figure shows the Poisson distribution for the mean value $\mu = 3$. For zero observed events $\mu = 3$ is the 95% upper confidence limit.

The Poisson confidence limits agree with the Bayesian credibility limits if background is absent but disagree in situation where background is expected but no background is observed. The likelihood function ignores predictions that are not realized but confidence limits do not satisfy the likelihood principle and depend on predictions that obviously are not realized. For instance, in an experiment where no event is observed the upper limit for the rate depends on the amount of predicted but obviously absent background. Coverage is a property of an ensemble of experiments and therefore results for individual measurements may contradict common sense.

8.5 Summary

- It is not possible to associate a probability to a parameter interval, unless a likelihood function and prior information are given. In many cases, however, the interval is so narrow, that the result is independent

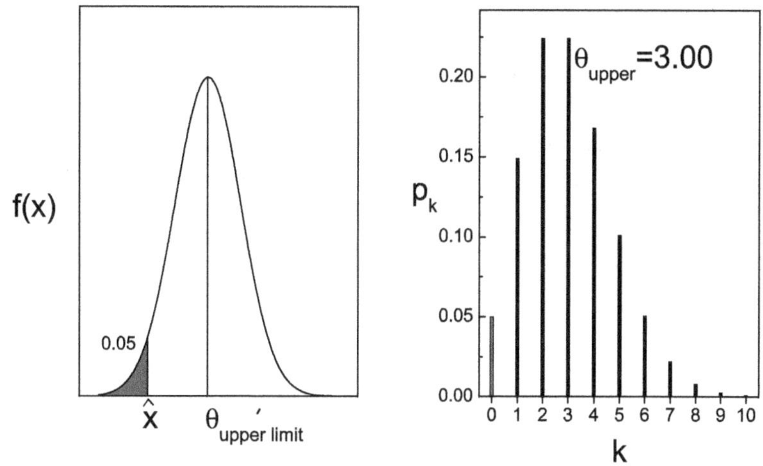

Fig. 8.10 95% Upper limits for the parameters of a continuous and a discrete distribution.

of the shape of sensible prior distributions and a uniform prior can be applied.

- If a likelihood function is available, the standard likelihood ratio limits should be used to represent the precision of the measurement. The limits may be asymmetric and multi-dimensional regions can be constructed. They are transformation invariant, permit to combine different measurement and summarize as far as possible the available information.
- If the log-likelihood function is parabolic and the prior can be approximated by a constant, e.g. the likelihood function is very narrow, the likelihood function is proportional to the p.d.f. of the parameter. Then the error limits represent one standard deviation and a 68.3 % probability interval. The standard error calculus is applicable.
- If the likelihood function is very asymmetric, due to low event numbers or due to constraints, the full likelihood function has to be retained to document the result.
- Nuisance parameters are eliminated by the methods described in Chapter 7.
- Error propagation is performed using the direct functional dependence of the parameters.

- Credibility and Confidence intervals should not be confounded with error intervals.
- Classical confidence intervals often neglect available information but are able to guarantee the well defined long term performance of procedures in many practical applications. In sciences where constants of nature are determined, they are of limited use. They depend on irrelevant information and neglect information contained in the likelihood function. Confidence limits of several experiments cannot be combined without loss of information.
- In particle physics it is common to represent upper and lower limits by confidence intervals. Preferable are credibility intervals, computed with a uniform (uninformative) prior.

Chapter 9

Unfolding

9.1 Introduction

In many experiments the measurements are deformed by limited acceptance, sensitivity, or resolution of the detectors. Knowing the properties of the detector, we are able to simulate these effects, but is it possible to invert this process, to reconstruct from a distorted event sample the original distribution from which the undistorted sample has been drawn?

There is no simple answer to this question. Apart from the unavoidable statistical uncertainties, the correction of losses is straight forward, but unfolding the effects, caused by the limited resolution, is difficult and feasible only by introducing a priory assumptions about the shape of the original distribution or by grouping the data in histogram bins. Therefore, we should ask ourselves, whether unfolding is really a necessary step of our analysis. If we want to verify a theoretical prediction for a distribution $f(\boldsymbol{x})$, it is much easier and more accurate to fold f with the known resolution and to compare then the smeared prediction and the experimental distributions with the methods discussed in Chapter 10. If a prediction contains interesting parameters, also those should be estimated by comparing the smeared distribution with the observed data. When we study, for instance, a sharp resonance peak on a slowly varying background, it will be very difficult, if not impossible, to determine the relevant parameters from an unfolded spectrum, while it is easy to fit them directly to the observed distribution, see Section 7.3 and Ref. [55]. However, in situations where a reliable theoretical description is missing, or where the measurement is to be compared with a distribution obtained in another experiment with different experimental conditions, unfolding of the data cannot be avoided. Examples are the determination of structure functions in deep inelastic scattering or transverse momentum distributions from the Large Hadron Collider at CERN where an obvious parametrization is missing.

The choice of the unfolding procedure depends on the goal, one is aiming for. We either can try to optimize the reconstruction of the distribution, with the typical trade-off between resolution and bias where we have a kind of probability density estimation (PDE) problem (see Chapter 12), or we can treat unfolding as an inference problem where the errors should contain the unknown result with a reasonable coverage probability.[1] The former approach dominates in most applications outside the natural sciences, for instance in picture unblurring, but is also adopted in particle physics and astronomy. We will follow both lines, the first provides the most likely shape of the distribution, but is not suited as a bases for a quantitative analysis, while the second permits to combine results and to compare them quantitatively to theoretical predictions. We will consider mainly histograms and spline approximations, but sketch also binning free methods which may become more popular with increased computer power.

General unfolding studies are found in Refs. [54, 56–58, 60, 61]. Specific methods are presented in Refs. [59, 62–67].

9.2 Discrete Inverse Problems and the Response Matrix

9.2.1 *Introduction and Definition*

Folding is described by the integral

$$g(x') = \int_{-\infty}^{\infty} h(x', x) f(x) dx. \tag{9.1}$$

The function $f(x)$ is folded with a response function[2] $h(x', x)$, resulting in the smeared function $g(x')$. We call $f(x)$ the *true distribution* and $g(x')$ the *smeared distribution* or the *observed distribution*. The three functions g, h, f can have discontinuities but of course the integral has to exist. The integral equation (9.1) is called Fredholm equation of the first kind with the kernel $h(x', x)$. If the function $h(x', x)$ is a function of the difference $x' - x$ only, (9.1) is denoted convolution integral, but often the terms convolution and folding are not distinguished. The relation (9.1) describes the

[1] We derive the errors from the shape of the likelihood function. In most cases with not too small event numbers, the definition coincides to a good approximation with the error definition derived from the coverage paradigm, Section 8.4. In this chapter arguing with coverage is more convenient than with the likelihood ratio.

[2] Sometimes it is also called folding function or transfer function

direct process of folding. We are interested in the *inverse problem*: Knowing g and h we want to infer $f(x)$. This inverse problem is classified by the mathematicians as *ill posed* because it has no unique solution. In the direct process high frequencies are washed out. The damping of strongly oscillating contributions in turn means that in mapping g to f high frequencies are amplified, and the higher the frequency, the stronger is the amplification. In fact, in practical applications we do not really know g, the information we have, consists only of a sample of observations with the unavoidable statistical fluctuations.[3] The fluctuations of g correspond to large perturbations of f and consequently to ambiguities.

The response function often, but not always, describes a simple resolution effect and then it is called *point spread function* (PSF). There exists also more complex situations, like in positron emission tomography (PET), where the relation between the observed distribution of two photons and the interesting distribution of their origin is more involved. In PET and many other applications the variables x and x' are multidimensional.

9.2.2 *The Histogram Representation*

9.2.2.1 *Discretization and the Response Matrix*

The disease of the inverse problem can partially be cured by discretization, which essentially means that we construct a parametric model. We usually replace the continuous functions by histograms which can be written as vectors $\boldsymbol{\theta}$ for the true histogram and $\boldsymbol{d}$ for the observed histogram The two histograms are connected by the response function, here by a matrix A. We get for the direct process:

$$\mathrm{E}(\boldsymbol{d}) = \mathsf{A}\boldsymbol{\theta}. \tag{9.2}$$

$$\mathrm{E}\begin{pmatrix} d_1 \\ d_2 \\ \cdot \\ \cdot \\ \cdot \\ d_N \end{pmatrix} = \begin{pmatrix} A_{11} & .. & A_{1M} \\ A_{21} & .. & A_{2M}\cdot \\ \cdot & .. & \cdot \\ \cdot & .. & \cdot \\ \cdot & .. & \cdot \\ A_{N1} & .. & A_{NM} \end{pmatrix} \cdot \begin{pmatrix} \theta_1 \\ \cdot \\ \cdot \\ \theta_M \end{pmatrix}.$$

Here d_i is the content of bin i of an *observed histogram*. $\mathrm{E}(\boldsymbol{d})$ is the expected value. A is called *response or folding matrix* and θ_j is the content

[3]In the statistical literature the fluctuations are called noise.

of bin j of the undistorted *true histogram* that we want to determine.

$$\begin{aligned} \theta_j &= \int_{bin\ j} f(x)dx \\ \mathrm{E}(d_i) &= \int_{bin\ i} dx' \int_{-\infty}^{\infty} h(x',x)f(x)dx \\ A_{ij} &= \mathrm{E}(d_{ij})/\theta_j \end{aligned} \tag{9.3}$$

where

$$\mathrm{E}(d_{ij}) = \int_{bin\ i} dx' \int_{bin\ j} h(x',x)f(x)dx$$

is the expected number observed in bin i that originates from true bin j. The value A_{ij} represents the probability that the detector registers an event in bin i that belongs to the true histogram bin j. This interpretation assumes that all elements of $\boldsymbol{d}$, A and $\boldsymbol{\theta}$ are positive. The number of columns M is the number of bins in the true histogram and the number of parameters that have to be determined. The number of rows N is the number of bins in the observed histogram. We do not want to have more parameters than measurements and require $N \geq M$. Normally we constrain the unknown true histogram, requiring $N > M$. With N bins of the observed histogram and M bins of the true histogram we have $N - M$ constraints. The relation between the various histograms is shown in Fig. 9.1. We follow the simpler right hand path where multinomial errors need not be handled.

We require that A is rank efficient which means that the rank is equal to the number of columns M. Formally, this means that all columns are linearly independent and at least M rows are linearly independent: No two bins of the true histogram should produce observed distributions that are proportional to each other. Unfolding would be ambiguous in this situation, but a simple solution of the problem is to combine the bins. More complex cases that lead to a rank deficiency never occur in practice. A more serious requirement is the following: By definition of A, the observed histogram must not contain events that originate from other sources than the M true bins. In other words, the range of the true histogram has to cover all observed events. This requirement often entails that only a small fraction of the events that are contained the border bins of the true histogram are found in the observed histogram. The correspondingly low efficiency leads to large errors of the reconstructed number of events in these bins. Published simulation studies often avoid this complication by restricting the range of the true variable.

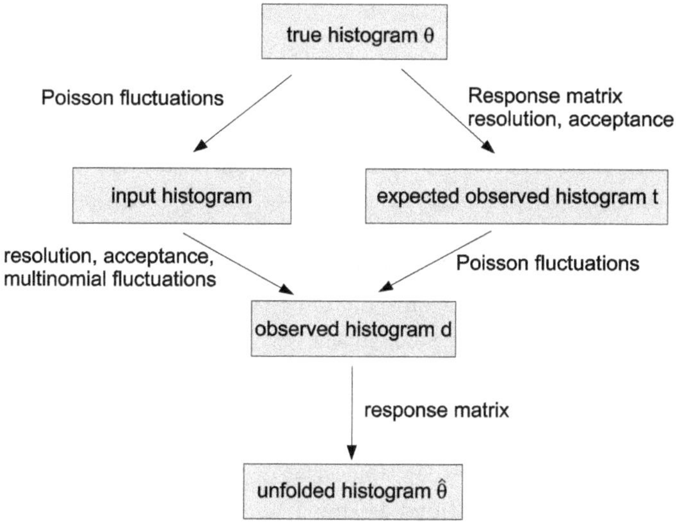

Fig. 9.1 Relations between the histograms involved in the unfolding process.

Some publications refer to a null space of the matrix A. The null space is spanned by vectors that fulfill $\mathsf{A}\boldsymbol{\theta} = 0$. With our definitions and the restrictions that we have imposed, the null space is empty.

In particle physics the experimental setups are mostly quite complex, and for this reason they are simulated with Monte Carlo programs. To construct the matrix A, we generate events following an assumed true distribution $f(x)$ characterized by the true variable x and a corresponding true bin j. The detector simulation produces the observed variable $x'$and the corresponding observed bin i. We will assume for the moment that we can generate an infinitely large amount of so-called Monte Carlo events such that we do not have to care about statistical fluctuations of the elements of A. The statistical fluctuations of the observed event numbers be described by the Poisson distribution.

There is another problem that we neglect but that we have to resume later: The matrix A depends to some extent on the true distribution which is not known in the Monte Carlo simulation. The dependence is small if the bins of the true distribution are narrow enough to neglect the fluctuations of $f(x)$ within a bin. This condition cannot always be maintained.

9.2.2.2 *The Need for Regularization*

The discrete model avoids the ambiguity of the continuous ill-posed problem but especially if the observed bins are narrow compared to the resolution, the matrix is badly conditioned which means that the inverse or pseudo-inverse of A contains large components. This is illustrated in Fig. 9.2 which shows two different original distributions and the corresponding distributions smeared with a Gaussian $\mathcal{N}(x - x'|0, 1)$. In spite of the extremely different original distributions, the smeared distributions of the samples are practically indistinguishable. This demonstrates the substantial information loss that is caused by the smearing, especially in the case of the distribution with four peaks. Sharp structures are washed out and can hardly be reconstructed. Given the observed histogram with some additional noise, it will be almost impossible to exclude one of the two candidates for the true distribution even with a huge amount of data. Naive unfolding by matrix inversion can produce oscillations as shown in Fig. 9.3.

If the matrix A is quadratic, we can simply invert (9.2) and get an estimate $\widehat{\boldsymbol{\theta}}$ of the true histogram.

$$\widehat{\boldsymbol{\theta}} = A^{-1}\boldsymbol{d}. \tag{9.4}$$

In practice this simple solution usually does not work because, as mentioned, our observations suffer from statistical fluctuations.

In Fig. 9.4, the result of a simple inversion of a data vector is depicted. The left hand plot is realized with 10 bins. It is clear that either fewer

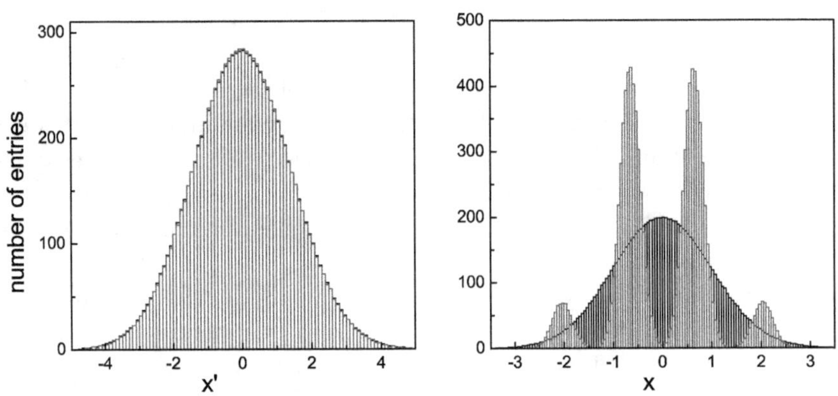

Fig. 9.2 Folded distributions (left) for two different distributions (right).

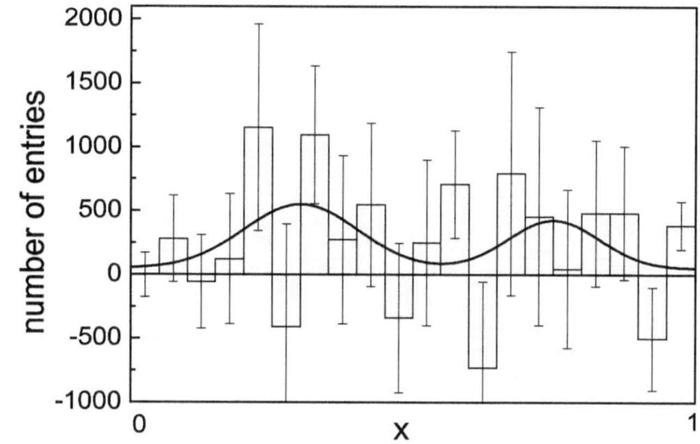

Fig. 9.3 Naive unfolding result obtained by matrix inversion. The curve corresponds to the true distribution.

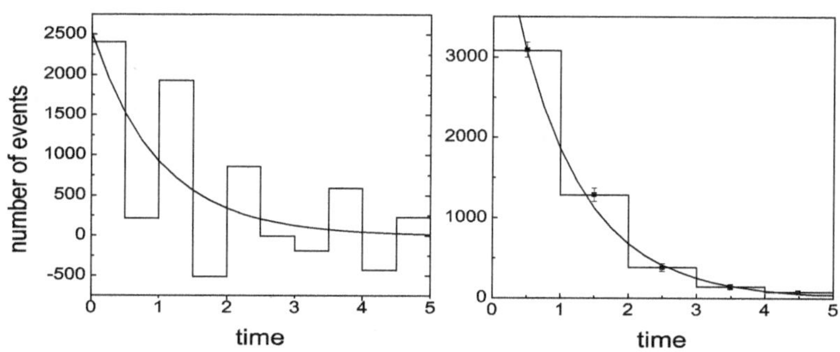

Fig. 9.4 Unfolding by matrix inversion with different binnings.

bins have to be chosen, see Fig. 9.4 right hand plot, or some smoothing (regularization) has to be applied.

9.2.3 *Expansion of the True Distribution*

Instead of representing the distribution $f(x)$ by a histogram, we can approximate it by an expansion into a finite sum of functions B_i. We choose normalized functions, $\int_{-\infty}^{\infty} B_i(x)dx = 1$.

$$f(x) \approx \sum_{j=1}^{M} \beta_j B_j(x). \tag{9.5}$$

We get

$$\begin{aligned} \mathrm{E}(d_i) &= \int_{bin\ i} dx' \sum_{j=1}^{M} \beta_j \int_{-\infty}^{\infty} h(x', x) B_j(x) dx \\ &= \sum_{j=1}^{M} A_{ij} \beta_j . \end{aligned} \tag{9.6}$$

with

$$A_{ij} = \int_{bin\ i} dx' \int_{-\infty}^{\infty} h(x', x) B_j(x) dx .$$

The response matrix element A_{ij} now is the probability to observe an event in bin i of the observed histogram that originates from the distribution B_j. In other words, the observed histogram is approximated by a superposition of the histograms produced by folding the functions B_j. Unfolding means to determine the amplitudes β_j of the functions B_j.

Below we will discuss the approximation of $f(x)$ by a superposition of basic spline functions (b-splines). For our applications the b-splines of order 2 (linear), 3 (quadratic) or 4 (cubic) are appropriate (see Appendix A.13). Unfolding then produces a smooth function which normally is closer to the true distribution than a histogram. The disadvantage of spline approximations compared to the histogram representation is that a quantitative comparison with predictions or the combination of several results is more difficult.

Remark: In probability density estimation (PDE) a histogram is considered as a first order spline function. The spline function corresponds to the line that limits the top of the histogram bins. The interpretation of a histogram in experimental sciences is different from that in PDE. Observations are collected in bins and then the content of the bin measures the integral of the function g over the bin and the bin content of the unfolded histogram is an estimate of the integral of f over that bin. A function can always be described correctly by a histogram,while the description by spline functions is an approximation. This has to be kept in mind when we compare the unfolding result to a prediction.

9.2.4 The Least Square Solution and the Eigenvector Decomposition

We have seen that high frequency contributions of the true distributions are strongly suppressed by the folding process. The decomposition of the response matrix allows to quantify the suppression as a function of the frequency.

9.2.4.1 The Least Square Solution

As mentioned, for a *square matrix* A, $M = N$, the solution of $\boldsymbol{\theta}$ is simply obtained by matrix inversion, $\widehat{\boldsymbol{\theta}} = \mathsf{A}^{-1}\boldsymbol{d}$. The error matrix $\mathsf{V}_\theta = \mathsf{A}^{-1}\mathsf{V}(\mathsf{A}^{-1})^T$ is derived by error propagation. We omit the calculation. In the limit where there is no smearing, A is diagonal and describes only acceptance losses.

The choice $M = N$ is not recommended. For $M \le N$ the least square function χ^2_{stat} is given by the following relation:

$$\chi^2_{stat} = \sum_{i=1}^{N} \frac{(t_i - d_i)^2}{t_i} = \sum_{i=1}^{N} \frac{(\sum_{k=1}^{M} A_{ik}\theta_k - d_i)^2}{\sum_{k=1}^{M} A_{ik}\theta_k} . \tag{9.7}$$

If the numbers d_i are not described by a simple Poisson distribution, we have to insert the weight matrix where $\mathsf{C} = \mathsf{V}^{-1}$ is the inverse of its error or covariance matrix V:

$$\chi^2_{stat} = \sum_{i,k=1}^{N} [(t_i - d_i)C_{ik}(t_k - d_k)] . \tag{9.8}$$

If the data follow a Poisson distribution where the statistics is high enough to approximate it by a normal distribution and where the denominator of (9.7) can be approximated by d_i, the least square minimum can be evaluated by a simple linear matrix calculus. (The linear LS solution is treated in Section 6.8.)

$$\chi^2_{stat} = \sum_{i=1}^{N} \frac{(t_i - d_i)^2}{d_i} = \sum_{i=1}^{N} \frac{(\sum_{k=1}^{M} A_{ik}\theta_k - d_i)^2}{d_i} , \tag{9.9}$$

We apply the transformations

$$\boldsymbol{d} \Rightarrow \boldsymbol{b} = \mathsf{A}^T\mathsf{C}\boldsymbol{d} , \tag{9.10}$$

$$\mathsf{A} \Rightarrow \mathsf{Q} = \mathsf{A}^T\mathsf{C}\mathsf{A} . \tag{9.11}$$

We call Q *least square matrix*. We get for the expected value of $\boldsymbol{b}$

$$\mathrm{E}(\boldsymbol{b}) = \mathsf{Q}\boldsymbol{\theta} \tag{9.12}$$

with the LS solution

$$\widehat{\boldsymbol{\theta}} = \mathsf{Q}^{-1}\boldsymbol{b} \tag{9.13}$$

and the error matrix V_θ of the solution

$$\mathsf{V}_\theta = \mathsf{Q}^{-1}.$$

We have simply replaced A by Q and $\boldsymbol{d}$ by $\boldsymbol{b}$. Both quantities are then known. Relation (9.13 replaces relation (9.2). The matrix Q is quadratic and can be inverted if the LS solution exists.

9.2.4.2 *Eigenvector Decomposition of the Least Square Matrix*

To understand better the origin of the fluctuations of the LS solution (9.13), we factorize the matrix Q in the following way: The matrix[4] $\mathsf{Q} = \mathsf{U}\Lambda\mathsf{U}^{-1}$ is composed of the diagonal matrix Λ which contains the eigenvalues of Q and the matrix U whose columns consist of the eigenvectors $\boldsymbol{u}_i$ of Q:

$$\mathsf{Q} = \left(\boldsymbol{u}_1\, \boldsymbol{u}_2\, ..\, \boldsymbol{u}_M\right) \begin{pmatrix} \lambda_1 & & & & \\ & \lambda_2 & 0 & & \\ & & . & & \\ & 0 & & . & \\ & & & & \lambda_M \end{pmatrix} \left(\boldsymbol{u}_1\, \boldsymbol{u}_2\, ..\, \boldsymbol{u}_M\right)^{-1}.$$

$$\mathsf{Q}\boldsymbol{u}_i = \lambda_i \boldsymbol{u}_i = \boldsymbol{v}_i,\ i = 1, ..., M. \tag{9.14}$$

Software to produce the eigenvector decomposition can be found in most mathematical computer libraries.

In case of eigenvalues that appear more than once, the eigenvectors are not uniquely defined. Linear orthogonal combinations can be created by rotations in the corresponding subspace but they produce the same LS solution.

The solution $\boldsymbol{\theta}$ can be expanded into the orthogonal unit eigenvectors $\boldsymbol{u}_i$:

$$\boldsymbol{\theta} = \sum_{i=1}^{M} a_i \boldsymbol{u}_i,\ \ \theta_k = \sum_{i=1}^{M} a_i u_{ik},$$

$$a_i = \boldsymbol{\theta} \cdot \boldsymbol{u}_i,\ \ a_i = \sum_{k=1}^{M} \theta_k u_{ik}.$$

---

[4]We require that the square $M \times M$ matrix Q has M linearly independent eigenvectors and that all eigenvalues are real and positive.

By construction, the amplitudes a_i are uncorrelated and the norm $||\theta||^2 = \Sigma\theta_i^2$ of the solution is given by

$$||\theta||^2 = \sum_{i=1}^{M} a_i^2 .$$

The transformed observed vector $\boldsymbol{b}$ is

$$\boldsymbol{b} = \sum_{i=1}^{M} a_i \lambda_i \boldsymbol{u}_i = \sum_{i=1}^{M} a_i \boldsymbol{v}_i .$$

In Fig. 9.5 we present an schematic example of a set of eigenvectors. A contribution $\boldsymbol{u}_i$ to the true histogram as shown on the left hand side will produce a contribution $\boldsymbol{v}_i$ to the observed histogram. It is of the same shape but reduced by the factor λ_i as shown on the right hand side. The eigenvalues decrease from top to bottom. Strongly oscillating components of the true histogram correspond to small eigenvalues. They are hardly visible in the observed data, and in turn, small contributions v_i to the observed

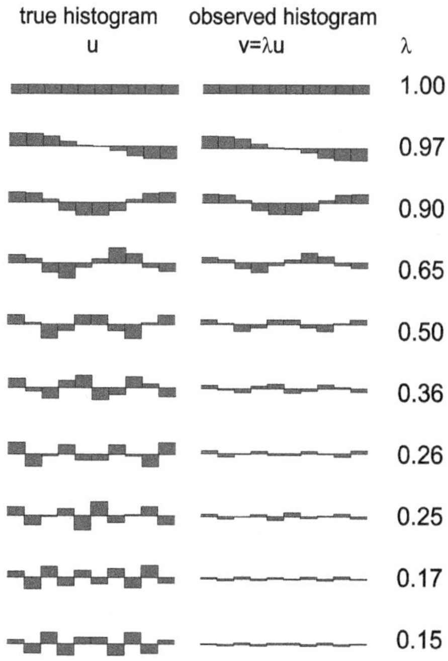

Fig. 9.5 Set of eigenvectors ordered according to decreasing eigenvalues. A contribution $\boldsymbol{u}_i$ in the true histogram corresponds to a contribution $\boldsymbol{v}_i$ to the observed histogram.

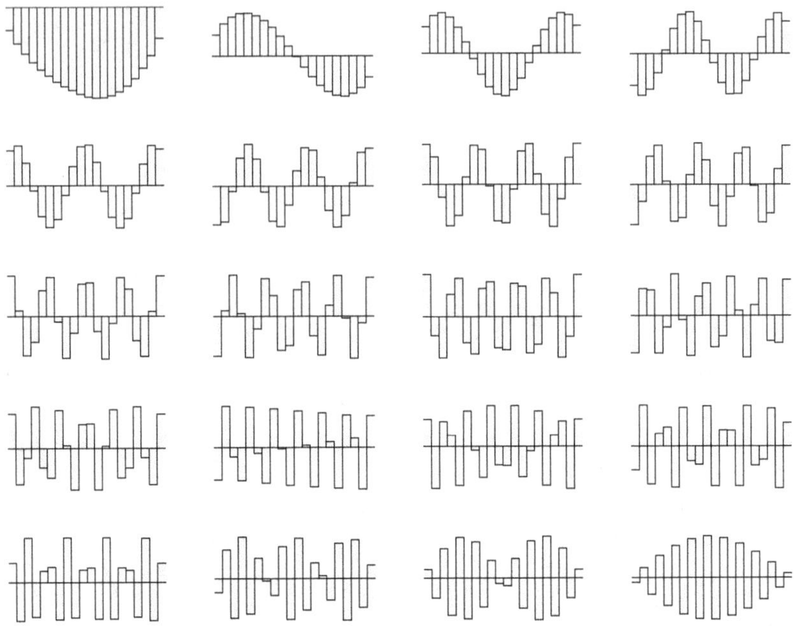

Fig. 9.6 Eigenvectors of the modified LS matrix ordered with decreasing eigenvalues.

data caused by statistical fluctuations can lead to rather large oscillating contributions $\boldsymbol{u}_i = \boldsymbol{v}_i/\lambda_i$ to the unfolded histogram if the eigenvalues are small. Eigenvector contributions with eigenvalues below a certain value cannot be reconstructed, because they cannot be distinguished from noise in the observed histogram.

The eigenvector decomposition is equivalent to the *singular value decomposition* (SVD). In the following we will often refer to the term SVD instead of the eigenvector decomposition, because the former is commonly used in the unfolding literature.

Example 9.1. Eigenvector decomposition of the LS matrix

In Fig. 9.6 the 20 eigenvectors of a LS matrix ordered with decreasing eigenvalue are displayed. The response matrix has 20 true and 40 observed bins. The graph is generated from a sample of 100 000 uniformly distributed events in the range of the observed and the true variables $0 < x, x' < 1$. The response function is a Gaussian with standard deviation $\sigma_s = 0.04$. The

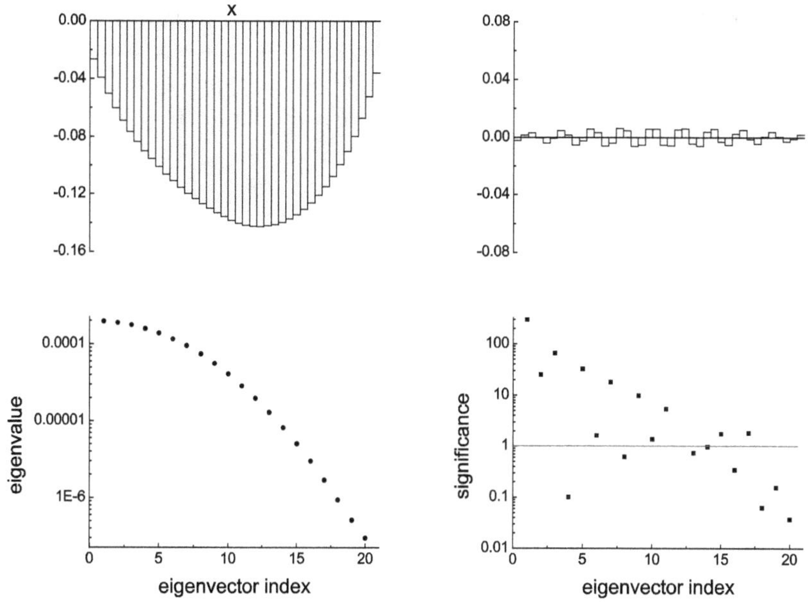

Fig. 9.7 Observed eigenvectors 1 (top left) and 20 (top right), eigenvalues (bottom left) and significance of eigenvector amplitudes (bottom right).

eigenvectors show an oscillatory behavior where the number of clusters corresponds roughly to the eigenvector index. In Fig. 9.7 top the eigenvectors 1 and 20 folded with the response matrix are shown. A contribution of eigenvector 20 to the observed histogram is similar to that of noise. The eigenvalues shown at the bottom left graph vary by about three orders in magnitude. This means that a contribution of the eigenvector 20 to the true distribution is suppressed in the observed distribution by a factor of 1000 with respect to a contribution of eigenvector 1. The bottom right hand graph shows the significance of the amplitudes that are attributed to the eigenvectors. Significance is defined as the absolute value of the amplitude divided by its error. As we have indicated above, the significance is expected to decreases with decreasing eigenvalue. Due to the symmetry of the problem, the amplitudes with even index should vanish. Statistical fluctuations in the simulation partially destroy the symmetry. Eigenvector contributions where the significance is below one, are compatible with being absent within one standard deviation.

The significance of of the amplitudes deteriorates strongly with increasing smearing. It is difficult to compensate a bad resolution of an experiment by increasing the statistics! We should always make an effort to avoid large smearing effects not only because large event numbers are required but also because the unfolding results then depend strongly on a precise knowledge of the response function and are less reliable.

9.2.4.3 *The Effective Number of Parameters*

When we unfold a histogram, the number of bins of the unfolded histogram is the number of free parameters in the fit. The previous example indicates that the number of independent parameters that we can determine in a given problem is rather limited. Below a certain eigenvalue λ_k, all parameters have a significance close to or below one. We can define an effective number of parameters $N_{eff} = k$ as the number of parameters with eigenvalues above or equal to this limit. For the example of Fig. 9.8 with a uniform distribution the effective number of parameters is $N_{eff} = 17$. There are also parameters left of index 17 that are compatible with being zero. We should not exclude the corresponding contributions, because the reason for the small values of the significance are not large errors, but small values of the fitted amplitudes as is indicated in the right hand graph. This graph shows that some amplitudes that correspond to small eigenvalues become

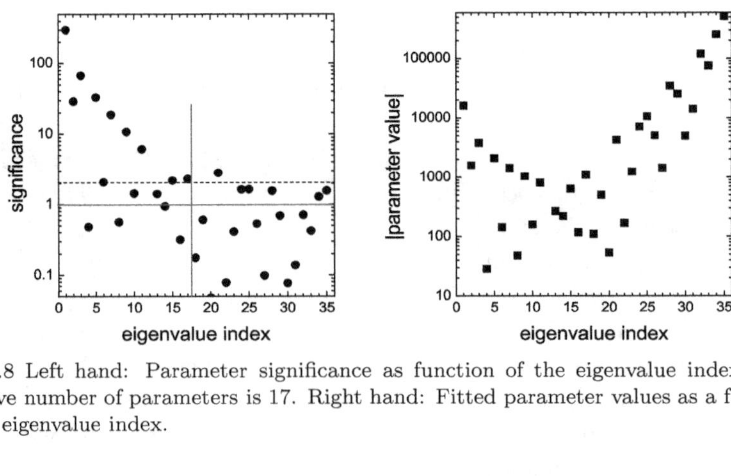

Fig. 9.8 Left hand: Parameter significance as function of the eigenvalue index. The effective number of parameters is 17. Right hand: Fitted parameter values as a function of the eigenvalue index.

rather large. This is due to the amplification of high frequency noise in the unfolding. The number of bins in the unfolded distribution should not be much larger than the effective number of parameters, because then we keep too much redundant information, but on the other hand it has to be large enough to represent the smallest significant eigenvector. A reasonable choice for the number of bins is about twice N_{eff}. The optimal number will also depend on the shape of the distribution.

SVD is not an optimal method but it provides insight in the origin of unfolding problems which can be summarized qualitatively as follows: The limited resolution suppresses high frequency contributions of the true histogram in the observed histogram where they cannot be distinguished from statistical fluctuations. In turn, these artificial fluctuations then produce necessarily high frequency contributions in the unfolded histogram. These fake oscillations have to be suppressed and thereby also the true oscillations vanish. The unfolding result is biased.

9.2.5 *The Maximum Likelihood Approach*

9.2.5.1 *Histogram Representation*

Whenever possible, we should apply a MLF instead of a LSF. With Poisson distributed event numbers d_i with expected values $E(d_i) = t_i = \sum_j A_{ij}\theta_j$ the probability to obtain d_i is

$$P(d_i) = \frac{e^{-t_i} t_i^{d_i}}{d_i!}$$

and the corresponding log-likelihood is up to an irrelevant constant

$$\begin{aligned} \ln L_{stat} &= \sum_{i=1}^{N} [d_i \ln t_i - t_i] \\ &= \sum_{i=1}^{N} \left[d_i \ln \sum_{j=1}^{M} A_{ij}\theta_j - \sum_{j=1}^{M} A_{ij}\theta_j \right] . \end{aligned} \tag{9.15}$$

Maximizing $\ln L_{stat}$ we obtain an estimate $\widehat{\boldsymbol{\theta}}$ of the true histogram.

Usually we have of the order of 20 bins and of course the same number of correlated parameters which have to be adjusted. In this situation the fit often does not converge very well. Instead of maximizing the log-likelihood with methods like Simplex, we can compute the solution iteratively with the expectation maximization (EM) method [68], see Appendix A.4.

The following alternating steps are executed:

- Folding step:

$$d_i^{(k)} = \sum_j A_{ij}\theta_j^{(k)}. \tag{9.16}$$

- Unfolding step:

$$\theta_j^{(k+1)} = \sum_{i=1}^{N} A_{ij}\theta_j^{(k)} \frac{d_i}{d_i^{(k)}} / \alpha_j \,. \tag{9.17}$$

Usually uniform starting values $\theta^{(0)} = 1$ are selected. The efficiency parameter $\boldsymbol{\alpha}$ corrects for acceptation losses, $\alpha_j = \sum_{i=1}^{M} A_{ij}$.

As will be discussed below, the iterative likelihood procedure can be used to produce a regularized unfolding solution by stopping the iteration.

Before the EM method had been invented, the iterative procedure had been introduced independently by Richardson and Lucy [69,70] specifically for the solution of unfolding problems. Later it was re-invented several times [62, 65, 71, 72]. That the result of the iteration converges to the maximum likelihood solution, is a general property of the EM method but was also proven by Vardi *et al.* [73] and later independently by Mülthei and Schorr [62]. For a discussion of the application to unfolding see [75].

9.2.5.2 *Spline Approximation*

Instead of the histogram representation other function sets can be used to approximate the true distribution. The coefficients can again be obtained with the iterative EM formalism. A smooth unfolding result is obtained by the Expansion into B-splines:

$$f(x) = \sum_{j=1}^{N} \theta_j B_j(x).$$

The relations (9.16) and (9.17) remain valid.

9.3 Unfolding with Explicit Regularization

9.3.1 *General Considerations*

The main field where professional unfolding methods are applied lies in image reconstruction. In medicine unblurring of tomographic pictures of arterial stenoses, of tumors or orthopedic objects are important. Other areas of interest are unblurring of images of astronomical surveys, of geographical maps, of tomographic pictures of tools or mechanical components like train wheels to detect damages. Also pattern recognition for example of fingerprints or the iris is an important field of interest. The goal of most applications is to dig out hidden or fuzzy structures from blurred images, to remove noise and to improve the contrast. Also in physics applications it is of interest to visualize hidden structures and to reconstruct distributions which may be used for instance to simulate experimental situations. We can profit from the fact that physics distributions are rather smooth. Often we can remove the roughness of an unfolding result without affecting very much the real structures of the true distribution.

To obtain a smooth distribution, several different regularization mechanisms are available.

In particle- and astrophysics the following methods are applied:

(1) *Truncation methods*: In the eigenvector decomposition of the LS matrix (equivalent to the SVD) low eigenvalue contributions to the unfolding solution are suppressed or eliminated.
(2) *Penalty methods*: A penalty term which is sensitive to unwanted fluctuations is introduced in the LS or ML fit of the unfolding solution. Typically, deviations from a uniform or a linear distribution are suppressed. Standard methods penalize curvature, low entropy or a large norm of the unfolding distribution.
(3) *Iterative fitting with early stopping*: A smooth distribution is iteratively modified and adjusted to the observation. The iteration process based on the EM method is stopped, before "unacceptable"oscillations emerge.
(4) *Implicit regularization*: The function $f(x)$ is expanded into a small number of smooth functions. A histogram with few bins or an expansion into a small number of spline functions is adjusted to the data, without explicit regularization.

A simple bin-by-bin correction method has been used in the past in some particle physics experiments. The ratio of the numbers in the observed and the true histogram in the simulation is used to correct the observed histogram. This approach generates a smooth distribution, but should be discarded because it often produces strongly biased results.

In the following, we first discuss some general properties of regularization approaches and then we describe the standard methods. We assume that the observed data contents of the bins follow Poisson distributions.

9.3.2 *Variable Dependence*

We have already stressed that smoothness criteria cannot be derived from basic principles. Smoothness is not well defined and the standard methods are not invariant against variable transformations.

Most unfolding methods have the convenient property that the unfolding result does not depend on the ordering of the bins in the unfolded distribution. This means that multi-dimensional distributions can be represented by one-dimensional histograms. An exception is regularization with a curvature penalty.

The dependence of the smoothness criteria on the chosen variable can be used to adapt the regularization approaches to specific problems. If, for instance, penalties favor a uniform distribution, we can choose a variable in which we expect that the true distribution is roughly uniform but in most cases it is better to adapt the penalty function, because then usually the resolution is approximately constant and equally sized bins are appropriate. On might for instance not penalize the deviations from uniformity for a nearly exponential distribution but the deviation from an exponential. The corresponding procedure in the iterative EM method is to select the starting distribution such that it corresponds to our expectation of the true distribution.

9.3.3 *Choice of the Regularization Strength*

A critical parameter in all unfolding procedures is the regularization strength which regulates the smoothness of the unfolding result and which determines bias and precision. The optimal value of the corresponding regularization parameter depends on the specific application. To fix it, we must have an idea about the shape of the true distribution. We might choose it differently for a structure function, a Drell-Yan distribution with possible spikes, a transverse momentum distribution and the distribution of the cosmic background radiation. From the data we can only derive an upper limit of the regularization parameter: The unfolded distribution has to be compatible within the statistical uncertainties with the observed histogram. Most unfolding methods try to approach a corresponding limit and eliminate fluctuations that are compatible with noise. There is no scientific justification for this pragmatic choice and one has to be aware of the fact that in this way real, interesting structures may be eliminated that could be resolved with higher statistics.

There exist many ways to fix the regularization parameter [54, 76]. We restrict the discussion to three common methods.

9.3.3.1 *Visual Inspection*

If we resign to the idea to use the unfolded distribution for parameter fits, it seems tolerable to apply subjective criteria for the choice of the regularization strength. By inspection of the unfolding results obtained with increasing regularization, we are to some extent able to distinguish fluctuations caused by noise from structures in the true distribution and to select a reasonable value of the regularization parameter. Probably, this method is in most cases as good as the following approaches which are partially quite involved.

9.3.3.2 *Truncation of the Eigenvector Solution*

We have seen that the unfolding result can be expanded into orthogonal components which are statistically independent.

We have studied above the eigenvector decomposition of the modified least square matrix and realized that the small eigenvalue components λ_i cause the unwanted oscillations. A smooth result is obtained by cutting all contributions with eigenvalues $\lambda_i, i = 1, ..., k)$ below a cut value λ_k. This procedure is called *truncated singular value decomposition* (TSVD).The value λ_k is chosen such that all eigenvectors with cut values greater than λ_k have statistically insignificant amplitudes. The truncation in the framework of the LSF has its equivalence in the ML method. We can order the eigenvectors of the covariance matrix of a MLF according to decreasing errors and retain only the dominant components.

9.3.3.3 *Minimization of the Integrated Square Error ISE*

A common measure of the agreement of an estimate of a probability density (PDE) with the true distribution is the integrated square error ISE. For a functions $f(x)$ and its PDE $\hat{f}(x)$ it is defined by

$$ISE = \int_{-\infty}^{\infty} \left[\hat{f}(x) - f(x)\right]^2 dx. \tag{9.18}$$

ISE is not defined for histograms in the way as physicists interpret them. To adapt the ISE concept to our needs, we modify the definition such that is measures the difference of the estimated content of the histogram $\hat{\theta}_i$ and the prediction θ_i. In addition we normalize it to the total number of events n and the number of true bins M.

$$ISE' = \sum_{i=1}^{M} \left(\hat{\theta}_i - \theta_i\right)^2 /n/M \tag{9.19}$$

ISE' depends mainly on the resolution, i.e. the response matrix and less on the shape of the true distribution. A crude guess of the latter can be used to estimate the regularization parameter r in a Monte Carlo simulation. (Here r is a generic name for the number of iterations in the EM method, the penalty strength or the cut in truncation approaches) The "‘true" distribution is folded and unfolded varying r and the value that minimizes ISE' is selected. The process can be iterated, but since the shape of the true distribution is not that critical, this will not be necessary in the majority of cases. The procedure consists of the following steps:

(1) Unfold $\boldsymbol{d}$ with varying regularization strength r and select the "best" value $\tilde{r}$ and $\tilde{\theta}^{(0)}$ by visual inspection of the unfolded histograms.
(2) Use $\tilde{\theta}^{(0)}$ as input for typically $n = 100$ simulations of "observed" distributions $\tilde{d}_i$, $i = 1, n$.
(3) Unfold each "observed"distribution with varying r and select the value $\tilde{r}_i$ that corresponds to the smallest value of ISE'. The value of ISE' is computed by comparing the unfolded histogram with $\tilde{\theta}^{(0)}$.
(4) Take the mean value $\bar{r}$ of the regularization strengths $\tilde{r}_i$, unfold the experimental distribution and obtain $\tilde{\theta}^{(1)}$. If necessary, go back to 2, replace $\tilde{\theta}^{(0)}$ by $\tilde{\theta}^{(1)}$ and iterate.

The procedure is independent of the regularization method.

9.3.4 *Error Assignment to Unfolded Distributions*

The regularization constrains the unfolded distribution. Therefore it introduces a bias and decreases the error δ_s obtained in the fit. The height of peaks is reduced, the width is increased, valleys are partially filled. The true uncertainties δ depend on the nominal error δ_s and the bias b, $\delta^2 = \delta_s^2 + b^2$. Increasing the regularization strength reduces δ_s but increases the bias b. Due to the unavoidable bias, the nominal error does not cover all distributions that are compatible with the observed data. The nominal errors depend on the selected regularization parameter and do not conform to the requirements stated in Section 4. The diagonal errors given in plots of the unfolded histogram are often misleading because the errors are correlated. Nevertheless, they may indicate qualitatively the range of acceptable true distributions.

9.3.4.1 *Calculation of the Nominal Error*

There are several way to calculate the errors:

(1) A common method is to apply error propagation starting from the observed data $\boldsymbol{d}$. To be consistent with the point estimate, the best estimate of the unfolded distribution $\hat{d}_i = \Sigma_k A_{ik} \hat{\theta}_k$ should be used instead. Error propagation is quite sensitive to non-linear relations which occur with low event numbers.
(2) The errors can be derived from the curvature matrix at the LS or ML estimates. This is the standard way in which symmetric error limits are computed in the common fitting programs. In principle also likelihood ratio or χ^2-contours can be computed. The parameter $\boldsymbol{\theta}$ is varied, the corresponding histogram $\boldsymbol{d}$ is computed and compared to $\widehat{\boldsymbol{d}} = \mathsf{A}\widehat{\boldsymbol{\theta}}$. Values of $\boldsymbol{\theta}$ that changes the difference $\ln L(\widehat{\boldsymbol{\theta}}) - \ln L(\boldsymbol{\theta})$ by $1/2$ fix the standard likelihood ratio error bounds. In the EM method with early stopping, $\boldsymbol{\theta}$ can be re-fitted starting from $\hat{d}$ and the errors can be provided by the fit program.
(3) We can use bootstrap re-sampling techniques [77], see Section 12.2. In short, the data sample is considered as representative of the true folded distribution. From the N observed events, N events are drawn with replacement. They form a bootstrap sample $\boldsymbol{d}^*$ which is histogrammed and unfolded. This procedure is repeated many times and in this way a set of unfolded distributions is generated. from which the fluctuations, confidence intervals and correlations can be extracted. For example, in a selected bin the standard 68 % confidence interval contains 68 % of the bootstrap results.

A more detailed discussion of the error estimation with bootstrap methods is presented in [78].

Usually the statistical uncertainty of the response matrix can be neglected. If this is not the case, the simplest way to include it, is to generate bootstrap samples of it.

9.3.5 *EM Unfolding with Early Stopping*

In a comparison [54] of the different regularization methods, the EM method performed better than competing approaches.

We have seen that the EM algorithm produces the MLE of the unfolded histogram. We start with a smooth distribution and suppress the

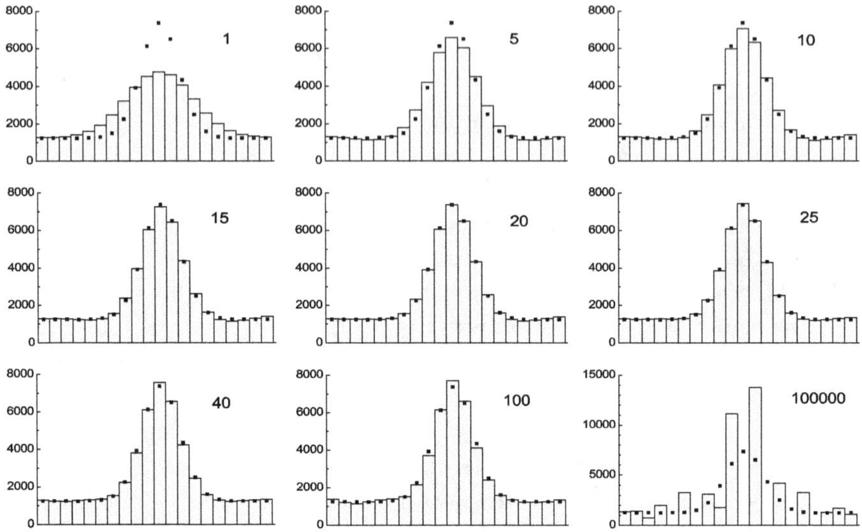

Fig. 9.9 Unfolding with the EM method for different numbers of iterations. The experimental resolution is $\sigma_S = 0.08$. The squares correspond to the true distribution.

fluctuations that emerge when we approach the MLE by stopping the iteration once the result is compatible with the data. We have to fix the starting distribution and the stopping condition.

Example 9.2. Unfolding with the EM method

In Fig. 9.10 the result is displayed of the iterative unfolding of the distorted superposition of a Gaussian with a uniform distribution. The response function is also of Gaussian shape with the same width as the peak. A sample of 50000 events has been generated. The iterative unfolding starts with a uniform distribution. The unfolded histograms are compared to the true histogram indicated by squares. The number of iterations is given in each plot. The last plot with extreme fluctuations corresponds to the MLE that is obtained with a large number of iterations. In Fig. 9.10 the test quantity ISE' is plotted as a function of the number of applied iterations.The optimal number of iterations that minimizes ISE' for the given data set is 25 but the result varies slowly with the number of iterations.

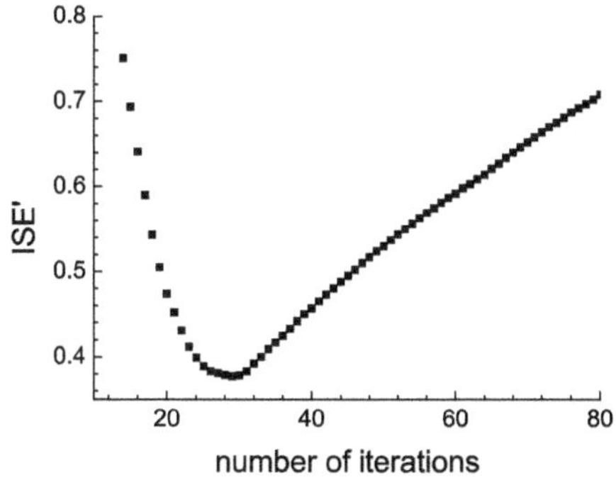

Fig. 9.10 ISE' as a function of the number of iterations.

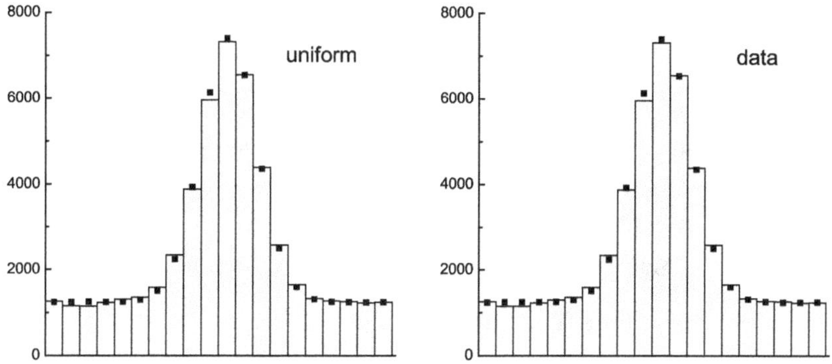

Fig. 9.11 Iterative unfolding with two different starting histograms, left uniform and right experimental.

Generally, the optimal number of iterations increases with the Gaussian smearing parameter σ_s and, of course, for $\sigma_s = 0$ no iteration would but necessary.

9.3.5.1 *Dependence on the starting distribution*

So far we have used a uniform starting distribution for the EM iteration. If there is prior knowledge of the approximate shape of the true distribution, for instance from previous experiments, then the uniform histogram can be replaced by a better estimate. Experience shows that the influence of the starting histogram on the unfolding result is usually rather weak.

Example 9.3. EM unfolding with different starting distributions

We repeat the unfolding of the distribution with 50000 events in the interval $[0 < x, x' < 1]$ and experimental resolution $\sigma_s = 0.08$ starting with the histogram of the observed events. (The choice of the example with a large number of events is less sensitive to statistical fluctuations than an example with low statistics and should indicate possible systematic effects.) The two results displayed in Fig. 9.11 are qualitatively indistinguishable. Starting with the uniform histogram, the lowest $ISE' = 0.0964$ is obtained after 40 iterations with $\chi^2 = 35.1$. With the observed histogram the values obtained after 38 iterations are $ISE' = 0.0940$ with the same value $\chi^2 = 35.1$. In a low statistics example with 500 events and resolution $\sigma_s = 0.04$ the minimum is reached already after 2 iterations with the $ISE' = 0.0488$ and 0.0487, respectively and values $\chi^2 = 36.0$ and 36.3.

The influence of the starting distribution on the unfolding result should be checked but in the majority of cases it is not necessary to deviate from a uniform histogram.

9.3.6 *Truncation of Small Eigenvector Contributions*

Descriptions of the method can be found in Refs. [66, 76].

9.3.6.1 *Truncated SVD*

The regularization is usually obtained by chopping off eigenvectors with low eigenvalues. The eigenvectors with the m largest eigenvalues are kept. The value m is chosen such that none of the remaining vectors contributes significantly to the decomposition. The significance, defined as the ratio of the amplitude and its error, $S_i = a_i/\delta_i$ of all eliminated vectors should be compatible with zero within one ore two standard deviations. Alternatively, the cutoff m can be chosen such that the ISE' is minimal.

Truncated SVD is mathematically attractive but suffers from drawbacks:

(1) The SVD solution is obtained by a linear LS fit. This implies that low event numbers in the observed histogram are not treated correctly. Combining bins with low event numbers can reduce the problem.

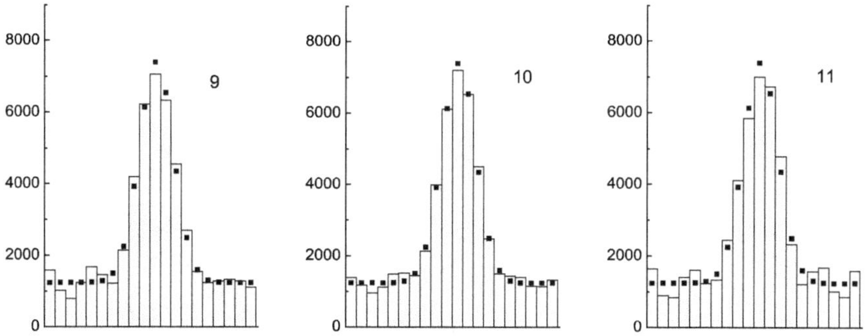

Fig. 9.12 Unfolding with truncated SVD

(2) The eigenvalue decomposition is mainly related to the properties of the response matrix and does not sufficiently take into account the shape of the true distribution. Small eigenvalues may correspond to significant structures in the true distribution and the corresponding eigenvectors may be eliminated by the truncation. The combination of the vectors belonging to several "insignificant" amplitudes may contribute significantly to the true distribution.

Figs. 9.12 show the unfolding results from TSVD for the same data that have been used to test the EM method. From the dependence of ISE' on the number of eigenvectors, Fig. 9.13 we find that the optimal number is 9. The agreement of the unfolded histogram with the true histogram is significantly worse than in the EM method. For low event numbers the method has the tendency to loose events in the unfolded histogram [54].

9.3.6.2 *Smooth Truncation*

It has been proposed [60, 76] to replace the brute force chopping off the noise dominated components by a smooth cut. This is accomplished by filter factors

$$\varphi(\lambda) = \frac{\lambda^2}{\lambda^2 + \lambda_0^2} \tag{9.20}$$

where λ_0 is the eigenvalue which fixes the degree of smoothing and λ is the eigenvalue corresponding to the coefficient which is to be multiplied by $\varphi(\lambda)$. The solution is then

$$\boldsymbol{\theta}_{reg} = \sum_{i=1}^{M} \varphi(\lambda_i) a_i \boldsymbol{u}_i .$$

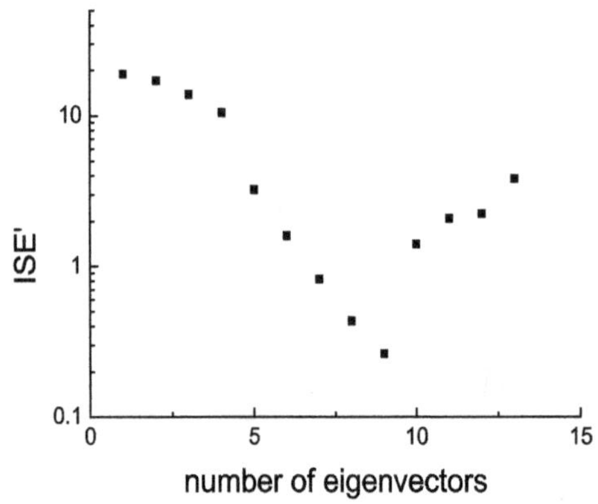

Fig. 9.13 Unfolding with TSVD: ISE' (arbitrary units) as a function of the number of eigenvectors.

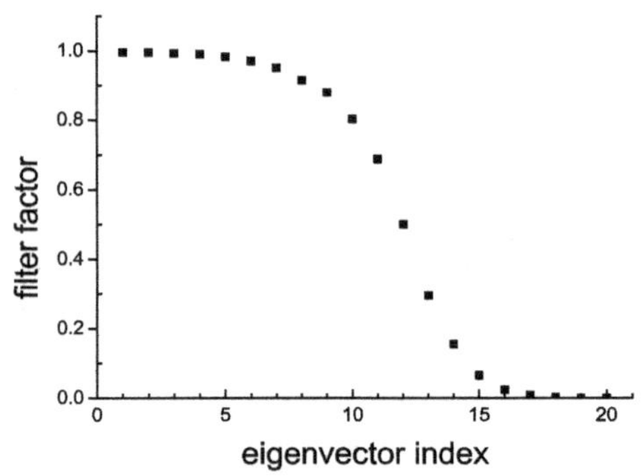

Fig. 9.14 Filter factor as a function of the eigenvector index.

The function 9.20 is displayed in Fig. 9.14. The amplitude of the eigenvector with eigenvalue $\lambda = \lambda_0$ is reduced by a factor 2. For large eigenvalues λ the filter factor is close to one and for small values it is close to zero. It is not obvious why a reduction of the amplitude of a component m and the inclusion of a fraction of the amplitude of a less significant component $n > m$ should improve the performance.

In [76] it is shown that the filtered SVD solution is equivalent to Tikhonov's norm regularization under the condition that the uncertainties of the observations correspond to white noise (normally distributed fluctuations with constant variance). We will come back to the norm regularization below.

9.3.7 *Penalty Regularization*

The EM and truncated SVD methods are very intuitive and general. If we have specific ideas about what we consider as smooth, we can penalize deviations from the wanted features by introduction of a penalty term R in the likelihood or LS expression:

$$\ln L = \ln L_{stat} - R, \tag{9.21}$$

$$\chi^2 = \chi^2_{stat} + R. \tag{9.22}$$

Here $\ln L_{stat}$ and χ^2_{stat} are the expressions given in (9.15) and (9.7).The sign of R is positive such that with increasing R the unfolded histogram becomes smoother. If we prefer a uniform distribution, R could be chosen proportional to the norm $||\theta||^2 = \sum_{1=1}^{N} \theta_i^2$. This is the simple Tikhonov regularization [81]. Popular are also the entropy regularization which again favors a uniform solution and the curvature regularization which prefers a linear distribution. All three methods have the tendency to reduce the height of peaks and to fill up valleys, a common feature of all regularization approaches. More sophisticated penalty functions can be invented if a priori knowledge about the true distribution is available. In particle physics, distributions often have a nearly exponential shape. Then one would select a penalty term which is sensitive to deviations from an exponential distribution.

9.3.7.1 *Curvature Regularization*

An often applied regularization function R is,

$$R(x) = r_c \left(\frac{\mathrm{d}^2 f}{\mathrm{d}x^2} \right)^2 . \tag{9.23}$$

It increases with the curvature of f and favors a linear unfolded distribution. The regularization constant r_c determines the power of the regularization.

For a histogram of M bins with constant bin width we approximate (9.23) by

$$R = r_c \sum_{i=2}^{M-1} \frac{(2\theta_i - \theta_{i-1} - \theta_{i+1})^2}{n^2} . \tag{9.24}$$

with n the total number of events and r_c the parameter that fixes the regularization strength.

The curvature penalty is a function of the content of three adjacent bins. It is not very efficient at the two border bins of the histogram. In the field of

PDE specific methods have been developed to avoid the problem [82]. More smoothing at the edges of the histogram can be achieved by increasing the bin size of the border bins or by increasing the penalty. The latter solution is adopted in [78].

9.3.7.2 *Entropy Regularization*

We borough the entropy concept from thermodynamics, where the entropy S measures the randomness of a state and the maximum of S corresponds to the equilibrium state which is the state with the highest probability. It has also been introduced into information theory and into Bayesian statistics to fix prior probabilities. However, there is no intuitive argument why the entropy should be especially suited to cure the fake fluctuations caused by the noise. It is probably the success of the entropy concept in other fields and its relation to probability which have been at the origin of its application in unfolding problems. We penalize a low entropy and thus favor a uniform distribution.

The entropy S of a discrete distribution with probabilities $p_i, i = 1, \dots, M$ is defined through the relation:

$$S = -\sum_{i=1}^{M} p_i \ln p_i .$$

For a random distribution the probability for one of the $n = \Sigma\theta_i$ events to fall into true bin i is given by θ_i/n. The maximum of the entropy corresponds to an uniform population of the bins, i.e. $\theta_i = const. = n/M$, and equals $S_{max} = \frac{1}{M} \ln M$, while its minimum $S_{min} = 0$ is found for the one-point distribution (all events in the same bin j) $\theta_i = n\delta_{i,j}$. We define the entropy regularization penalty with the regularization strength r_e of the distribution by

$$R = r_e \sum_{i=1}^{M} \frac{\theta_i}{n} \ln \frac{\theta_i}{n} . \tag{9.25}$$

Adding R to χ^2 or subtracting it from $\ln L$ can be used to smoothen a distribution.

A draw-back of a regularization based on the entropy or the norm is that distant bins are related, while smearing is a usually a local effect. Entropy regularization is popular in astronomy [83, 84] and has been adopted in particle physics [63, 85].

9.3.7.3 *Tikhonov or Norm Regularization*

The most obvious and simplest way to regularize unfolding results is to penalize a large value of the norm squared $||\boldsymbol{\theta}||^2$ of the solution:

$$R = \frac{r_n}{n^2}\sum_{i=1}^{M}\theta_i^2. \tag{9.26}$$

The norm regularization has first been proposed by Tikhonov [81]. Minimizing the norm implies a bias towards a small number of events in the unfolded distribution. To reduce this effect, contrary to the originally proposed penalty, we normalize the norm to the number of events squared n^2.

9.3.8 *Comparison of the Methods*

To compare the performance of the five methods we take an example from [54] where a more detailed comparison is presented.

Example 9.4. Comparison of Unfolding Methods

We simulate the distribution

$$f(x) = 0.2\mathcal{N}(-2,1) + 0.5\mathcal{N}(2,1) + 0.3\mathcal{U}$$

defined in the interval $[-7, 7]$ with smearing $\sigma_s = 1$ which has been used in [78]. The function and its smeared version are displayed in Fig. 9.15 top left. The unfolded distributions obtained with the EM, the truncated SVD and three penalty methods for the first of 10 samples with 5000 events are depicted in the same figure. The optical inspection does not reveal large differences between the results. The mean values of ISE' from the 10 samples are presented in Fig. 9.16. They indicates that the EM method and entropy regularization perform better than the other approaches. From a repetition with 100 simulated experiments, we find that the mean value of ISE' is smaller for the EM method compared to the entropy regularization by a factor of 2.00 ± 0.16.

Similar results have been obtained with other distributions [54].

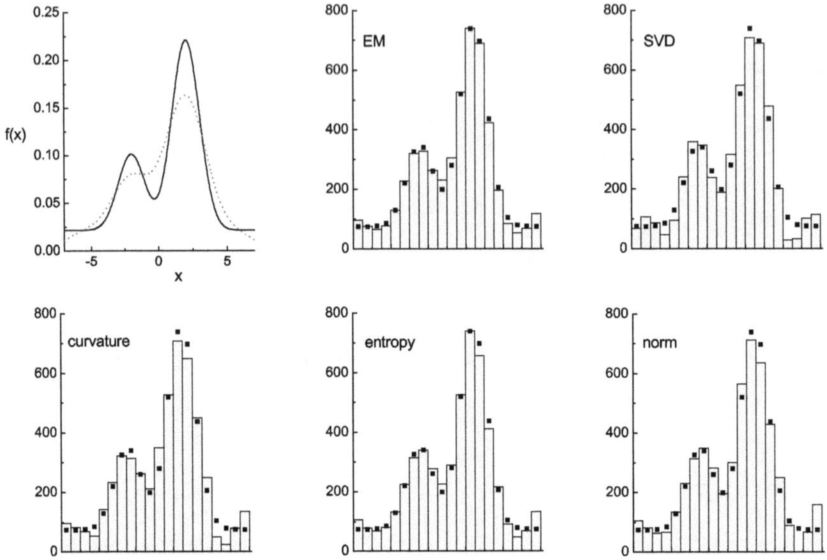

Fig. 9.15 Unfolding results from different methods. The top left hand plot shows the true distribution and its smeared version. The squares correspond to the true distribution.

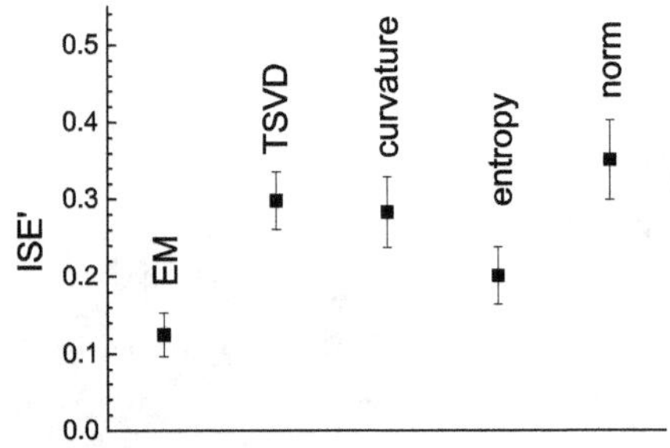

Fig. 9.16 ISE' averaged over 10 experiments.

9.3.9 *Spline Approximation*

Simulations of particle experiments often are based on PDEs. For instance unfolded proton structure functions are required to predict cross sections in proton proton collisions at the Large Hadron Collider at CERN. For these kind of simulations coarse binned histograms are not optimal and smooth unfolding results are preferred which can be obtained with spline approximations [57, 78, 85]. Spline approximations are sketched in Section 11.2.4 and formulas are given in Appendix A.13. Monotone distributions like transverse momentum distributions in particle physics can be described by linear splines. Distributions with bumps are better approximated with quadratic or cubic splines. The higher the order of the spline approximation, the more difficult it is to adjust the spline function at the borders of the distribution.

The representation of the unfolded function by a superposition of spline functions reduces the dependence of the unfolding result on the function used in the simulation of the response matrix. In the methods with penalty regularization, the construction of a response matrix and the dependence of the unfolding result on the distribution used in the Monte Carlo simulation of the response matrix can be avoided altogether with the parameter estimation method explained in Chapter 6

It has to be noted that independent of the regularization a systematic error is introduced by the fact that the true distribution is approximated by the spline curve. It can happen that this approximation is poor but normally it is excellent within the statistical uncertainties.

Once the elements of the response matrix have been computed, the unfolding proceeds in the same way as with histograms. The coefficients β_i are fitted to the observed data vector $\boldsymbol{d}$ in the same way as in the histogram representation.

Example 9.5. Unfolding to a spline curve

We apply the EM method to three data samples of our standard one-peak example with 5000 events and smearing resolution $\sigma_s = 0.08$. Twenty cubic B-splines are fitted to the data. Each time the number of iterations is selected which minimizes ISE' In Fig. 9.17 the results are depicted. The convergence of the log-likelihood is displayed in Fig. 9.18. The convergence is initially fast and then the residual value decays exponentially.

More detailed studies are necessary to establish the promising performance of the EM-unfolding into a superposition of b-splines.

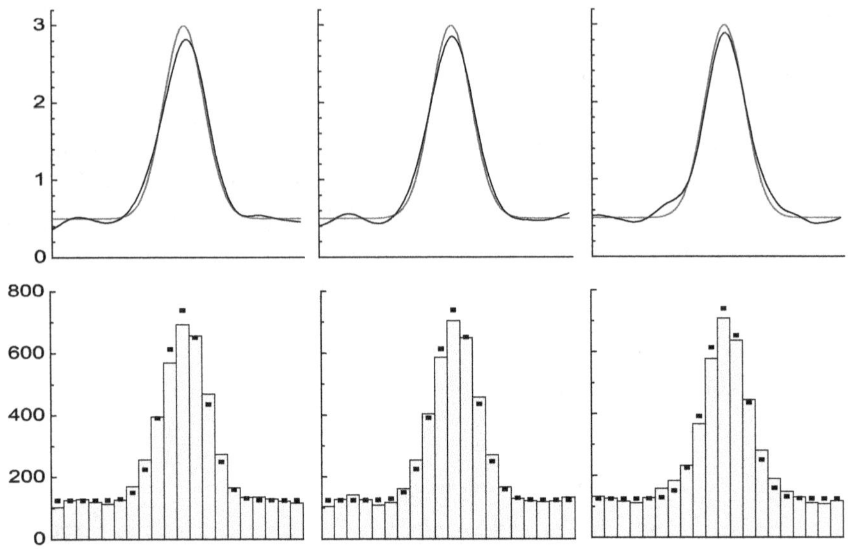

Fig. 9.17 EM unfolding to a spline function. Three examples are presented. The top graphs show the true distribution together with the unfolding results. The bottom graphs contain the corresponding histograms.

9.3.10 *Statistical and Systematic Uncertainties of the Response Matrix*

Until now we have assumed that we know exactly the probability A_{ij} to observe an event in bin i which is located in bin j of the true histogram. But A depends on the shape of the distribution f used in the simulation unless the smearing is large compared to the bin width in the true histogram. On the other hand, small bins mean strong oscillations and correlations between neighboring bins. We have two contradicting conditions: To be independent of the shape of $f(x)$ we would like to choose small bins, to avoid strong correlations, we prefer wide bins. A way out in situations where the statistics is relatively large, could be to unfold with narrow bins and to combine bins after the unfolding. With little statistics this procedure is difficult to follow [54] because then the errors are asymmetric and the linear error propagation used in combining bins is a bad approximation. Iteration of the Monte Carlo input distribution can improve the precision of the response matrix and in most cases leads to satisfactory results. A possible solution is to proceed in two steps: In step 1 the response matrix

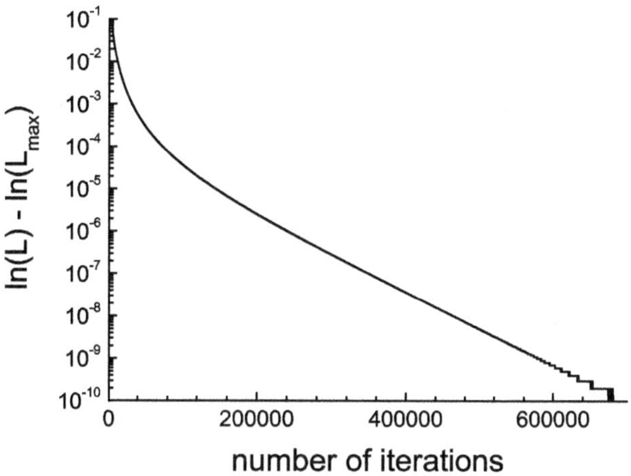

Fig. 9.18 Convergence of the ML spline fit with the number of iterations.

is generated in the usual way. In step 2 the data are unfolded to a superposition of splines and the result is used to generate an improved response matrix.

Statistical uncertainties of the response matrix can be estimated with the bootstrap method (see Chapter 11). A systematic error is introduced, if the experimental resolution is not known exactly. In nuclear and particle physics the sample size is often the limiting factor, in other fields like optics, the uncertainty of the resolution could dominate.

Example 9.6. Effect of a biased resolution to an unfolding result

A normal distribution is folded with a normal resolution function with resolution σ_f and unfolded with σ'_f. The relative error is

$$\delta = \frac{\left|\sigma_f - \sigma'_f\right|}{\sigma_f},$$

We obtain an artificial broadening of a Gaussian line after unfolding by

$$\sigma_{art}^2 = \left|\sigma_f^2 - \sigma_f'^2\right|,$$
$$\sigma_{art} = \sigma_f(2\delta - \delta^2)^{1/2} \approx \sqrt{2\delta}\sigma_f,$$

where σ_{art}^2 has to be added to the squared width of the original line. Thus a Dirac δ-function becomes a normal distribution of width σ_{art}. Even small deviations in the resolution can lead to a substantial artificial broadening of narrow peaks if the width of the smearing function is larger than that of the peaks in the true distribution.

9.4 Unfolding with Implicit Regularization

If we want to document the experimental information in such a way that it is conserved for a comparison with a theory that might be developed in future or if we want to compare or combine the date with those of another experiment, we must avoid the bias introduced by the explicit regularization. This can be achieved by keeping the distorted data together with the resolution function. Such a procedure is optimal because no information is wasted but it has severe drawbacks: Two large datasets, the experimental data and the Monte Carlo sample would have to be published and the whole analysis work would be left to the scientist who wants to use the data.

A simple and and efficient approach is to suppress oscillations without smoothing by choosing wide bins in the unfolded histograms. The unfolded histogram is obtained by matrix inversion, a least square or a maximum likelihood fit as described in Section 7.1. We call this procedure *implicit regularization.* An example is presented in Fig. 9.4. The histogram together with the error matrix allows to estimate parameters of future theoretical functions. If the function exists already when the raw data exist, the function parameters can be adjusted to the data with the method of Section 7.3 without a loss of information. The bin width of the unfolded histogram has to be chosen wide enough to reduce the correlations and the corresponding diagonal errors sufficiently to permit a linear error calculus. We have to accept that some information is lost. In most experiments the experimental smearing is small and the loss is negligible.

As discussed above, wide bins have the disadvantage that the response matrix depends on the distribution that is used to generate it. To cure the problem, this distribution can be approximated by the unfolding result obtained with explicit regularization to a spline approximation. The remaining bias is usually negligible.

9.5 Treatment of Background

We distinguish two different situations.

Background is most cases generated by a Poisson random process. We can unfold the observed histogram as usual and subtract the background from the unfolded histogram. Either its shape and amount is known, then it can be subtracted directly, or the background has to be evaluated from the histogram. It has to be parametrized and its amount and the parameters have to be fitted in regions of the histogram, where the background dominates. Background subtraction is treated in Section 7.4.

If the background is due to some malfunction of the detector and not Poisson distributed, then it has to be estimated and subtracted in the observed histogram. Iterative unfolding is no longer possible because it relies on the Poisson distribution of the number of events in the observed histogram. A LS fit has to be applied with penalty regularization.

9.6 Summary and Recommendations for the Unfolding of Histograms

Let us summarize our conclusions:

- Whenever an existing theory has to be verified, or parameters of it have to be estimated, the prediction should be folded and compared to the observed data. The results then are independent of the distribution used in the Monte Carlo simulation and the unavoidable information losses of the unfolding procedures are avoided.
- If this is not the case, the experimental results have to be published in a way that unknown biases are excluded. This is achieved by unfolding with bins large enough to avoid excessive oscillations.
- It should be attempted to generate enough Monte Carlo events such that the statistical uncertainty introduced by the response matrix is negligible. If this is not possible, its contribution to the error of the unfolded distribution can be estimated by bootstrap techniques or by variation of the Monte Carlo statistics.
- Uncertainties in the shape of the distribution used to generate the response matrix have to be estimated and taken into account by adding a systematic error. To keep the uncertainties small, the distribution can be approximated by a spline function that is determined by a preliminary unfolding the data with explicit regularization.

To visualize the true distribution, it is useful to unfold with explicit regularization.

- The preferred choice is EM iterative unfolding. It is technically very simple. The standard starting distribution is uniform, but it can be adapted to specific problems. The method is independent of the dimension of the histogram, the size and the ordering of the bins.
- The eigenvector decomposition provides insight in the origin of the unfolding problematic. Regularization by truncation of low eigenvalue components (TSVD) in many cases is not competitive.
- Regularization with roughness penalties produce similar, but often slightly worse results as the EM method. Curvature penalties are difficult to apply in three or more dimensions. In typical examples entropy regularization performs better than norm and curvature regularization.
- The regularization biases the results. It excludes distributions with high frequency contributions that are compatible with the data and underestimates the errors. In a strictly frequentist view, the unfolded distributions have to be considered only as qualitative results.
- The regularization is necessary to avoid narrow local fluctuations but it affects also global structures, filling valleys and reducing the height of peaks. These biases can partially be estimated by Monte Carlo simulations and corrected.

Algorithms of available unfolding computer programs are often based on the experience from only a single or few simple examples. The results and especially the quoted error estimates should be used with great care.

9.7 Binning-Free Methods

Binning free methods are able to unfold very narrow structures like points and lines that cannot be represented by histograms. They are especially suited for low statistics distributions in high dimensional spaces where histogram bins would suffer from too few events. A further advantage of binning free unfolding is the possibility to construct arbitrary histograms of the unfolded points with various selection criteria. A draw back of these methods lies in the absence of a simple error handling which includes correlations. So far, there is little experience with binning-free unfolding. The methods that are presented here are still being developed.

9.7.1 *Iterative Unfolding with Local Density Approximation*

We can realize the EM method also in a binning free way [64].

We start with a Monte Carlo sample of events, each event being defined by the true coordinate x and the observation x'. During the iteration process we modify at each step a weight which we associate to the events such that the densities in the observation space of simulated and real events approach each other. Initially all weights are equal to one. At the end of the procedure we have a sample of weighted events which corresponds to the unfolded distribution.

To this end, we estimate a local density $d'(x_i')$ in the vicinity of any point x_i' in the observation space. (For simplicity, we restrict ourselves again to a one-dimensional space since the generalization to several dimensions is trivial.) The following density estimation methods (see Chapter 12) lend themselves:

(1) The density is taken as the number of observations within a certain fixed region around x_i', divided by the length of the region. The length should correspond roughly to the resolution, if the region contains a sufficient number of entries.
(2) The density is chosen proportional to the inverse length of that interval which contains the K nearest neighbors, where K should be not less than about 10 and should be adjusted by the user to the available resolution and statistics.

We denote by $t(x)$ the simulated density in the true space at location x, by $t'(x')$ the folded simulated density at x' and the corresponding data density be $d'(x')$. The density $d'(x')$ is estimated from the length of the interval containing K events, $t'(x')$ from the number of simulated events $M(x')$ in the same interval. The simulated densities are updated in each iteration step k. We associate a preliminary weight

$$w_i^{'(1)} = \frac{d'(x_i')}{t'^{(0)}(x_i')} = \frac{K}{M(x')}$$

to the Monte Carlo event i. The weighted events in the vicinity of x represent a new density $t^{(1)}(x)$ in the true space. We now associate a true weight w_i to the event which is just the average over the preliminary weights of all K events in the neighborhood of x_i, $w_i = \sum_j w_j'/K$. With the smoothed weight w_i a new observed simulated density $t'^{(1)}$is computed. In the k-th iteration the preliminary weight is given by

$$w_i'^{(k+1)} = \frac{d'(x_i')}{t'^{(k)}(x_i')} w_i^{(k)} .$$

The weight will remain constant once the densities t' and d' agree. As result we obtain a discrete distribution of coordinates x_i with appropriate weights w_i, which represents the unfolded distribution. The degree of regularization depends on the parameters K used for the density estimation.

The method is obviously not restricted to one-dimensional distributions, and is indeed useful in multidimensional cases, where histogram bins suffer from small numbers of entries. We have to replace x_i, x_i' by $\boldsymbol{x}_i$, $\boldsymbol{x}_i'$, and the regions for the density estimation are multi-dimensional.

9.7.2 *The Migration Method*

We start again with an experimental sample of N observed point locations $\{\boldsymbol{x}_1', \dots, \boldsymbol{x}_N'\}$ and a transfer function $t(\boldsymbol{x}_i', \boldsymbol{x}_i)$ that describes the probability observation $\boldsymbol{x}_i'$ given the true locations $\boldsymbol{x}_i$. The p.d.f. for the observation $\boldsymbol{x}_k'$ is

$$f_k = \boldsymbol{x}_k' | \boldsymbol{x}_1, \dots, \boldsymbol{x}_N) = \sum_{i=1}^{N} t(\boldsymbol{x}_k', \boldsymbol{x}_i)$$

The likelihood function for the unknown locations $\boldsymbol{x_i}$ is

$$L = \sum_{k=1}^{N} f_k / N$$

In principle the MLE of the parameters that correspond to the unfolded sample could be obtained with standard methods. However, there are two obstacles: 1. The transfer function is often not known in analytic form and 2. standard maximizing programs can hardly cope with the large number of parameters.

The basic idea of a possible solution [87] of the problem is the following: We generate a Monte Carlo sample of the same size as the experimental sample $\{\boldsymbol{y}_1', \dots, \boldsymbol{y}_N'\}$. If the analytic form of the transfer function is known, we compute the likelihood $L(\boldsymbol{y}_1, \dots, \boldsymbol{y}_N)$ for the parameters according to (reflikemig). We select randomly a Monte Carlo event and let it migrate by a random amount Δy in a random direction $\boldsymbol{y_i} \to \boldsymbol{y_i} + \Delta \boldsymbol{y_i}$. If the likelihood increases, we keep the move, if not, we reject it. The initial location of the $\boldsymbol{y_i}$ can be set on top of the observations.

Of course maximizing the likelihood leads to the same artifacts as observed in the histogram based methods. The true points form clusters which, eventually, degenerate into discrete distributions. A smooth result is obtained by stopping the maximizing process before the maximum has been reached. For definiteness, similar to the case of histogram unfolding, a fixed difference of the likelihood from its maximum value should be chosen to stop the maximization process. Similarly to the histogram case, this difference should be of the order of $\Delta L \approx \sqrt{NDF/2}$ where the number of degrees of freedom NDF is equal to the number of points times the dimension of the space.

There may be applications, for instance in astronomy, where we are interested to find point sources and their intensity. Then the described unfolding procedure could be used without regularization.

Example 9.7. Deconvolution by fitting the true event locations

Fig. 9.20 top shows 2000 points randomly generated according to the superposition of two normal distributions denoted as $\mathcal{N}(x', y'|\mu_x, \mu_y, \sigma_x, \sigma_y)$:

$$f(x', y') = 0.6\mathcal{N}(x', y'| -2, 0, 1, 1) + 0.4\mathcal{N}(x', y'| +2, 0, 1, 1).$$

The response function again is a normal distribution centered at the true points with symmetric standard deviations of one unit. It is used to convolute the original distribution with the result shown at the center of Fig. 9.20. The starting values of the parameters $\hat{x}_i, \hat{y}_i$ are set equal to the observed locations x'_i, y'_i. Randomly selected points are then moved within squares of size 4×4 units and moves that improve the likelihood are kept. After 5000 successful moves the procedure is stopped to avoid clustering of the true points. The result is shown in the lower plot of Fig. 9.20. On the right hand side of the same figure the projections of the distribution onto the x axis in form of histograms are presented.

The resolution of the response function t is normally not a simple analytic function, but only numerically available through a Monte Carlo simulation. Thus we associate to each true Monte Carlo point i a set of K generated observations $\{\boldsymbol{y}'_{i1}, \ldots, \boldsymbol{y}'_{iK}\}$ which we call satellites [87] and which move together with $\boldsymbol{y}_i$. The test quantity ϕ that replaces the likelihood function is now a function of the N experimental positions and the $N \times K$

smeared Monte Carlo positions. Choices of the test statistic ϕ are presented in Chapter 10. We recommend to use the variable *energy*.

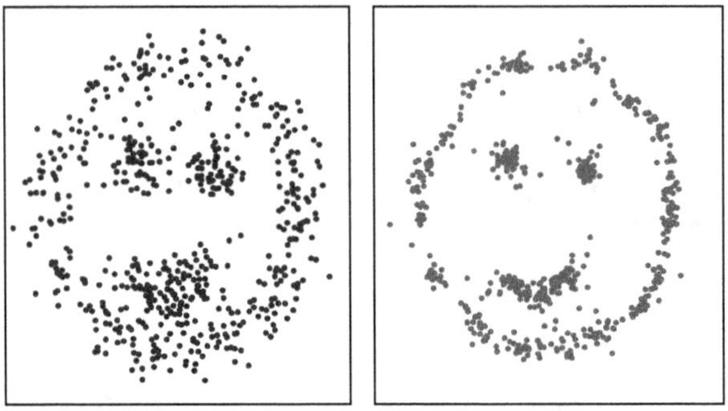

Fig. 9.19 Deconvolution of a blurred picture with the satellite method.

The migration distances $\Delta \boldsymbol{y}_i$ should be taken from a distribution with a width somewhat larger than the measurement resolution, while the exact shape of the distribution is not relevant. We therefore recommend to use a uniform distribution, for which the generation of random numbers is faster than for a normal or other distributions. The unfolding result is independent from these choices, but the number of iteration steps can raise appreciably for a bad choice of parameters.

Example 9.8. Deconvolution of a blurred picture

Figure 9.19 shows a two-dimensional application. The observed picture consisted of lines and points which are convoluted with a two-dimensional normal distribution. In the Monte Carlo simulation for each true point $K = 25$ satellites have been generated. The energy ϕ is minimized. The resolution of the lines in the deconvoluted figure on the right hand side is restricted by the low experimental statistics. For the eyes the restriction is predominantly due to the low Monte Carlo factor K. Each eye has $N = 60$ points. The maximal resolution for a point measured N times is obtained for measurement error σ_f as

$$\begin{aligned} \Delta x &= \sigma_f \sqrt{\frac{1}{N} + \frac{1}{K}} \\ &= \sigma_f \sqrt{\frac{1}{60} + \frac{1}{25}} = 0.24\,\sigma_f . \end{aligned}$$

Measurement resolution and acceptance should stay approximately constant in the region in which the events migrate. When we start with a reasonably good approximation of the true distribution, this condition is usually satisfied. In exceptional cases it would be necessary to update the distribution of the satellites y'_{ik} after each move, i.e. to simulate or correct them once again. It is more efficient, however, to update all elements together periodically after a certain number of migration steps.

The number K determines the maximal resolution of the unfolded distribution, it has therefore a regularization effect; e.g. for a measurement resolution σ_f and $K = 16$ the minimal sampling interval is $\sigma_T = \sigma_f/\sqrt{K} = \sigma_f/4$.

If the true p.d.f. has several maxima, we may find several relative minima of the energy. In this case a new stochastic element has to be introduced in the minimization (see Section 5.2.7.1). In this case a move towards a position with smaller energy is not performed automatically, but only preferred statistically.

We have not yet explained how acceptance losses are taken into account. The simplest possibility is the following: If there are acceptance losses, we need $K_{i0} > K$ trials to generate the K satellites of the event y_i. Consequently, we relate a weight $w_i = K_{0i}/K$ to the element y_i. At the end of the unfolding procedure we obtain a weighted sample.

9.7.3 *Summary of Binning-Free Methods*

The advantage of binning-free methods is that there the information loss caused by binning is avoided. Narrow structures can be resolved. Unfolding produces again points in the observation space which can be subjected to selection criteria and collected into arbitrary histograms, while methods working with histograms have to decide on the corresponding parameters before the unfolding is performed.

The binning-free, iterative method based on PDE has the disadvantage that the user has to choose some parameters. It requires sufficiently high statistics in all regions of the observation space.

The satellite method is well suited for small samples and multidimensional distributions, where other methods have difficulties. Structures with narrow peaks, curves and points that cannot be described by histograms can be unfolded. For large samples it is rather slow, even on large computers.

The binning-free likelihood method requires an analytic response function. It is much faster than the satellite method.

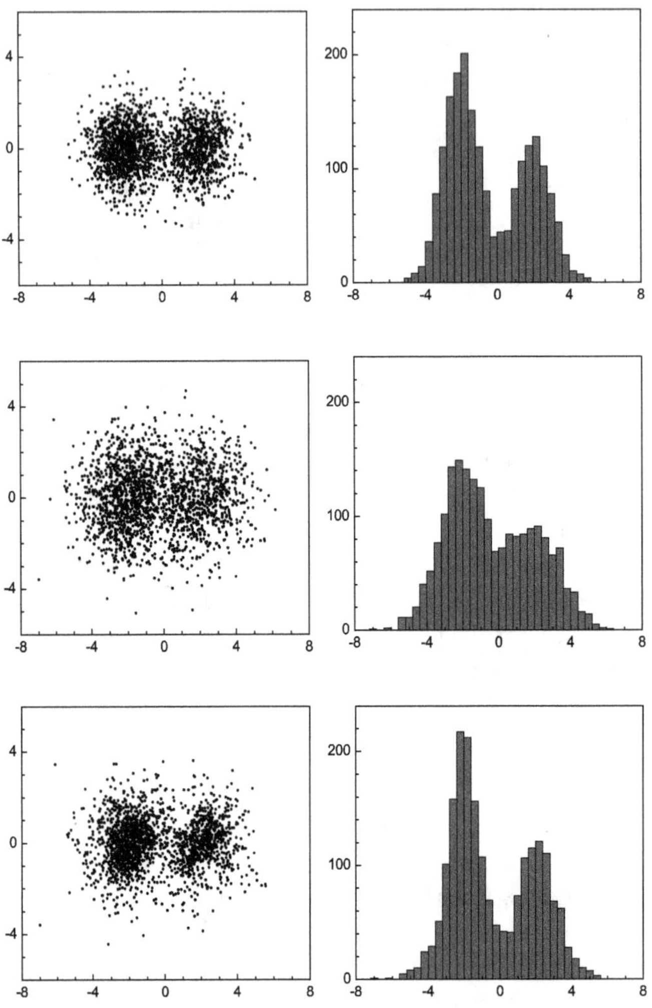

Fig. 9.20 Deconvolution of point locations. The central plot on the left hand side is deconvoluted and shown in the bottom plot. The true point distribution is given in the top plot. The right hand side shows the corresponding projections onto the x axis in form of histograms.

Chapter 10

Hypothesis Tests and Significance of Signals

10.1 Introduction

So far we treated problems where a data sample was used to discriminate between completely fixed competing hypotheses or to estimate free parameters of a given distribution. Now we turn to the task to quantify the compatibility of observed data with a given hypothesis. We distinguish between the following topics:

(a) Classification, for example event selection in particle reactions.
(b) Testing the compatibility of a distribution with a theoretical prediction, i.e. goodness-of-fit tests.
(c) Testing whether two samples originate from the same population, two-sample tests.
(d) Quantification of the significance of signals, like the Higgs signal.

The hypothesis that we intend to test is called the *null hypothesis* H_0. In most tests the alternative hypothesis H_1 is simply "H_0 is false". Often, additional characteristics of H_1 are very vague and cannot be quantified, but a test makes sense only if we have an idea about H_1. Formally, a test is associated with a decision: *accept* or *reject*. This is obvious for classification problems, while in the other cases we are mostly satisfied with the quotation of the so-called p-value, introduced by R. Fisher, which measures the compatibility of a given data sample with the null hypothesis.

There is some confusion about the terms *significance test* and *hypothesis test* which has its origin in a controversy between R. Fisher on one side and J. Neyman[1] and E. Pearson[2] on the other side. Fisher had a more pragmatic

[1] Jerzy Neyman (1894–1981), Polish mathematician
[2] Egon Sharpe Pearson (1895–1980), English statistician

view while Neyman–Pearson emphasized a strictly formal treatment with prefixed criteria whether to accept or reject the hypothesis. We will not distinguish between the two terms but use the term *significance* mainly for the analysis of small signals. We will talk about tests even when we do not decide on the acceptance of H_0.

The test procedure has to be fixed before looking at the data.[3] To base the selection of a test and its parameters on the data which we want to analyze, to optimize a test on the bases of the data or to terminate the running time of an experiment as a function of the output of a test would bias the result.

Goodness-of-fit (GOF) tests and two-sample tests are closely related. Goodness-of-fit test are often applied after a parameter of a distribution has been adjusted to the observed data. In this case the hypothesis depends on one or several free parameters. We have a composite hypothesis and a composite test. Two sample test are applied if data are to be compared to a prediction that is modeled by a Monte Carlo sample. Sometimes it is of interest to check whether experimental conditions have changed. To test the hypothesis that this is not the case, samples taken at different times are compared.

At the end of this chapter we will treat another case in which we have a partially specified alternative and which plays an important role in physics. There the goal is to investigate whether a small signal is significant or explainable by a fluctuation of a background distribution. We call this procedure *signal test.*

10.2 Some Definitions

Before addressing goodness-of-fit tests, we introduce some notations used in the statistical literature.

10.2.1 *Single and Composite Hypotheses*

We distinguish between *simple* and *composite* hypotheses. The former fix the population uniquely. Thus H_0: "The sample is drawn from a normal distribution with mean zero and variance one, i.e. $\mathcal{N}(0,1)$." is a simple hypothesis. If the alternative is also simple, e.g. H_1 : "$\mathcal{N}(5,1)$", then we have the task to decide between two simple hypotheses which we have

[3] Scientists often call this a *blind* analysis.

already treated in Chapter 6, Section 6.3. In this case there exists an optimal test, the likelihood ratio test.

Composite hypotheses are characterized by free parameters, like H_0: "The sample is drawn from a normal distribution."The user will adjust mean and variance of the normal distribution and test with a goodness-of-fit comparison whether the adjusted Gaussian is compatible with the data.

10.2.2 *Test Statistic, Critical Region and Significance Level*

After we have fixed the null hypothesis and the admitted alternative H_1, we must choose a *test statistic* $t(\boldsymbol{x})$, which is a function of the sample values, $\boldsymbol{x} \equiv \{x_1, \dots, x_N\}$, that discriminates between $f_0(t|H_0)$ and the distribution of H_1.

When we test, for instance, the hypothesis that a coordinate is distributed according to $\mathcal{N}(0, 1)$, then for a sample consisting of a single measurement x, a reasonable test statistic is the absolute value $|x|$. We assume that if H_0 is wrong then $|x|$ would be large. A typical test statistic is the χ^2 deviation of a histogram from a prediction. Large values of χ^2 indicate that something might be wrong with the prediction or with the experimental data.

Before we apply the test that is related to a decision, we have to fix a *critical region* K which leads to the rejection of H_0 if t is located inside of it. Under the condition that H_0 is true, the probability of rejecting H_0 is $P\{t \in K|H_0\} = \alpha$ where $0 \leq \alpha \leq 1$ normally is a small quantity (e.g. 5%). It is called *significance level*or *size of the test* . For a test based on the χ^2-statistic, the critical region is defined by $\chi^2 > \chi^2_{\max}(\alpha)$ where the critical value $\chi^2_{\max}$ is a function of the significance level α. It fixes the range of the critical region, $\chi^2 > \chi^2_{\max}$.

To compute rejection probabilities, we have to compute the p.d.f. $g(t)$ of the test statistic for H_0. In some cases it is known as we will see below, but in other cases it has to be obtained by a Monte Carlo simulation. The distribution g has to include all experimental conditions under which t is determined like the acceptance and the measurement uncertainty of t.

Example 10.1. Test of a predicted counting rate

A theory H_0 predicts $n_0 = 100$ rare events in an experiment. Observed are $\hat{n} = 130$. We test whether there is an excess of events due to a process not considered in H_0. The significance level set is $\alpha = 0.01$ which means

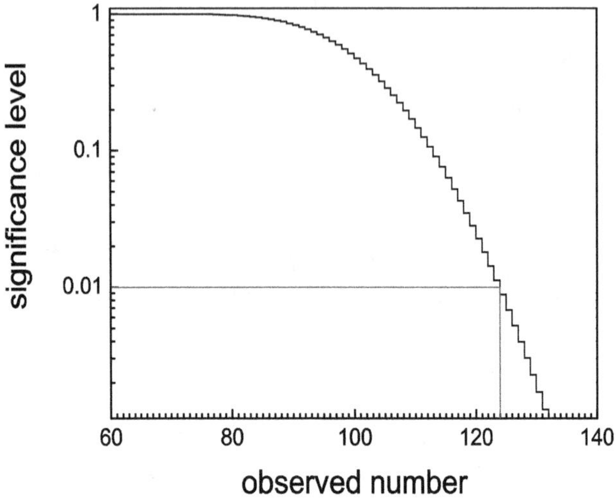

Fig. 10.1 Relation between the critical value n of a Poisson experiment with mean 100 and the significance level. Observation $n > 124$ are excluded at a significance level of 1%.

that with 1% probability H_0 will be excluded if it is true. Figure 10.1 shows the significance level $\alpha = 1 - F(n)$,with $F(n)$ the distribution function, as a function of n. The critical region $n \geq n_c$ starts at $n_c = 125$ and extends to infinity and H_0 will be excluded at a significance level of 1%. (The p-value that will be defined below is 0.0023).

Example 10.2. Particle selection based on the invariant mass

In a experiment with a magnetic spectrometer $K^0 \to \pi^+\pi^-$ events are selected. An event is accepted if the mass $m_{\pi\pi}$ reconstructed from the decay particles agrees with the mass m_k of the kaon within 2 standard deviations. The measurement errors are normally distributed. The null hypothesis is that $x = (m_{\pi\pi} - m_k)/\delta$ follows a normal distribution with mean zero and standard deviation one, $x \sim \mathcal{N}(0,1)$. The test statistic is $|x|$, the critical region is $|x| \geq 2$ and the size of the test is $\alpha = \frac{2}{\sqrt{2\pi}} \int_2^\infty \exp(-x^2/2)dx = 0.0455$.

10.2.2.1 *Errors of the First and Second Kind, Power of a Test*

After the test parameters are selected, we can apply the test to our data. If the actually obtained value of t is outside the critical region, $t \notin K$, then we accept H_0, otherwise we reject it. This procedure implies four different outcomes with the following a priori probabilities:

(1) $H_0 \cap t \in K$, $P\{t \in K|H_0\} = \alpha$: *error of the first kind.* (H_0 is true but rejected),
(2) $H_0 \cap t \notin K$, $P\{t \notin K|H_0\} = 1 - \alpha$ (H_0 is true and accepted),
(3) $H_1 \cap t \in K$, $P\{t \in K|H_1\} = 1 - \beta$ (H_0 is false and rejected),
(4) $H_1 \cap t \notin K$, $P\{t \notin K|H_1\} = \beta$: *error of the second kind* (H_0 is false but accepted).

When we apply the test to a large number of data sets or events, then the rate α, the error of the first kind, is the inefficiency in the selection of H_0 events, while the rate β, the error of the second kind, represents the background with which the selected events are contaminated with H_1 events. Of course, for α given, we would like to have β as small as possible. Given the rejection region K which depends on α, also β is fixed. However, we usually have only a vague idea about the properties of H_1 and cannot compute β. For a reasonable test we expect that β is monotonically decreasing with α increasing. The more we restrict the region where H_0 is accepted, the more the background should be reduced. With $\alpha \to 0$ also the critical region K is shrinking, while the *power* $1 - \beta$ must decrease, and the background is less suppressed. For fixed α, the power indicates the quality of a test, i.e. how well alternatives to H_0 can be rejected.

The power is a function of the significance level α. Tests which provide maximum power $1 - \beta$ with respect to H_1 for all values of α are called *uniformly most powerful (UMP) tests.* Only in rare cases where H_1 is restricted in some way, there exists an optimum, i.e. a UMP test. If both hypotheses are simple, then as already mentioned in Chapter 6, Section 6.3, according to a lemma of J. Neyman and E. S. Pearson, the likelihood ratio can be used as test statistic to discriminate between H_0 and H_1 and provides a uniformly most powerful test.

The interpretation of α and β as error rates makes sense when many experiments or data sets of the same type are investigated. In a search experiment where we want to find out whether a certain physical process or a phenomenon exists or in an isolated GOF test they refer to virtual experiments and it is not obvious which conclusions we can draw from their values.

10.2.3 *Consistency and Bias of Tests*

A test is called *consistent* if its power tends to unity as the sample size tends to infinity. In other words: If we have an infinitely large data sample, we should always be able to decide between H_0 and the alternative H_1.

We also want that independent of α the rejection probability for H_1 is higher than for H_0, i.e. $\alpha < 1 - \beta$. Tests that violate this condition are called *biased*. Consistent tests are asymptotically unbiased.

When H_1 represents a family of distributions, consistency and unbiasedness are valid only if they apply to all members of the family. Thus in case that the alternative H_1 is not specified, a test is biased if there is an arbitrary hypothesis different from H_0 with rejection probability less than α and it is inconsistent if we can find a hypothesis different from H_0 which is not rejected with power unity in the large sample limit.

Example 10.3. Bias and inconsistency of a test

Assume, we select in an experiment events of the type $K^0 \to \pi^+\pi^-$. The invariant mass $m_{\pi\pi}$ of the pion pairs has to match the K^0 mass. Due to the finite experimental resolution the experimental masses of the pairs are normally distributed around the kaon mass m_K with variance σ^2. With the null hypothesis H_0 that we observe only $K^0 \to \pi^+\pi^-$ decays, we may apply to our sample a test with the test quantity $t = \overline{(m_{\pi\pi} - m_K)^2}/\sigma^2$, the normalized mean quadratic difference between the observed masses of N pairs and the nominal K^0 mass. Our sample is accepted if it satisfies $t < t_0$ where t_0 is the critical quantity which determines the error of the first kind α and the acceptance $1 - \alpha$. The distribution of Nt under H_0 is a χ^2-distribution with N degrees of freedom. Clearly, the test is biased, because we can imagine mass distributions with acceptance larger than $1 - \alpha$, for instance a uniform distribution in the range $t \leq t_0$. This test is also inconsistent, because it would favor this specific realization of H_1 also for infinitely large samples. Nevertheless it is not unreasonable for very small samples in the considered case and for $N = 1$ there is no alternative. The situation is different for large samples where more powerful tests exist which take into account the Gaussian shape of the expected distribution under H_0.

While consistency is a necessary condition for a sensible test, bias of a test applied to a small sample cannot always be avoided and is tolerable under certain circumstances.

The formal definitions of this section are important in many applications, outside physics, like in drug or fertilizer tests. Physicists talk about efficiency, purity or contamination. The use of the terms *error of the first and second kind* and *size of test* is rather an exception.

10.2.4 *The p-Value*

10.2.4.1 *Definition*

Strictly speaking, the result of a test is that a hypothesis is "accepted" or "rejected". In most practical situations it is useful to replace this digital answer by a continuous parameter, the so called *p-value* which is a monotonic function $p(t)$ of the test statistic t and which measures the compatibility of the sample with the null hypothesis, a small value of p casting some doubt on the validity of H_0. To illustrate the meaning of p-values, we return to the example of a normally distributed measurement x. Here for a measurement $\hat{x}$ the p-value is equal to the probability to observe $|x| \geq |\hat{x}|$:

$$p = \frac{2}{\sqrt{2\pi}} \int_{\hat{x}}^{\infty} e^{-x^2/2} dx .$$

For a measurement $x = 0$ where the observation coincides exactly with the prediction of H_0 we get $p = 1$ and for $x = \infty$ we obtain $p = 0$. The p-value of the counting rate example above with a predicted Poisson rate of 100 and 130 observed counts is $p = 0.0023$, see Fig. 10.1.

To simplify the general definition of the test statistic definition, we assume that the test statistic t is confined to values between zero and infinity with a critical region $t > t_c$.[4] Its distribution under H_0 be $f_0(t)$. Then we have

$$p(t) = 1 - \int_0^t f_0(t')\mathrm{d}t' = 1 - F_0(t), \tag{10.1}$$

with F_0 the distribution function. Since p is a unique monotonic function of t, we can consider p as a normalized test statistic which is completely equivalent to t.

The relationship between the different quantities which we have introduced is shown in Fig. 10.2. The upper graph represents the p.d.f. of

---

[4]This condition can always be realized for one-sided tests by a variable transformation. For two-sided tests, p-values cannot be defined.

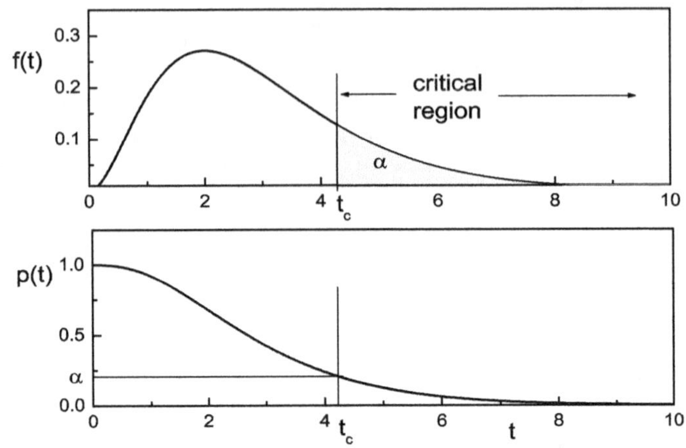

Fig. 10.2 Distribution of a test statistic and corresponding p-value curve.

the test statistic under H_0. The critical region extends from t_c to infinity. The a priori rejection probability for a sample under H_0 is α, equal to the integral of the distribution of the test statistic over the critical region. The lower graph shows the p-value function. It starts at one and is continuously decreasing to zero at infinity. The smaller the test statistic is – think of χ^2 – the higher is the p-value. At $t = t_c$ the p-value is equal to the significance level α. The condition $p < \alpha$ leads to rejection of H_0 in classifications. Due to its construction, the p.d.f. of the p-value under H_0 is uniform.

10.2.4.2 *Interpretation and Use of p-Values*

Since the distribution of p under H_0 is uniform in the interval $[0, 1]$, all values of p in this interval are equally probable. When we reject a hypothesis under the condition $p < 0.1$ we have a probability of 10% to reject H_0. The rejection probability would be the same for a rejection region $p > 0.9$. The reason for cutting at low p-values is the expectation that distributions of H_1 would produce low p-values.

The name p-value is derived from the word probability, *but the p-value is not the probability that the hypothesis under test is true.* It is the probability under H_0 to obtain a value of the test statistic t that is larger than the value that is actually observed or, equivalently, the probability to obtain a p-value which is smaller than the observed one. A p-value between zero and p is expected to occur in the fraction p of experiments if H_0 is true.

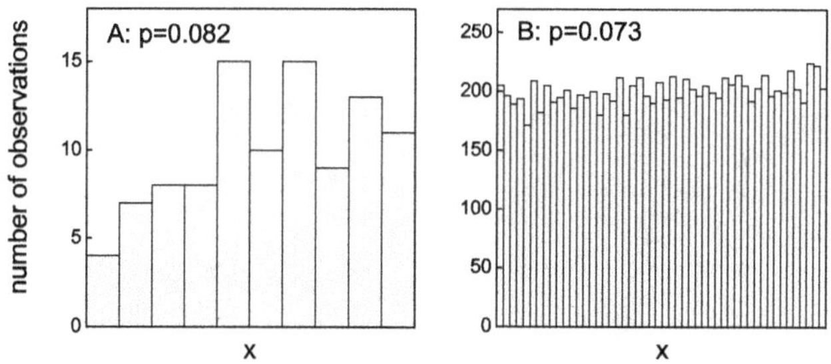

Fig. 10.3 Comparison of two experimental histograms to a uniform distribution.

Example 10.4. The p-value and the probability of a hypothesis

In Fig. 10.3 we have histogrammed two distributions from two simulated experiments A and B. Are these uniform distributions? For experiment B with 10000 observations this is conceivable, while for experiment A with only 100 observations it is difficult to guess the shape of the distribution. Alternatives like strongly rising distributions are excluded in B but not in A. We would therefore attribute a higher probability for the validity of the hypothesis of a uniform distribution for B than for A, but the p-values based on the χ^2-test are very similar in both cases, namely $p \approx 0.08$.

We learn from this example that the p-value is more sensitive to deviations from H_0 in large samples than in small samples. Since in practice small unknown systematic errors can rarely be excluded, we should not be astonished that in high statistics experiments often small p-values occur. The systematic uncertainties which usually are not considered in the null hypothesis then dominate the purely statistical fluctuation.

Even though we cannot transform significant deviations into probabilities for the validity of a hypothesis, they provide useful hints for hidden measurement errors or a contamination with background. In our example a linearly rising distribution is added to uniform distributions. The fractions are 45% in experiment A and 5% in experiment B.

In classification problems we are able to compare many replicates of measurements to the same hypothesis. In particle physics experiments usually a huge number of tracks has to be reconstructed. The track parameters are adjusted by a χ^2-fit to measured points assuming normally distributed uncertainties. The χ^2 value of each fit can be used as a test statistic and

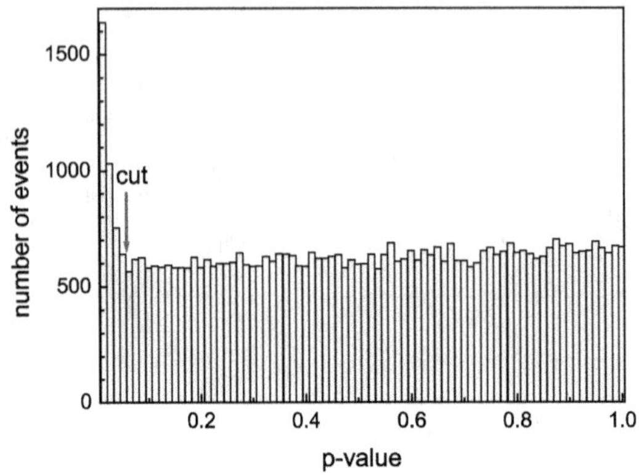

Fig. 10.4 Experimental distribution of p-values.

transformed into a p-value, often called χ^2-probability. Histograms of p-values obtained in such a way are very instructive. They often look like the one shown in Fig. 10.4. The plot has two interesting features: It is slightly rising with increasing p-value which indicates that the errors have been slightly overestimated. The peak at low p-values is due to fake tracks which do not correspond to particle trajectories and which we would eliminate almost completely by a cut[5] at about $p_c = 0.05$. We would have to pay for it by a loss of good tracks of slightly less than 5%. A more precise estimate of the loss can be obtained by an extrapolation of the smooth part of the p-value distribution to $p = 0$.

10.2.4.3 *Combination of p-Values*

If two p-values p_1, p_2 which have been derived from independent test statistics t_1, t_2 are available, we would like to combine them to a single p-value p. The at first sight obvious idea to set $p = p_1 p_2$ suffers from the fact that the distribution of p will not be uniform. A popular but arbitrary choice is

$$p = p_1 p_2 \left[1 - \ln(p_1 p_2)\right] \tag{10.2}$$

which can be shown to be uniformly distributed [90]. This choice has the unpleasant feature that the combination of the p-values is not associative,

[5] In this example the requirement to fix the critical region before looking at the data is violated, but the tolerated bias is small.

i.e. $p[(p_1, p_2), p_3] \neq p[p_1, (p_2, p_3)]$. There is no satisfactory way to combine p-values.

We propose, if possible, not to use (10.2) but to go back to the original test statistics and construct from them a combined statistic t and the corresponding p distribution. For instance, the obvious combination of two χ^2-statistics would be $t = \chi_1^2 + \chi_2^2$.

10.3 Classification

In classification problems we decide whether to accept or reject hypotheses as a result of a test. Examples are event selection (e.g. B quark production), particle track selection on the bases of the quality of reconstruction and particle identification, (e.g. electron identification based on calorimeter or Cerenkov information). Typical for these examples is that we examine a number of similar objects and accept a certain error rate α. The goal is to find the optimal test statistic. Its critical value determines the acceptance of the selected events and its contamination by background and indirectly the statistical and the systematic uncertainty of the result. The choice depends on the physics goal and on how well we can estimate the amount of background. For instance, when we select B particle decays, to determine the production rate, we will probably allow for a higher contamination than if we want to determine the lifetime of the B mesons. Often it is useful to transform the test statistic into a p-value where we know that it should be uniformly distributed under H_0.

Sophisticated classification methods have been developed in the last few decades along with the increased computing power. We will discuss some of them in Chapter 11 where we introduce artificial neural networks and decision trees. The goodness-of-fit-tests that will be treated in the following section can also be applied to classification problems.

10.4 Goodness-of-Fit Tests

10.4.1 *General Remarks*

Goodness-of-fit (GOF) tests check whether a sample is compatible with a given distribution. Scientists usually have a quite good feeling for deviations between two distributions just by looking at a plot. For instance, when we examine the statistical distribution of Fig. 10.5, we will realize that its description by an exponential distribution is rather unsatisfactory. The

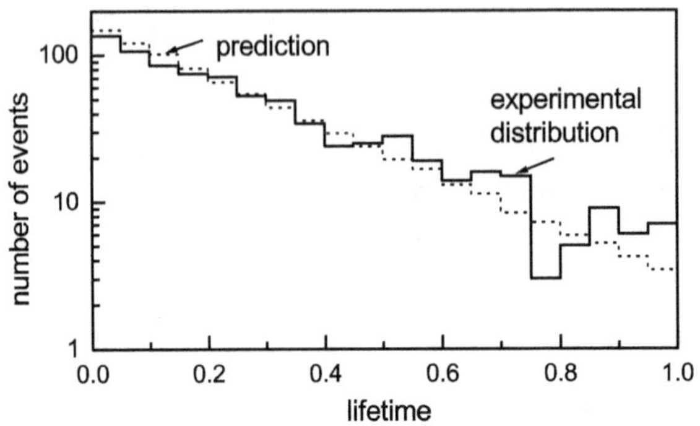

Fig. 10.5 Comparison of an experimental distribution to a prediction.

question is: How can we quantify the disagreement? Without an idea about a possible alternative description it is difficult to select an efficient test procedure.

To test whether a roulette is behaving correctly, we check whether all numbers occur with equal probability. We could construct a roulette which is producing all numbers sequentially. It would pass the test but not the requirement. However this behavior is not what we imagine for a standard roulette, we would rather expect that possibly some numbers occur more often than others and this is what the test should exclude. When we test a random number generator we would be interested, for example, in a periodicity of the results or a correlation between subsequent numbers and we would choose a different test.

GOF tests are not only used to check the validity of a hypothesis but also serve to detect unknown systematic errors in experimental results. When we measure the mean life of an unstable particle, we know that the lifetime distribution is exponential but to apply a GOF test is informative, because a low p-value may indicate a contamination of the events by background, an unsatisfactory simulation of the detector properties or problems with the experimental equipment.

A typical test quantity is the variable χ^2 which we have introduced to adjust parameters of functions to experimental histograms or measured points with known error distributions. In the least square method of parameter inference, see Chapter 6, the parameters are fixed such that the sum χ^2 of the normalized quadratic deviations is minimum. Deviating parameter values produce larger values of χ^2, consequently we expect the same effect when we compare the data to a wrong hypothesis. If χ^2 is abnormally large, it is likely that the null hypothesis is not correct.

Physicists use almost exclusively the χ^2-test, even though for many applications more powerful tests are available. Scientists also often overestimate the significance of the χ^2-test results. Other tests like the *Kolmogorov–Smirnov Test* and tests of the *Cramer–von Mises family* avoid the always somewhat arbitrary binning of histograms in the χ^2-test. These tests are restricted to univariate distributions, however.

If events of a data sample are characterized by several discriminating features, multi-dimensional tests are more powerful than separate one-dimensional tests where the features are tested independently. These tests should be binning-free to avoid bins with low event numbers.

Other binning-free methods can also be applied to multivariate distributions.

Sometimes students think that the likelihood L_0 of the null hypothesis is a powerful test statistic, e.g. for H_0 with single event distribution $f_0(x)$ the product $\Pi_i f_0(x_i)$. That this is not a good idea is demonstrated in Fig. 10.6 where the null hypothesis is represented by a fully specified normal distribution. From the two samples, the narrow one clearly fits the distribution worse but it has the higher likelihood. A sample where all observations are located at the center would per definition maximize the likelihood but such a sample would certainly not support the null hypothesis but rather a narrow Gaussian.

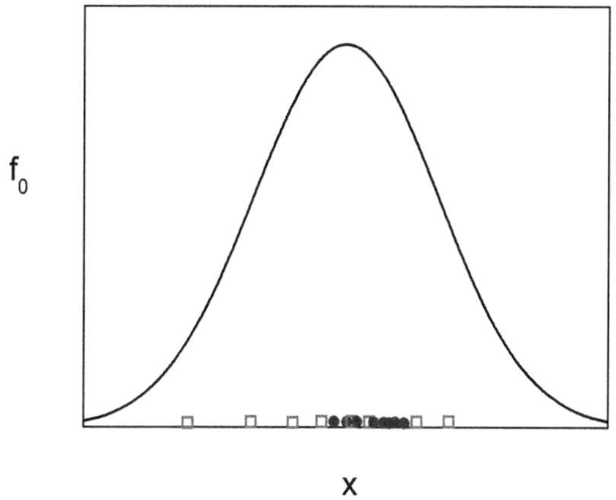

Fig. 10.6 Two different samples and a hypothesis.

While the indicated methods are *distribution-free*, i.e. applicable to arbitrary distributions specified by H_0, there are procedures to check the agreement of data with specific distributions like normal, uniform or exponential distributions. These methods are of inferior importance for physics applications. We will deal only with *distribution-free* methods.

We will not discuss tests related to *order statistics*. These tests are mainly used in connection with time series and are not very powerful in most of our applications.

At the end of this section we want to remind that parameter inference with a valid hypothesis and GOF test which doubt the validity of a hypothesis touch two completely different problems. Whenever possible deviations can be parametrized, it is always appropriate to determine the likelihood function of the parameter and use the likelihood ratio to discriminate between different parameter values.

A good review of GOF tests can be found in [89], in which, however, recent developments are missing.

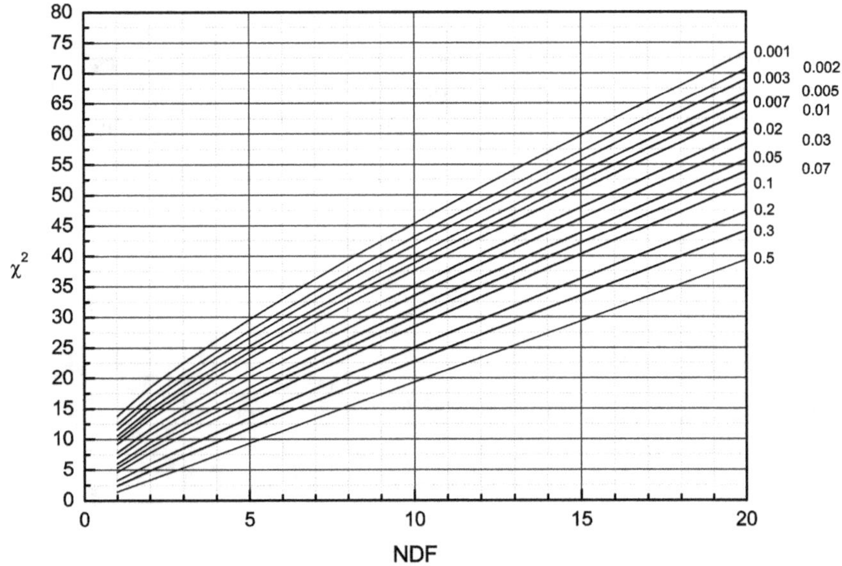

Fig. 10.7 Critical χ^2-values as a function of the number of degrees of freedom with the p-value as parameter.

10.4.2 *The χ^2-Test in Generalized Form*

In Section 7.1.1 we have compared a histogram with Poisson distributed data in B bins to a prediction t:

$$\chi^2 = \sum_{i=1}^{B} \frac{(d_i - t_i)^2}{t_i}. \tag{10.3}$$

The value of χ^2 can be used to test the compatibility of the data with the theoretical prediction t. Large values of χ^2 indicate that the description of the data by the prediction is doubtful.

Very small values are equally unlikely under H_0, but alternative hypotheses rather produce high values of χ^2. Therefore the χ^2-test is one-sided.[6] To transform χ^2 to a p-value, we have to know the distribution of χ^2. For not too low numbers d_i the Poisson distribution can be approximated by a normal distribution and then χ^2 follows a χ^2 distribution with $f = B$ degrees of freedom.

If the predictions t_i depend on parameters, that have been adjusted to the data, we have to deal with a composite hypothesis. With additional Z adjusted parameters we remain with $f = B - Z$ degrees of freedom. In the Appendix A.8 it is shown, that the test quantity can still be described by the χ^2 distribution, if two further conditions are fulfilled: The predictions t_i are linear functions of the parameters and the dependence of the denominators t_i on the parameters is negligible.

The relation between the observed value χ^2 and the p-value of the one-sided χ^2-test is displayed in Fig. 10.7 as function of the number of the degrees of freedom.

When the number of entries per histogram bin is small, the approximation that the variations are normally distributed is not justified. In this case or or if there are sizable violation of the stated conditions above, the χ^2-distribution should no longer be used to calculate the p-value.

Nevertheless we can still use in this situation the sum of quadratic deviations χ^2 as test statistic. The distribution $f_0(\chi^2)$ has then to be determined by a Monte Carlo simulation.

The χ^2-test can also be applied to the result of a least square adjustment of measurements y_i to a function, see Section 6.8:

$$\chi^2 = \sum_{i=1}^{N} \frac{(y_i - t(x_i, \boldsymbol{\theta}))^2}{\delta_i^2}. \tag{10.4}$$

---

[6]The χ^2-test is also called *Pearson test* after the British mathematician and statistician Karl Pearson (1857–1980), who has introduced the test already in 1900.

Again similar conditions as those stated above have to be satisfied at least approximately: linear dependence of the prediction on the parameters, normally distributed errors δ_i and independence of the errors of the parameters. A further generalization is given in Appendix A.8 where weighted events and statistical errors of the theoretical predictions, resulting from the usual Monte Carlo calculation, are considered.

The value of χ^2 is independent of the signs of the deviations. However, if several adjacent bins show an excess (or lack) of events like in the left hand histogram of Fig. 10.8 this indicates a systematic discrepancy which one would not expect at the same level for the central histogram which produces the same value for χ^2. Because correlations between neighboring bins do not enter in the test, a visual inspection is often more effective than the mathematical test. Sometimes it is helpful to present for every bin the value of χ^2 multiplied by the sign of the deviation either graphically or in form of a table.

Example 10.5. χ^2-comparison for a two-dimensional histogram

In the following table for a two-dimensional histogram the values of χ^2 accompanied with the sign are presented. The absolute values are well confined in the range of our expectation but near the right hand border we observe an accumulation of positive deviations which points to a systematic effect.

i \ j	1	2	3	4	5	6	7	8
1	0.1	-0.5	1.3	-0.3	1.6	-1.1	2.0	1.2
2		-1.9	0.5	-0.4	0.1	-1.2	1.3	1.5
3			-1.2	-0.8	0.2	0.1	1.3	1.9
4				0.2	0.7	-0.6	1.1	2.2

Remark: The quantity δ^2 has to be calculated under the assumption that the theoretical description which is to be tested is correct. For example, instead of ascribing to a measured quantity an error δ'_k which is proportional to its value y_k, a corrected error

$$\delta_k = \delta'_k \frac{t_k}{y_k}$$

should be used.

Sometimes extremely small values of χ^2 are presented. The reason is in most cases an overestimation of the errors.

The variable χ^2 is frequently used to separate signal events from background. In this case it is not required that χ^2 follows the χ^2-distribution. It is only necessary that it is a discriminative test statistic. The separation could be performed on the bases of a histogram like the one presented in Fig.10.4.

10.4.2.1 *The χ^2-Test for Small Samples*

When the number of entries per histogram bin is small, the approximation that the variations are normally distributed is not justified. Consequently, the χ^2-distribution should no longer be used to calculate the p-value.

Nevertheless we can use in this situation the sum of quadratic deviations χ^2 as test statistic. The distribution $f_0(\chi^2)$ has then to be determined by a Monte Carlo simulation. The method still works pretty well.

10.4.2.2 *Warning*

The assumption that the distribution of the test statistic under H_0 is described by a χ^2-distribution relies on the following assumptions: 1. The entries in all bins of the histogram are normality distributed. 2. The expected number of entries depends linearly on the free parameters in the considered parameter range. An indication for a non-linearity are asymmetric errors of the adjusted parameters. 3. The estimated uncertainties δ_i in the denominators of the summands of χ^2 are independent of the parameters. Deviations from these conditions affect mostly the distribution at large values of χ^2 and thus the estimation of small p-values. Corresponding conditions have to be satisfied when we test the GOF of a curve to measured points. Whenever we are not convinced about their validity we have to generate the distribution of χ^2 by a Monte Carlo simulation.

10.4.2.3 *The Choice of Binning*

For Poisson distributed, binned data the χ^2-test looses significance when the number of bins becomes too large.

To estimate the effect of fine binning for a smooth deviation, we consider a systematic deviation which is constant over a certain region with a total number of entries N_0 and which produces an excess of εN_0 events. Partitioning the region into B bins would add to the statistical χ^2 in each single bin the contribution:

$$\chi_s^2 = \frac{(\varepsilon N_0/B)^2}{N_0/B} = \frac{\varepsilon^2 N_0}{B}.$$

For B bins we increase χ^2 by $\varepsilon^2 N_0$ which is to be compared to the purely statistical contribution χ_0^2 which is in average equal to B. The significance S, i.e. the systematic deviation in units of the expected fluctuation $\sqrt{2B}$ is

$$S = \varepsilon^2 \frac{N_0}{\sqrt{2B}}.$$

It decreases with the square root of the number of bins.

We recommend a fine binning only if deviations are considered which are restricted to narrow regions. This could be for instance pick-up spikes. These are pretty rare in our applications. Systematic deviations are rather produced by a non-linearity of measurement devices or by background that extends over a large region. Then wide intervals are to be preferred.

In [91] it is proposed to choose the number of bins according to the formula $B = 2N^{2/5}$ as a function of the sample size N.

Example 10.6. Dependence of the power of the test on the bin width of a histogram an exponential distribution

In Fig. 10.5 a histogrammed sample is compared to an exponential. The sample contains, besides observations following this distribution, a small contribution of uniformly distributed events. From Table 10.1 we recognize that this defect expresses itself by small p-values and that the corresponding decrease becomes more pronounced with decreasing number of bins.

Some statisticians propose to adjust the bin parameters such that the number of events is the same in all bins. In our table this partitioning is denoted by e.p. (equal probability). In the present example this does not improve the significance.

10.4.3 *The Likelihood Ratio Test*

10.4.3.1 *General Form*

The likelihood ratio test compares H_0 to a parameter dependent alternative H_1 which includes H_0 as a special case. The two hypothesis are defined

Table 10.1 p-values for χ^2 and EDF statistic.

test	p value
χ^2, 50 Bins	0.10
χ^2, 50 Bins, e.p.	0.05
χ^2, 20 Bins	0.08
χ^2, 20 Bins, e.p.	0.07
χ^2, 10 Bins	0.06
χ^2, 10 Bins, e.p.	0.11
χ^2, 5 Bins	0.004
χ^2, 5 Bins, e.p.	0.01
D_{max}	0.005
W^2	0.001
A^2	0.0005

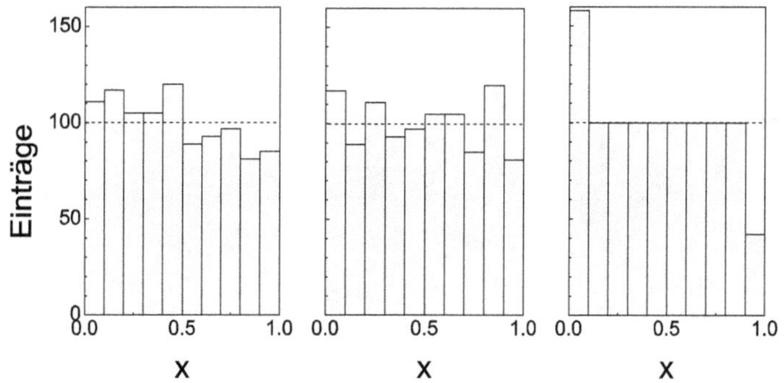

Fig. 10.8 The left-hand and the central histogram produce the same χ^2 p-value, the left-hand and the right-hand histograms produce the same Kolmogorov p-value.

through the p.d.f.s $f(\boldsymbol{x}|\boldsymbol{\theta})$ and $f(\boldsymbol{x}|\boldsymbol{\theta}_0)$ where the parameter set $\boldsymbol{\theta}_0$ is a subset of $\boldsymbol{\theta}$, often just a fixed value of $\boldsymbol{\theta}$. The test statistic is the likelihood ratio λ, the ratio of the likelihood of H_0 and the likelihood of H_1 where the parameters are chosen such that they maximize the likelihoods for the given observations $\boldsymbol{x}$. It is given by the expression

$$\lambda = \frac{\sup L(\boldsymbol{\theta}_0|\boldsymbol{x})}{\sup L(\boldsymbol{\theta}|\boldsymbol{x})}, \tag{10.5}$$

or equivalently by

$$\ln \lambda = \sup \ln L(\boldsymbol{\theta}_0|\boldsymbol{x}) - \sup \ln L(\boldsymbol{\theta}|\boldsymbol{x}).$$

If $\boldsymbol{\theta}_0$ is a fixed value, this expression simplifies to $\ln \lambda = \ln L(\boldsymbol{\theta}_0|\boldsymbol{x}) - \sup \ln L(\boldsymbol{\theta}|\boldsymbol{x})$.

From the definition (10.5) follows that λ always obeys $\lambda \leq 1$.

Example 10.7. Likelihood ratio test for a Poisson count

Let us assume that H_0 predicts $\mu_0 = 10$ decays in an hour, observed are 8. The likelihood to observe 8 for the Poisson distribution is $L_0 = \mathcal{P}_{10}(8) = e^{-10}10^8/8!$. The likelihood is maximal for $\mu = 8$, it is $L = \mathcal{P}_8(8) = e^{-8}8^8/8!$ Thus the likelihood ratio is $\lambda = \mathcal{P}_{10}(8)/\mathcal{P}_8(8) = e^{-2}(5/4)^8 = 0.807$. The probability P to observe a ratio smaller than or equal to 0.807 is

$$P = \sum_k \mathcal{P}_{10}(k) \quad \text{for } k \text{ with} \quad \mathcal{P}_{10}(k) \leq 0.807\, \mathcal{P}_{10}(10).$$

Relevant numerical values of $\lambda(k, \mu_0) = \mathcal{P}_{\mu_0}(k)/\mathcal{P}_k(k)$ and $\mathcal{P}_{\mu_0}(k)$ for $\mu_0 = 10$ are given in the following table.

k	8	9	10	11	12	13
λ	0.807	0.950	1.000	0.953	0.829	0.663
$\mathcal{P}$	0.113	0.125	0.125	0.114	0.095	0.073

It is seen, that the sum over k runs over all k, except $k = 9, 10, 11, 12$: $p = \Sigma_{k=0}^{8}\mathcal{P}_{10}(k) + \Sigma_{k=13}^{\infty}\mathcal{P}_{10}(k) = 1 - \Sigma_{k=9}^{12}\mathcal{P}_{10}(k) = 0.541$ which is certainly acceptable.

The likelihood ratio test in this general form is useful to discriminate between a specific and a more general hypothesis, a problem which we will study in Section 10.6.2. To apply it as a goodness-of-fit test, we have to histogram the data.

10.4.3.2 *The Likelihood Ratio Test for Histograms*

We have shown that the likelihood $L_0 = \prod_i f_0(x_i)$ of a sample cannot be used as a test statistic, but when we combine the data into bins, a likelihood ratio can be defined for the histogram and used as test quantity. The test variable is the ratio of the likelihood for the hypothesis that the bin content is predicted by H_0 and the likelihood for the hypothesis that maximizes the likelihood for the given sample. The latter is the likelihood for the hypothesis where the prediction for the bin coincides with its content. If H_0 is not simple, we take the ratio of the maximum likelihood allowed by H_0 and the unconstrained maximum of L.

For a bin with content d, prediction t and p.d.f. $f(d|t)$ this ratio is $\lambda = f(d|t)/f(d|d)$ since at $t = d$ the likelihood is maximal. For the histogram we have to multiply the ratios of the B individual bins. Instead we change to the log-likelihoods and use as test statistic

$$V = \ln \lambda = \sum_{i=1}^{B} [\ln f(d_i|t_i) - \ln f(d_i|d_i)].$$

If the bin content follows the Poisson statistics we get (see Chapter 6, Section 7.1)

$$\begin{aligned} V &= \sum_{i=1}^{B} [-t_i + d_i \ln t_i - \ln(d_i!) + d_i - d_i \ln d_i + \ln(d_i!)] \\ &= \sum_{i=1}^{B} [d_i - t_i + d_i \ln(t_i/d_i)]. \end{aligned}$$

The distribution of the test statistic V is not universal, i.e. not independent of the distribution to be tested as in the case of χ^2. It has to be determined by a Monte Carlo simulation. In case parameters of the prediction have been adjusted to data, the parameter adjustment has to be included in the simulation.

The method can be extended to weighted events and to the case of Monte Carlo generated predictions with corresponding statistical errors, see Appendix A.8.

Asymptotically, $N \to \infty$, the test statistic V approaches $-\chi^2/2$ as is seen from the expansion of the logarithm, $\ln(1 + x) \approx x - x^2/2$. After introducing $x_i = (d_i - t_i)/t_i$ which, according to the law of large numbers, becomes small for large d_i, we find

$$\begin{aligned} V &= \sum_{i=1}^{B} [t_i x_i - t_i(1 + x_i) \ln(1 + x_i)] \\ &\approx \sum_{i=1}^{B} t_i \left[x_i - (1 + x_i)(x_i - \frac{1}{2}x_i^2)\right] \\ &\approx \sum_{i=1}^{B} t_i \left(-\frac{1}{2}x_i^2\right) = -\frac{1}{2}\sum_{i=1}^{B} \left(\frac{(d_i - t_i)^2}{t_i}\right) = -\frac{1}{2}\chi_B^2, \end{aligned}$$

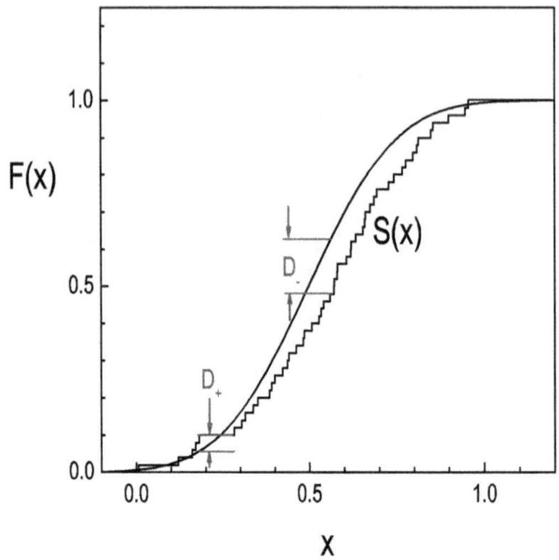

Fig. 10.9 Comparison of the empirical distribution function$S(x)$ with the theoretical distribution function $F(x)$.

and thus $-2V$ is distributed according to a χ^2-distribution with B degrees of freedom, but then we may also use directly the χ^2-test.

If the prediction is normalized to the data, we have to replace the Poisson distribution by the multinomial distribution. We omit the calculation and present the result:

$$V = \sum_{i=1}^{B} d_i \ln(t_i/d_i).$$

In this case, V approaches asymptotically the χ^2-distribution with $B-1$ degrees of freedom.

10.4.4 *The Kolmogorov–Smirnov Test*

The subdivision of a sample into intervals is arbitrary and thus subjective. Unfortunately some experimenters use the freedom to choose histogram bins such that the data agree as well as possible with the theoretical description in which they believe. This problem is excluded in binning-free tests which have the additional advantage that they are also applicable to small samples.

The Kolmogorov–Smirnov test compares the distribution function

$$F_0(x) = \int_{-\infty}^{x} f_0(x)\,\mathrm{d}x$$

to the corresponding experimental quantity S,

$$S(x) = \frac{\text{Number of observations with } x_i < x}{\text{Total number}}.$$

The test statistic is the maximum difference D between the two functions:

$$\begin{aligned} D &= \sup |F(x) - S(x)| \\ &= \sup(D_+, D_-). \end{aligned}$$

The quantities D_+, D_- denote the maximum positive and negative difference, respectively. $S(x)$ is a step function, an experimental approximation of the distribution function and is called *Empirical Distribution Function* (*EDF*). It is depicted in Fig. 10.9 for an example and compared to the distribution function $F(x)$ of H_0. To calculate $S(x)$ we sort all N elements in ascending order of their values, $x_i < x_{i+1}$ and add $1/N$ at each location x_i to $S(x)$. Then $S(x_i)$ is the fraction of observations with x values less or equal to x_i,

$$\begin{aligned} S(x_i) &= \frac{i}{N}, \\ S(x_N) &= 1. \end{aligned}$$

As in the χ^2-test we can determine the expected distribution of D, which will depend on N and transform the experimental value of D into a p-value. To get rid of the N dependence of the theoretical D distribution we use $D^* = \sqrt{N}D$. Its distribution under H_0 is for not too small N ($N >\approx 100$) independent of N and available in form of tables and graphs. For event numbers larger than 20 the approximation $D^* = D(\sqrt{N}+0.12+0.11/\sqrt{N})$ is still a very good approximation.[7] The function $p(D^*)$ is displayed in Fig. 10.10.

The Kolmogorov–Smirnov test emphasizes more the center of the distribution than the tails because there the distribution function is tied to the values *zero* and *one* and thus is little sensitive to deviations at the borders. Since it is based on the distribution function, deviations are integrated over a certain range. Therefore it is not very sensitive to deviations which are localized in a narrow region. In Fig. 10.8 the left-hand and the right-hand histograms have the same excess of entries in the region left of the center.

[7] D does not scale exactly with $\sqrt{N}$ because S increases in discrete steps.

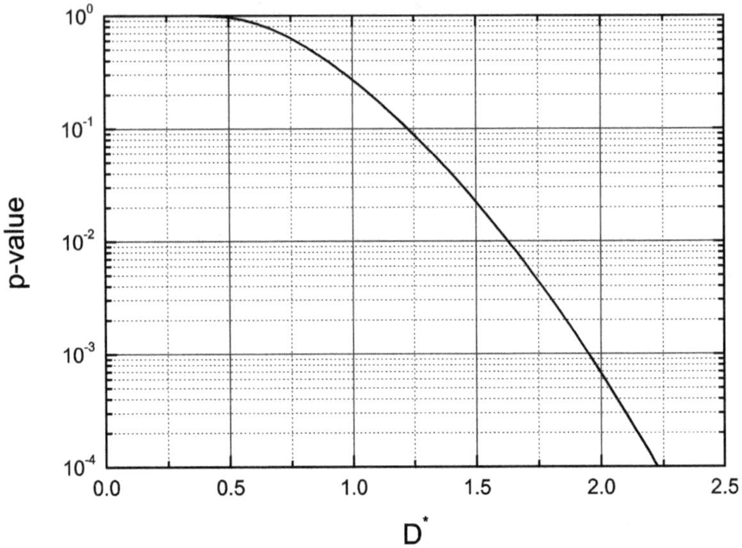

Fig. 10.10 p-value as a function of the Kolmogorov test statistic D^* .

The Kolmogorov–Smirnov test produces in both cases approximately the same value of the test statistic, even though we would think that the distribution of the right-hand histogram is harder to explain by a statistical fluctuation of a uniform distribution. This shows again, that the power of a test depends strongly on the alternatives to H_0. The deviations of the left-hand histogram are well detected by the Kolmogorov–Smirnov test, those of the right-hand histogram much better by the Anderson–Darling test which we will present below.

There exist other EDF tests [89], which in most situations are more effective than the simple Kolmogorov–Smirnov test.

10.4.5 *Tests of the Kolmogorov–Smirnov and Cramer–von Mises Families*

In the *Kuiper test* the test statistic V is the sum of the two deviations of the empirical distribution function S from F, $V = D_+ + D_-$. This quantity is designed for distributions "on the circle". This are distributions where the beginning and the end of the distributed quantity are arbitrary, like the distribution of the azimuthal angle which can be presented with equal justification in all intervals $[\varphi_0, \varphi_0 + 2\pi]$ with arbitrary φ_0.

The tests of the Cramer–von Mises family are based on the quadratic difference between F and S. The simple *Cramer–von Mises test* employs the test statistic

$$W^2 = \int_{-\infty}^{\infty} [(F(x) - S(x)]^2 \, \mathrm{d}F.$$

In most situations the *Anderson–Darling test* with the test statistic A^2 and the test of *Watson* with the test statistic U^2

$$A^2 = N \int_{-\infty}^{\infty} \frac{[S(x) - F(x)]^2}{F(x)\,[1 - F(x)]} \mathrm{d}F,$$

$$U^2 = N \int_{-\infty}^{\infty} \left\{ S(x) - F(x) - \int_{-\infty}^{\infty} [S(x) - F(x)] \, \mathrm{d}F \right\}^2 \mathrm{d}F,$$

are superior to the Kolmogorov–Smirnow test.

The test of Anderson emphasizes especially the tails of the distribution while Watson's test has been developed for distributions on the circle. The formulas above look quite complicated at first sight. They simplify considerably when we perform a *probability integral transformation* (*PIT*). This term stands for a simple transformation of the variate x into the variate $z = F_0(x)$, which is uniformly distributed in the interval $[0, 1]$ and which has the simple distribution function $H_0(z) = z$. With the transformed step distribution $S^*(z)$ of the sample we get

$$A^2 = N \int_{-\infty}^{\infty} \frac{[S^*(z) - z]^2}{z(1 - z)} \mathrm{d}z,$$

$$U^2 = N \int_{-\infty}^{\infty} \left\{ S^*(z) - z - \int_{-\infty}^{\infty} [S^*(z) - z] \, \mathrm{d}z \right\}^2 \mathrm{d}z.$$

In the Appendix A.6 it is shown how to compute the test statistics. There also the asymptotic distributions are collected.

10.4.6 *Neyman's Smooth Test*

This test [94] is different from those discussed so far in that it parametrizes the alternative hypothesis. Neyman introduced the smooth test in 1937 (for a discussion by E. S. Pearson see [95]) as an alternative to the χ^2-test, in that it is insensitive to deviations from H_0 which are positive (or negative) in several consecutive bins. He insisted that in hypothesis testing the investigator has to bear in mind which departures from H_0 are possible and thus to fix partially the p.d.f. of the alternative. The test is called "smooth"because, contrary to the χ^2-test, the alternative hypothesis approaches H_0 smoothly for vanishing parameter values. The hypothesis under test H_0 is again that the sample after the PIT, $z_i = F_0(x_i)$, follows a uniform distribution in the interval $[0, 1]$.

The smooth test excludes alternative distributions of the form

$$g_k(z) = \sum_{i=0}^{k} \theta_i \pi_i(z), \tag{10.6}$$

where θ_i are parameters and the functions $\pi_i(z)$ are modified orthogonal Legendre polynomials that are normalized to the interval $[0, 1]$ and symmetric or antisymmetric with respect to $z = 1/2$:

$$\begin{aligned}
\pi_0(z) &\equiv 1, \\
\pi_1(z) &= \sqrt{3}(2z - 1), \\
\pi_i(z) &= \sqrt{2i + 1} P_i(2z - 1).
\end{aligned}$$

Here $P_i(x)$ is the Legendre polynomial in the usual form. The first parameter θ_0 is fixed, $\theta_0 = 1$, and the other parameters are restricted such that g_k is positive. The user has to choose the parameter k which limits the degree of the polynomials. If the alternative hypothesis is suspected to contain narrow structures, we have to admit large k. The test with $k = 1$ rejects a linear contribution, $k = 2$ in addition a quadratic component and so on. Obviously, the null hypothesis H_0 corresponds to $\theta_1 = \cdots = \theta_k = 0$, or equivalently to $\sum_{i=1}^{k} \theta_i^2 = 0$. We have to look for a test statistic which increases with the value of this sum.

For a sample of size N the test statistic proposed by Neyman is

$$r_k^2 = \frac{1}{N} \sum_{i=1}^{k} t_i^2 = \frac{1}{N} \sum_{i=1}^{k} \left(\sum_{j=1}^{N} \pi_i(z_j) \right)^2 . \tag{10.7}$$

This choice is plausible, because a large absolute value of t_i is due to a strong contribution of the polynomial π_i to the observed distribution and thus also to a large value of θ_i^2, while under H_0 we have for $i \geq 1$

$$\langle t_i \rangle = N \langle \pi_i(z) \rangle = 0,$$

because

$$\int_0^1 \pi_i(z) \, \mathrm{d}z = 0.$$

Asymptotically, $N \to \infty$, under H_0 the test statistic r_k^2 follows a χ^2-distribution with k degrees of freedom (see 3.6.7). This is a consequence of the orthonormality of the polynomials π_i and the central limit theorem: We have

$$\mathrm{var}(t_i) = \langle t_i^2 \rangle = N \int_0^1 \pi_i^2(z)\,\mathrm{d}z = N$$

and as a sum of N random variables the statistic $t_i/\sqrt{N}$ is normally distributed for large N, with expectation value zero and variance one. Due to the orthogonality of the π_i, the t_i are uncorrelated. For small N the distribution of the test statistic r_k^2 has to be obtained by a Monte Carlo simulation.

In any case, large values of r_k^2 indicate bad agreement of the data with H_0, but for a fixed value of k the smooth test is not consistent.[8] Its power approaches unity for $N \to \infty$ only for the class of alternatives H_k having a PIT which is represented by an expansion in Legendre polynomials up to order k. Hence with respect to these, while usually uninteresting, restricted alternatives it is consistent. Thus for large samples and especially for the exclusion of narrow structures k should not be chosen too small. The value of k in the smooth test corresponds roughly to the number of bins in the χ^2-test.

The smooth test is in most cases superior to the χ^2-test. This can be understood in the following way: The smooth test scrutinizes not only for structures of a fixed frequency but for all frequencies up to k while the χ^2-test with $B \gg 1$ bins is rather insensitive to low frequency variations.

Remark: The alternative distribution quoted by Neyman was the exponential

$$g_k(z) = C \exp\left(\sum_{i=0}^{k} \theta_i \pi_i(z)\right) \tag{10.8}$$

where $C(\boldsymbol{\theta})$ is a normalization constant. Neyman probably chose the exponential form, because it guaranties positivity without further restrictions of the parameters θ_i. Moreover, with this class of alternatives, it has been shown by E. S. Pearson [95] that the smooth test can be interpreted as a likelihood ratio test. Anyway, (10.6) or (10.8) serve only as a motivation to choose the test statistic (10.7) which is the relevant quantity.

10.4.7 *The L_2 Test*

The binning-free tests discussed so far are restricted to one dimension, i.e. to univariate distributions. We now turn to multivariate tests.

[8] For $k = 1$, for instance, the test cannot exclude distributions concentrated near $z = 1/2$.

A very obvious way to express the difference between two distributions f and f_0 is the integrated quadratic difference

$$L_2 = \int [f(\boldsymbol{r}) - f_0(\boldsymbol{r})]^2 \, \mathrm{d}\boldsymbol{r}. \tag{10.9}$$

Unfortunately, we cannot use this expression for the comparison of a sample $\{\boldsymbol{r}_1, \dots, \boldsymbol{r}_N\}$ with a continuous function f_0, but we can try to derive from our sample an approximation of f. Such a procedure is called *probability density estimation* (*PDE*). A common approach (see Chapter 12) is the Gaussian smearing or smoothing. The N discrete observations at the locations $\boldsymbol{r}_i$ are transformed into the function

$$f_G(\boldsymbol{r}) = \frac{1}{N} \sum e^{-\alpha(\boldsymbol{r}_i - \boldsymbol{r})^2}.$$

The smearing produces a broadening which has also to be applied to f_0:

$$f_{0G}(\boldsymbol{r}) = \int f_0(\boldsymbol{r}') e^{-\alpha(\boldsymbol{r} - \boldsymbol{r}')^2} \mathrm{d}\boldsymbol{r}'.$$

We now obtain a useful test statistic L_{2G},

$$L_{2G} = \int [f_G(\boldsymbol{r}) - f_{0G}(\boldsymbol{r})]^2 \, \mathrm{d}\boldsymbol{r}.$$

So far the L_2 test [96] has not found as much attention as it deserves because the calculation of the integral is tedious. However its Monte Carlo version is pretty simple. It offers the possibility to adjust the width of the smearing function to the density f_0. Where we expect large distances of observations, the Gaussian width should be large, $\alpha \sim f_0^2$.

A more sophisticated version of the L_2 test is presented in [96]. The Monte Carlo version is included in Section 10.4.10, see below.

10.4.8 *Comparing a Data Sample to a Monte Carlo Sample and the Metric*

We know turn to tests where we compare our sample not to an analytic distribution but to a Monte Carlo simulation of f_0. This is not a serious restriction because anyhow acceptance and resolution effects have to be taken into account in the majority of all experiments. Thus the null hypothesis is usually represented by a simulation sample.

To compare two samples, we have to construct a relation between observations of the samples which in the multidimensional case has to depend in some way on the distance between them. We can define the distance in the usual way using the standard euclidean metric but since the different dimensions often represent completely different physical quantities, e.g. spatial distance, time, mass etc., we have considerable freedom in the choice of the metric and we will try to adjust the metric such that the test is powerful.

We usually want that all coordinates enter with about equal weight into the test. If, for example, the distribution is very narrow in x but wide in y, then the distance r of points is almost independent of y and it is reasonable to stretch the distribution in the x direction before we apply a test. Therefore we propose for the general case to scale linearly all coordinates such that the empirical variances of the sample are the same in all dimensions. In addition we may want to get rid of correlations when, for instance, a distribution is concentrated in a narrow band along the x-y diagonal.

Instead of transforming the coordinates we can use the *Mahalanobis distance*[9] in order to normalize distances between observations $(\boldsymbol{x}_1, \ldots, \boldsymbol{x}_N\}$ with sample mean $\overline{\boldsymbol{x}}$. (The bold-face symbols here denote P-dimensional vectors describing different features measured on each of the N sampled objects.)

The Mahalanobis distance d_M of two observations $\boldsymbol{x}$ and $\boldsymbol{x}'$ is

$$d_M = \sqrt{(\boldsymbol{x} - \boldsymbol{x}')^T \mathsf{V}^{-1} (\boldsymbol{x} - \boldsymbol{x}')},$$

with

$$V_{ij} = \sum_{n=1}^{N} (x_{ni} - \overline{x}_i)(x_{nj} - \overline{x}_j)/N.$$

It is equivalent to the euclidean distance after a linear transformation of the vector components which produces a sample with unity covariance matrix. If the covariance matrix is diagonal, then the resulting distance is the normalized Euclidean distance in the P-dimensional space:

$$d_M = \sqrt{\sum_{p=1}^{P} \frac{(x_p - x'_p)^2}{\sigma_p^2}}.$$

In the following tests the choice of the metric is up to the user. In many situations it is reasonable to use the Mahalanobis distance, even though moderate variations of the metric normally have little influence on the power of a test.

[9]This is a distance measure introduced by P. C. Mahalanobis in 1936.

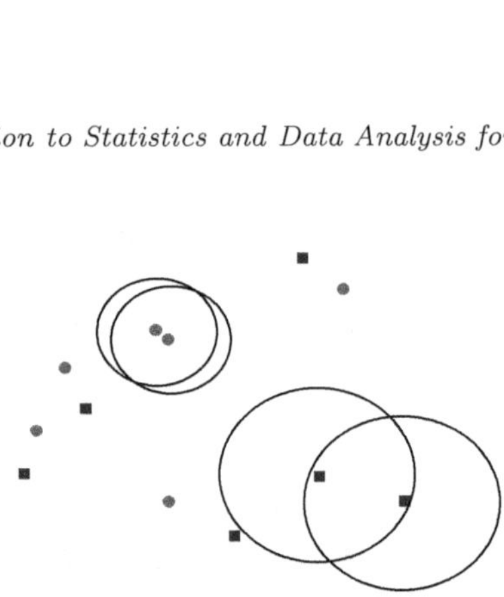

Fig. 10.11 K-nearest-neighbor test.

10.4.9 *The k-Nearest-Neighbor Test*

We consider two samples, one generated by a Monte Carlo simulation of a null distribution f_0 and the experimental sample. The test statistic is the number $n(k)$ of observations of the mixed sample where all of its k nearest neighbors belong to the same sample as the observation itself. This is illustrated in Fig. 10.11 for an unrealistically simple configuration. We find $n(1) = 4$ and $n(2) = 4$. The parameter k is a small number to be chosen by the user, in most cases it is one, two or three.

Of course we expect n to be large if the two parent distributions are very different. The k-nearest-neighbor test is very popular and quite powerful but has a severe caveat: We would like to have the number M of Monte Carlo observations much larger than the number N of experimental observations. In the situation with $M \gg N$ each observation tends to have as next neighbor a Monte Carlo observation and the test becomes less significant.

10.4.10 *The Energy Test*

The relatively new energy test has first be introduced by Szekely [98] and has been re-invented and extended by Aslan and Zech [97].

The energy test is inspired by physics. In electrostatics the energy of a mixture of positive and negative charges of one unit, is minimal if the positive and the negative charges are equally distributed, i.e. the charge is zero everywhere. If the number N of positive and the number M of negative charges M are different, a minimum of energy is still obtained if the distributions are the same. The energy has three components, the energies Φ_1 and Φ_2 caused by the repulsive force of the charges of the same sign, positive or negative, and Φ_3 by the attractive force of the charges of opposite sign.

$$\Phi = \Phi_1 + \Phi_2 + \Phi_3$$

For a sample of N positive charges $1/N$ located at positions $\boldsymbol{r}_i$ and a sample of M negative charges $1/M$ located at positions $\boldsymbol{x}_{0,j}$, the energy is

$$\Phi_1 = \frac{1}{N(N-1)} \sum_i \sum_{j>i} R(|\boldsymbol{r}_i - \boldsymbol{r}_j|) \tag{10.10}$$

$$\Phi_2 = \frac{1}{NM} \sum_i \sum_j R(|\boldsymbol{r}_i - \boldsymbol{r}_{0j}|) \tag{10.11}$$

$$\Phi_3 = \frac{1}{M(M-1)} \sum_i \sum_{j>i} R(|\boldsymbol{r}_{0,i} - \boldsymbol{r}_{0,j}|) \tag{10.12}$$

In its statistical interpretation Φ is a test statistic for the assumption that two samples are compatible. (Two sample tests are discussed in the next section. The distance function R is not necessarily the function $R(r) = 1/r$ of electrostatics. Other functions are often more powerful. With the following distance functions the energy test is consistent, i.e. in the limit of $N, M \to \infty$ the energy is minimal if both samples originate from the same distribution.

The distance function R relates sample points and simulated points of the null hypothesis to each other. To be consistent, the energy has to be minimal if $f(x)$ and $f_0(x)$ are identical. This condition is realized for the following distance functions:

$$\begin{aligned}
&\text{powerlaw:} && R(r) = r^k, \quad (-2 < k < 2,\ k \neq 0), \\
&\text{logarithm:} && R(r) = -\ln(r+\epsilon)\ , \\
&\text{normal:} && R(r) = \exp(-cr^2)\ , \quad (c > 0).
\end{aligned}$$

For our applications the logarithmic function $R(r) = -\ln(r+\epsilon)$ and the normal function $R(r) \sim \exp(-cr^2)$ are more suitable than $1/r$. The small positive constant ε suppresses the pole of the logarithmic distance function. Its value should be chosen approximately equal to the experimental resolution[10] but variations of ε by a large factor have no sizable influence on the result of tests.

Here we are interested to test whether the positive sample is compatible with the p.d.f. f_0 from which the negative Monte Carlo sample has been

[10] Distances between two points that are smaller than the resolution are accidental and thus insignificant.

drawn. This means that the number M Monte Carlo charges should be large compared to the N charges of the experimental sample. The distribution of the test statistic under H_0 can be obtained by generating a large number of Monte Carlo samples of size M. If M is large and the significance level α is small, the computation time may become long. A faster alternative is to generate a large number of sets by a resampling technique:

A data set of $2M$ observations is generated, which by splitting it, allows us to obtain two simulated values of ϕ. Then we shuffle the elements and compute again two additional values of ϕ. The shuffling is repeated as long as needed to get the required statistics. An efficient shuffling technique invented by Fisher and Yates and improved by Durstenfeld is described in the Appendix A.7. The values of ϕ are correlated, but the correlation is mostly negligible. In case of doubts, the shuffle should be repeated with several independent $2N$ sets. From the fluctuation of the p-values its error can be derived. Useful is also the scaling property of the distribution of the test statistic $g(\phi_{N,M})$. For samples with not too small values N, M and N', M' that differ in size by a positive scaling factor s, $N' = sN$, $M' = sM$ we get $g(\phi_{N,M}) = g(s\phi_{N',M'})$ [99]. This property permits to derive the distribution for large samples from that of small samples.

The energy test is quite powerful in many applications and has the advantage that it is not required to sort the sample elements. With Gaussian distance function it is equivalent with the L_2 test. It is more general than the latter, in that it allows to use various distance functions.

10.4.11 *Tests Designed for Specific Problems*

The power of tests depends on the alternatives. If we have an idea of it, even if it is crude, we can design a GOF test which is especially sensitive to the deviations from H_0 which we have in mind. The distribution of the test statistic has to be produced by a Monte Carlo simulation.

Example 10.8. Designed test: three region test

Experimental distributions often show a local excess of observations which are either just a statistical fluctuation or originate from a physical process. To check whether an experimental sample is compatible with the absence of a bump caused by a physical process, we may use the following *three region test.* We subdivide the domain of the variable in three regions with expected numbers of observations n_{10}, n_{20}, n_{30} and look for differences

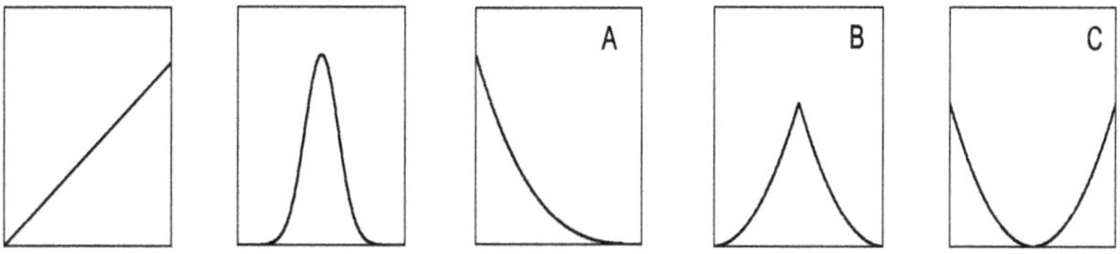

Fig. 10.12 Different admixtures to a uniform distribution.

to the corresponding experimental numbers n_1, n_2, n_3. The subdivision is chosen such that the sum of the differences is maximum. The test statistic R_3 is

$$R_3 = \sup_{n_1,n_2} \left[(n_1 - n_{10}) + (n_2 - n_{20}) + (n_3 - n_{30})\right].$$

Notice, that $n_3 = N - n_1 - n_2$ is a function of n_1 and n_2. The generalization to more than three regions is trivial. Like in the χ^2-test, we could also divide the individual squared differences by their expected value:

$$R_3' = \sup_{n_1,n_2} \left[\frac{(n_1 - n_{10})^2}{n_{10}} + \frac{(n_2 - n_{20})^2}{n_{20}} + \frac{(n_3 - n_{30})^2}{n_{30}}\right].$$

10.4.12 *Comparison of Tests*

10.4.12.1 *Univariate Distributions*

Whether a test is able to detect deviations from H_0 depends on the distribution f_0 and on the kind of distortion. Thus there is no test which is most powerful in all situations.

To get an idea of the power of different tests, we consider six different admixtures to a uniform distribution and compute the fraction of cases in which the distortion of the uniform distribution is detected at a significance level of 5%. For each distribution constructed in this way, we generate 1000 random mixtures with 100 observations each. The distributions which we add are depicted in Fig. 10.12. One of them is linear, two are normal with different widths, and three are parabolic. The χ^2-test was performed with 12 bins following the prescription of Ref. [91], the parameter of Neyman's

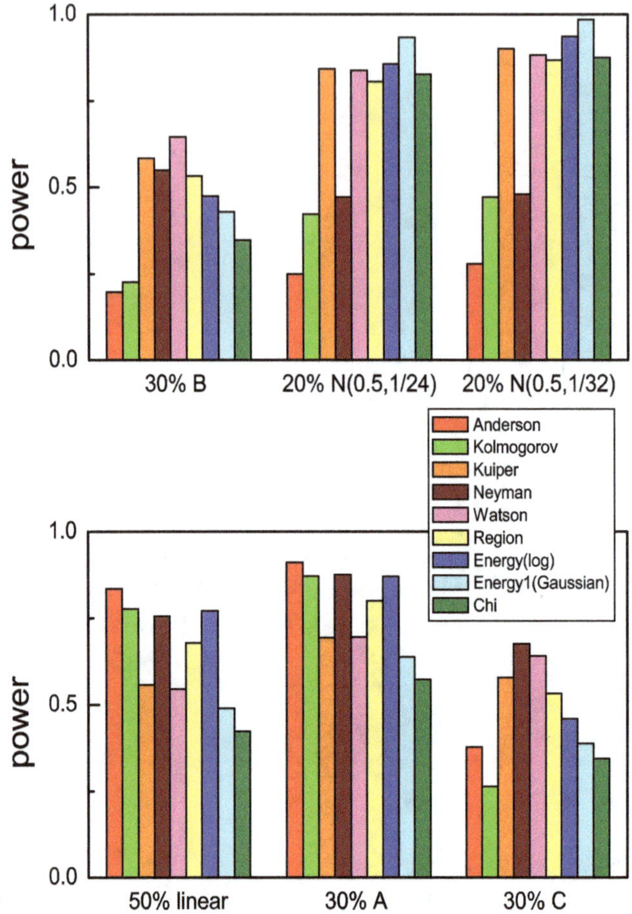

Fig. 10.13 Power (fraction of identified distortions) of different tests.

smooth test was $k = 2$ and the width of the Gaussian of the energy test was $s = 1/8$. The sensitivity of different tests is presented in Fig. 10.13.

The histogram of Fig. 10.13 shows that none of the tests is optimum in all cases. The χ^2-test performs only mediocrely. Probably a lower bin number would improve the result. The tests of Neyman, Anderson–Darling and Kolmogorov–Smirnov are sensitive to a shift of the mean value while the Anderson–Darling test reacts especially to changes at the borders of the distribution. The tests of Watson and Kuiper detect preferentially variations of the variance. Neyman's test and the energy test with logarithmic distance function are rather efficient in most cases.

10.4.12.2 *Multivariate Distributions*

The goodness-of-fit of multivariate distributions cannot be tested very well with simple tests. The χ^2-test often suffers from the small number of entries per bin. Here the k-nearest-neighbor test and the energy test with the long range logarithmic distance function are much more efficient.

Example 10.9. GOF test of a two-dimensional sample

Figure 10.14 shows a comparison of a sample H_1 with a two-dimensional normal distribution (H_0). H_1 corresponds to the distribution of H_0 but contains an admixture of a linear distribution. The p-value of the energy test is 2%. With a χ^2-test with 9 bins we obtain a p-value of 71%. It is unable to identify the deformation of f_0.

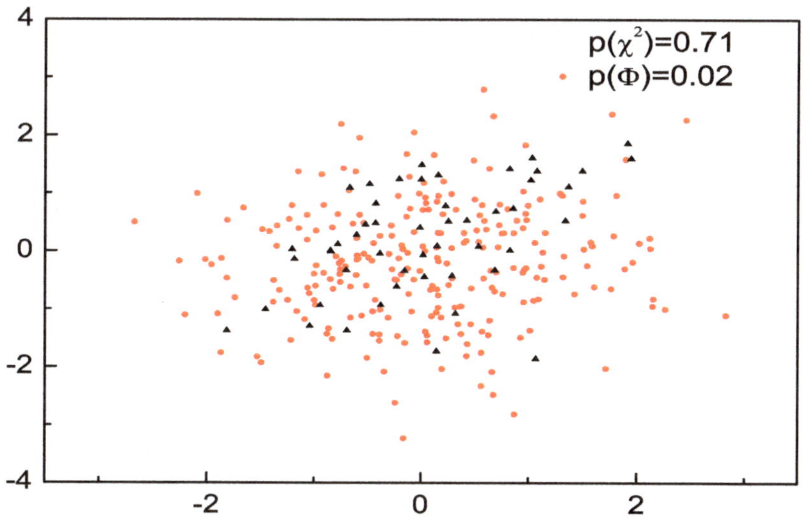

Fig. 10.14 Comparison of a normally distributed sample (red) from H_0 with a linear admixture (triangles) with the normal distribution of H_0.

10.5 Two-Sample Tests

10.5.1 *The Problem*

A standard situation in particle physics is that H_0 cannot be compared directly to the data but has first to be transformed to a Monte Carlo sample, to take into account acceptance losses and resolution effects. We have to compare two samples, a procedure which we had already applied in the *energy test*. Here the distribution of the test statistic needed to compute p-values can be generated by a simple Monte Carlo program.

In other sciences, a frequently occurring problem is that the effectiveness of two or several procedures have to be compared. This may concern drugs, animal feed or the quality of fabrication methods. A similar problem is to test whether a certain process is stable or whether its results have changed during time. Also in the natural sciences we frequently come across the problem that we observe an interesting phenomenon in one data sample which apparently has disappeared in another sample taken at a later time. It is important to investigate whether the two data samples are compatible with each other. Sometimes it is also of interest to investigate whether a Monte Carlo sample and an experimental sample are compatible. Thus we are interested in a statistical procedure which tells us whether two samples A and B are compatible, i.e. drawn from the same parent distribution. Thereby we assume that the parent distribution itself is unknown.

For some one-dimensional distributions it is often possible to find a sensible test statistic and to derive its H_0 distribution f_0. If this is not the case, we use the bootstrap resampling method:

To generate the distribution of the test statistic V, we combine the two samples to a new sample with $M + N$ elements and form new pairs of samples, with M and N elements. We draw M random elements from the combined sample and associate them to A and the remaining elements to B. Computationally this is done by reshuffling as described above in the section dealing with the standard energy test. This is easier than to use systematically all individual possibilities. For each generated pair i we determine the statistic V_i. This procedure is repeated many times and the values V_i form the reference distribution. Our experimental p-value is equal to the fraction of generated V_i which are larger than V_{AB}:

$$p = \frac{\text{Number of permutations with } V_i > V_{AB}}{\text{Total number of permutations}} .$$

10.5.2 *The χ^2-Test*

To test whether two samples are compatible, we can apply the χ^2-test or the Kolmogorov–Smirnov test with minor modifications.

When we calculate the χ^2-statistic we have to normalize the two samples A and B of sizes N and M to each other. For a_i and b_i entries in bin i, $a_i/N - b_i/M$ should be compatible with zero. With the usual error propagation we obtain an estimate $a_i/N^2 + b_i/M^2$ of the quadratic error of this quantity and

$$\chi^2 = \sum_{i=1}^{B} \frac{(a_i/N - b_i/M)^2}{(a_i/N^2 + b_i/M^2)}. \qquad (10.13)$$

It follows approximately a χ^2-distribution of $B - 1$ degrees of freedom, but not exactly, as we had to replace the expected values by the observed numbers in the error estimation. We have to be careful if the number of observations per bin is small.

10.5.3 *The Likelihood Ratio Test*

The likelihood ratio test is less vulnerable to low event numbers than the χ^2-test.

Setting $r = M/N$ we compute the likelihood that we observe in a single bin a entries with expectation λ and b entries with expectation $\rho\lambda$, where the hypothesis H_0 is characterized by $\rho = r$:

$$L(\lambda, \rho|a, b) = \frac{\mathrm{e}^{-\lambda}\lambda^a}{a!}\frac{\mathrm{e}^{-\rho\lambda}(\rho\lambda)^b}{b!}.$$

Leaving out constant factors the log-likelihood is

$$\ln L = -\lambda(1+\rho) + (a+b)\ln\lambda + b\ln\rho.$$

We determine the conditional maximum likelihood value of λ under $\rho = r$ and the corresponding log-likelihood:

$$1 + r = (a+b)\frac{1}{\hat{\lambda}_c},$$

$$\hat{\lambda}_c = \frac{a+b}{1+r},$$

$$\ln L_{cmax} = (a+b)\left[-1 + \ln\frac{a+b}{1+r}\right] + b\ln r.$$

The unconditional maximum of the likelihood is found for $\hat{\lambda} = a$ and $\hat{\rho} = b/a$:

$$\ln L_{umax} = -(a+b) + a\ln a + b\ln b.$$

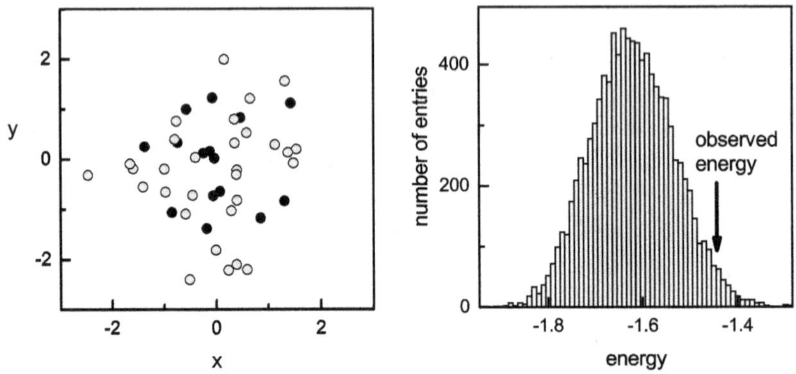

Fig. 10.15 Two-sample energy test. Left hand: the samples which are to be compared. Right hand: distribution of test statistic and actual value.

Our test statistic is V_{AB}, the logarithm of the likelihood ratio, now summed over all bins:

$$\begin{aligned} V_{AB} &= \ln L_{cmax} - \ln L_{umax} \\ &= \sum_i \left[(a_i + b_i) \ln \frac{a_i + b_i}{1 + r} - a_i \ln a_i - b_i \ln b_i + b_i \ln r \right]. \end{aligned}$$

Note that $V_{AB}(r) = V_{BA}(1/r)$, as it should.

Its distribution under H_0 is generated as explained above by resampling.

10.5.4 *The Kolmogorov–Smirnov Test*

Also the Kolmogorov–Smirnov test can easily be adapted to a comparison of two samples. We construct the test statistic in an analogous way as above. The test statistic is $D^* = D\sqrt{N_{eff}}$, where D is the maximum difference between the two empirical distribution functions S_A, S_B, and N_{eff} is the effective or equivalent number of events, which is computed from the relation:

$$\frac{1}{N_{eff}} = \frac{1}{N} + \frac{1}{M}.$$

In a similar way other EDF multi-dimensional tests which we have discussed above can be adjusted.

10.5.5 *The Energy Test*

For a binning-free comparison of two samples A and B with M and N observations we can again use the energy test [97] which in the multi-dimensional case has only few competitors.

We compute the energy ϕ_{AB} in the same way as above, replacing the Monte Carlo sample by one of the experimental samples. The expected distribution of the test statistic ϕ_{AB} is computed in the same way as for the likelihood ratio test from the combined sample using the permutation technique. Our experimental p-value is equal to the fraction of generated ϕ_i from the bootstrap sample which are larger than ϕ_{AB}:

$$p = \frac{\text{Number of permutations with } \phi_i > \phi_{AB}}{\text{Total number of permutations}} .$$

Example 10.10. Comparison of two samples

We compare two two-dimensional samples with 15 and 30 observations with the energy test. The two samples are depicted in a scatter plot at the left hand side of Fig. 10.15. The energy of the system is $\phi_{AB} = -1.480$ (The negative value arises because we have omitted the term ϕ_3). From the mixed sample 10000 sample combinations have been selected at random. Its energy distribution is shown as a histogram in the figure. The arrow indicates the location of ϕ_{AB}. It corresponds to a p-value of 0.06. We can estimate the error of the p-value computing it from many permutation sets each with a smaller number of permutations. From the variation of p from 100 times 100 permutations we find $\delta p = 0.02$. The p-value is small, indicating that the samples belong to different populations. Indeed they have been drawn from different distributions, a uniform distribution, $-1.5 < x, y < 1.5$ and a normal distribution with standard deviations $\sigma_x = \sigma_y = 1$.

10.5.6 *The k-Nearest-Neighbor Test*

The k-nearest-neighbor test is per construction a two-sample test. The distribution of the test statistic is obtained in exactly the same way as in the two-sample energy test which we have discussed in the previous section.

The performance of the k-nearest neighbors test is similar to that of the energy test. The energy test (and the L^2 test which is automatically included in the former) is more flexible than the k-nearest neighbors test and includes all observation of the sample in the continuous distance function. The k-nearest neighbors test on the other hand is less sensitive to variations of the density which are problematic for the energy test with the Gaussian distance function of constant width.

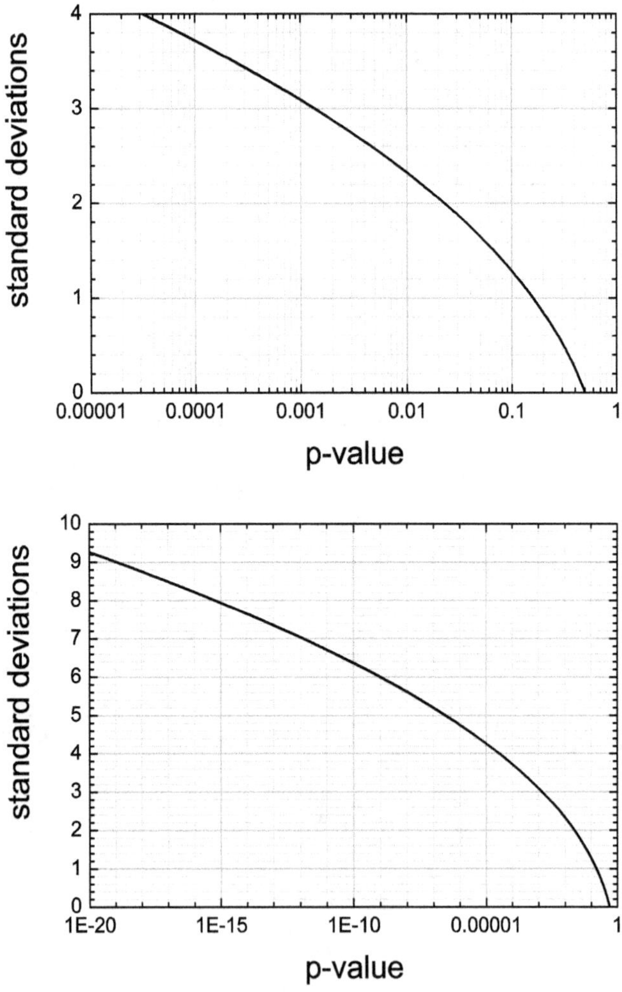

Fig. 10.16 Transformation of p-values to one-sided number of standard deviations.

10.6 Significance of Signals

10.6.1 *Introduction*

Tests for signals are closely related to goodness-of-fit tests but their aim is different. We are not interested to verify that H_0 is compatible with a sample but we intend to quantify the evidence of signals which are possibly present in a sample which consists mainly of uninteresting background. Here not only the distribution of the background has to be known but in addition we must be able to parametrize the alternative which we search for. The null hypothesis H_0 corresponds to the absence of deviations from the background. The alternative H_s is not fully specified, otherwise it would be sufficient to compute the simple likelihood ratio which we have discussed in Chapter 6.

Signal tests are applied when we search for rare decays or reactions like neutrino oscillations. Another frequently occurring problem is that we want to interpret a line in a spectrum as indication for a resonance or a new particle. To establish the evidence of a signal, we usually require a very significant deviation from the null hypothesis, i.e. the sum of background and signal has to describe the data much better than the background alone because particle physicists look in hundreds of histograms for more or less wide lines and thus always find candidates[11] which in most cases are just background fluctuations. For this reason, signals are only accepted by the community if they have a significance of at least four or five standard deviations. In cases where the existence of new phenomena is not unlikely, a smaller significance may be sufficient. A high significance for a signal corresponds to a low p-value of the null hypothesis.

To quote the p-value instead of the significance as expressed by the number of standard deviations by which the signal exceeds the background expectation is to be preferred because it is a measure which is independent of the form of the distribution. However, the standard deviation scale is better suited to indicate the significance than the p-values scale where very small values dominate. For this reason it has become customary to transform the p-value p into the number of Gaussian standard deviations

[11]This is the so-called *look-else-where effect.*

s_G which are related through

$$p = 1/\sqrt{2\pi} \int_{s_G}^{\infty} \exp(-x^2/2)dx \tag{10.14}$$

$$= \left[1 - \mathrm{erf}(s_G/\sqrt{2})\right]/2. \tag{10.15}$$

The function $s_G(p)$ is given in Fig. 10.16. Relations (10.14), (10.15) refer to one-sided tests. For two-sided tests, p has to be multiplied by a factor two.

When we require very low p-values for H_0 to establish signals, we have to be especially careful in modeling the distribution of the test statistic. Often the distribution corresponding to H_0 is approximated for instance by a polynomial with some uncertainties in the parameters and assumptions which are difficult to implement in the test procedure. It is better to underestimate the significance of a signal than to present evidence for a new phenomenon based on a doubtful number.

To illustrate this problem we return to our standard example where we search for a line in a one-dimensional spectrum. Usually, the background under an observed bump is estimated from the number of events outside but near the bump in the so-called side bands. If the side bands are chosen too close to the signal they are affected by the tails of the signal, if they are chosen too far away, the extrapolation into the signal region is sensitive to the assumed shape of the background distribution which often is approximated by a linear or quadratic function. This makes it difficult to estimate the size and the uncertainty of the expected background with sufficient accuracy to establish the p-value for a large (>4 st. dev.) signal. As numerical example let us consider an expectation of 1000 background events which is estimated by the experimenter too low by 2%, i.e. equal to 980. Then a 4.3 st. dev. excess would be claimed by him as a 5 standard deviation effect and he would find a p-value that is too low by a factor of 28. We also have to be careful with numerical approximations, for instance when we approximate a Poisson distribution by a Gaussian. These uncertainties have to be included in the simulation of the distribution of the test statistic.

Usually, the likelihood ratio, i.e. the ratio of the likelihood which maximizes H_s and the maximum likelihood for H_0 is the most powerful test statistic. In some situations a relevant parameter which characterizes the signal strength is more informative.

10.6.2 *The Likelihood Ratio Test*

10.6.2.1 *Definition*

An obvious candidate for the test statistic is the likelihood ratio (LR) which we have introduced and used in Section 10.4 to test goodness-of-fit of histograms, and in Section 10.5 as a two-sample test. We repeat here its general definition:

$$\lambda = \frac{\sup\left[L_0(\boldsymbol{\theta}_0|\boldsymbol{x})\right]}{\sup\left[L_s(\boldsymbol{\theta}_s|\boldsymbol{x})\right]},$$
$$\ln\lambda = \ln\sup\left[L_0(\boldsymbol{\theta}_0|\boldsymbol{x})\right] - \ln\sup\left[L_s(\boldsymbol{\theta}_s|\boldsymbol{x})\right]$$

where L_0, L_s are the likelihoods under the null hypothesis and the signal hypothesis, respectively. The supremum is to be evaluated relative to the parameters, i.e. the likelihoods are to be taken at the MLEs of the parameters. The vector $\boldsymbol{x}$ represents the sample of the N observations $x_1, \ldots, x_N$ of a one-dimensional geometric space. The extension to a multi-dimensional space is trivial but complicates the writing of the formulas. The parameter space of H_0 is assumed to be a subset of that of H_s. Therefore λ will be smaller or equal to *one*.

For example, we may want to find out whether a background distribution is described significantly better by a cubic than by a linear distribution:

$$f_0 = \alpha_0 + \alpha_1 x, \tag{10.16}$$
$$f_s = \alpha_0 + \alpha_1 x + \alpha_2 x^2 + \alpha_3 x^3. \tag{10.17}$$

We would fit separately the parameters of the two functions to the observed data and then take the ratio of the corresponding maximized likelihoods.

Frequently the data sample is so large that we better analyze it in form of a histogram. Then the distribution of the number of events y_i in bin i, $i = 1, \ldots, B$ can be approximated by normal distributions around the parameter dependent predictions $t_i(\boldsymbol{\theta})$. As we have seen in Chapter 7, Section 7.1 we then get the log-likelihood

$$\ln L = -\frac{1}{2}\sum_{i=1}^{B}\frac{\left[y_i - t_i\right]^2}{t_i} + const.$$

which is equivalent to the χ^2-statistic, $\chi^2 \approx -2\ln L$. In this limit the likelihood ratio statistic is equivalent to the χ^2-difference, $\Delta\chi^2 = \min\chi_0^2 - \min\chi_s^2$, of the χ^2-deviations, $\min\chi_0^2$ with the parameters adjusted to the null hypothesis H_0, and $\min \chi_s^2$ with its parameters adjusted to the alternative hypothesis H_s, background plus signal:

$$\ln \lambda = \ln \sup \left[L_0(\boldsymbol{\theta}_0|\boldsymbol{y}) \right] - \ln \sup \left[L_s(\boldsymbol{\theta}_s|\boldsymbol{y}) \right] \tag{10.18}$$

$$\approx -\frac{1}{2}(\min \chi_0^2 - \min \chi_s^2). \tag{10.19}$$

The p-value derived from the LR statistic does not take into account that a simple hypothesis is a priori more attractive than a composite one which contains free parameters. Another point of criticism is that the LR is evaluated only at the parameters that maximize the likelihood while the parameters suffer from uncertainties. Thus conclusions should not be based on the p-value only.

A Bayesian approach applies so-called *Bayes factors* to correct for the mentioned effects but it is not very popular because it has other caveats. Its essentials are presented in the Appendix A.15

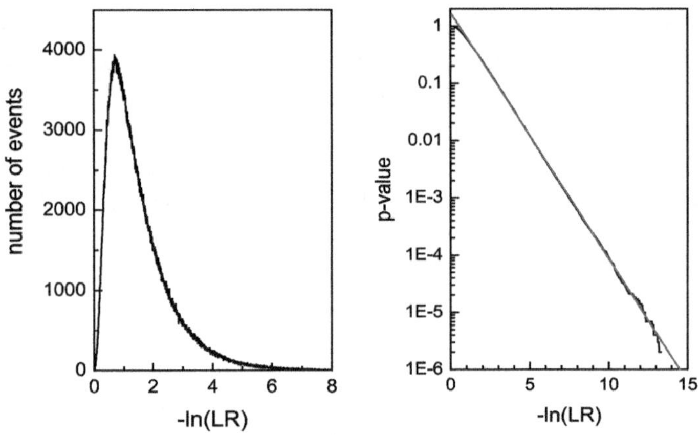

Fig. 10.17 Distributions of the test statistic under H_0 and p-value as a function of the test statistic.

10.6.2.2 *Distribution of the Test Statistic*

The distribution of λ under H_0 in the general case is not known analytically; however, if the approximation (10.19) is justified, the distribution of $-2\ln\lambda$ under certain additional regularity conditions and the conditions mentioned at the end of Section 10.4.2 will be described by a χ^2-distribution. In the example corresponding to relations (10.16) (10.17) this

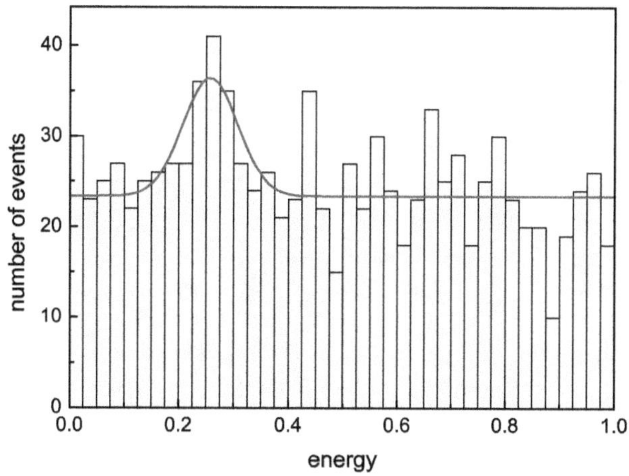

Fig. 10.18 Histogram of the event sample used for the likelihood ratio test. The curve is an unbinned likelihood fit to the data.

would be a χ^2-distribution of 2 degrees of freedom since f_s compared to f_0 has 2 additional free parameters. Knowing the distribution of the test statistic reduces the computational effort required for the numerical evaluation of p-values considerably.

Let us look at a specific problem: We want to check whether an observed bump above a continuous background can be described by a fluctuation or whether it corresponds to a resonance. The two hypotheses may be described by the distributions

$$f_0 = \alpha_0 + \alpha_1 x + \alpha_2 x^2, \tag{10.20}$$

$$f_s = \alpha_0 + \alpha_1 x + \alpha_2 x^2 + \alpha_3 \mathcal{N}(x|\mu, \sigma), \tag{10.21}$$

and we can again use $\ln \lambda$ or $\Delta\chi^2$ as test statistic. Since we have to define the test before looking at the data, μ and σ will be free parameters in the fit of f_s to the data. Unfortunately, now $\Delta\chi^2$ no longer follows a χ^2-distribution of 3 degrees of freedom and has a significantly larger expectation value than expected from the χ^2-distribution. The reason for this dilemma is that for $\alpha_3 = 0$ which corresponds to H_0, the other parameters μ and σ are undefined, and thus part of the χ^2-fluctuation in the fit to f_s is unrelated to the difference between f_s and f_0. More generally, only if the following conditions are satisfied, $\Delta\chi^2$ follows in the large number limit a χ^2-distribution with the number of degrees of freedom given by the difference of the number of free parameters of the null and the alternative hypotheses:

(1) The distribution f_0 of H_0 has to be a special realization of the distribution f_s of H_s.
(2) The fitted parameters have to be inside the region, i.e. off the boundary, allowed by the hypotheses. For example, the MLE of the location of a Gaussian should not be outside the range covered by the data.
(3) All parameters of H_s have to be defined under H_0.

If one of these conditions is not satisfied, the distribution of the test statistic has to be obtained via a Monte Carlo simulation. This means that we generate many fictive experiments of H_0 and count how many of those have values of the test statistic that exceed the one which has actually been observed. The corresponding fraction is the p-value for H_0. This is a fairly involved procedure because each simulation includes fitting of the free parameters of the two hypotheses. In Ref. [90] it is shown that the asymptotic behavior of the distribution can be described by an analytical function. In this way the amount of simulation can be reduced.

Example 10.11. Distribution of the likelihood ratio statistic

We consider a uniform distribution (H_0) of 1000 events in the interval $[0,1]$ and as alternative a resonance with Gaussian width, $\sigma = 0.05$, and arbitrary location μ in the range $0.2 \leq \mu \leq 0.8$ superposed to a uniform distribution. The free parameters are ε, the fraction of resonance events and μ. The logarithm of the likelihood ratio statistic is

$$\begin{aligned}
\ln\lambda &= \ln\sup\left[L_0(\boldsymbol{\theta}_0|\boldsymbol{x})\right] - \ln\sup\left[L_s(\boldsymbol{\theta}_s|\boldsymbol{x})\right] \\
&= \sum_{i=1}^{1000}\ln(1) - \sum_{i=1}^{1000}\ln\left[1-\hat{\varepsilon}+\frac{\hat{\varepsilon}}{\sqrt{2\pi}\sigma}\exp\left(-\frac{(x_i-\hat{\mu})^2}{2\sigma^2}\right)\right] \\
&= -\sum_{i=1}^{1000}\ln\left[1-\hat{\varepsilon}+\frac{\hat{\varepsilon}}{\sqrt{2\pi}\sigma}\exp\left(-\frac{(x_i-\hat{\mu})^2}{2\sigma^2}\right)\right],
\end{aligned}$$

essentially the negative logarithm of the likelihood of the MLE. Fig. 10.17 shows the results from a million simulated experiments. The distribution of $-\ln\lambda$ under H_0 has a mean value of -1.502 which corresponds to $\langle\Delta\chi^2\rangle = 3.004$. The p-value as a function of $-\ln\lambda$ follows asymptotically an exponential as is illustrated in the right hand plot of Fig. 10.17.

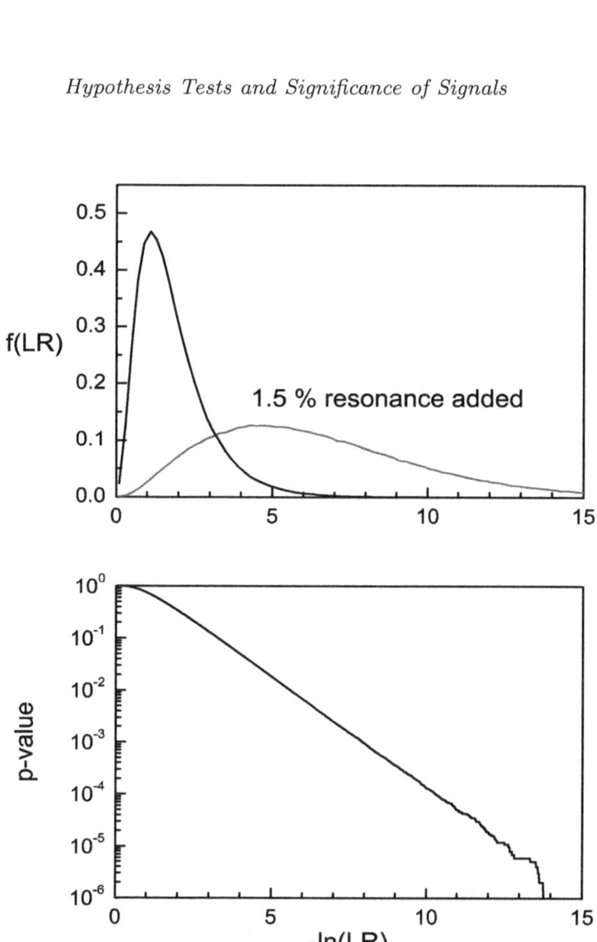

Fig. 10.19 Distributions of the test statistic for H_0 and for experiments with a 1.5% resonance contribution. In the lower graph the p-value for H_0 is given as a function of the test statistic.

Thus it is possible to extrapolate the function to smaller p-values which is necessary to claim large effects. Figure 10.18 displays the result of an experiment where a likelihood fit finds a resonance at the energy 0.257. It contains a fraction of 0.0653 of the events. The logarithm of the likelihood ratio is 9.277. The corresponding p-value for H_0 is $p_{LR} = 1.8 \cdot 10^{-4}$. Hence it is likely that the observed bump is a resonance. In fact it had been generated as a 7 % contribution of a Gaussian distribution $\mathcal{N}(x|0.25, 0.05)$ to a uniform distribution.

We have to remember though that the p-value is not the probability that H_0 is true, it is the probability that H_0 simulates the resonance of the size seen in the data or larger. In a Bayesian treatment, see Appendix A.15, we find betting odds in favor of H_0 of about 2% which is much less impressive. The two numbers refer to different issues but nonetheless we have to face the fact that the two different statistical approaches lead to different conclusions about how evident the existence of a bump really is.

In experiments with a large number of events, the computation of the p-value distribution based on the unbinned likelihood ratio becomes excessively slow and we have to turn to histograms and to compute the likelihood ratio of H_0 and H_s from the histogram. Figure 10.19 displays some results from the simulation of 10^6 experiments of the same type as above but with 10000 events distributed over 100 bins.

In the figure the distributions of the LR for a signal for H_0 and for experiments with 1.5% resonance added is shown. The large spread of the signal distributions reflects the fact that identical experiments by chance may observe a very significant signal or just a slight indication of a resonance.

10.6.2.3 *General Multichannel Case*

We now extend the likelihood ratio test to the multichannel case. We assume that the observations $\boldsymbol{x}_k$ of the channels $k = 1, \dots, K$ are independent of each other. The overall likelihood is the product of the individual likelihoods. For the log-likelihood ratio we then have to replace (10.18) by

$$\ln \lambda = \sum_{k=1}^{K} \left\{ \ln \sup \left[L_{0k}(\boldsymbol{\theta}_{0k}|\boldsymbol{x}_k) \right] - \ln \sup \left[L_{sk}(\boldsymbol{\theta}_{sk}|\boldsymbol{x}_k) \right] \right\} .$$

As an example, we consider an experiment where we observe bumps at the same mass in K different decay channels, bumps which are associated to the same phenomenon, i.e. a particle decaying into different secondaries.

When we denote the decay contribution into channel k by ε_k, the p.d.f. of the decay distribution by $f_k(\boldsymbol{x}_k|\boldsymbol{\theta}_k)$ and the corresponding background distributions by $f_{0k}(\boldsymbol{x}_k|\boldsymbol{\theta}_{0k})$, the distribution under H_0 is

$$f_0(\boldsymbol{x}_1, \dots, \boldsymbol{x}_K|\boldsymbol{\theta}_{01}, \dots, \boldsymbol{\theta}_{0K}) = \prod_{k=1}^{K} f_{0k}(\boldsymbol{x}_k|\boldsymbol{\theta}_{0k})$$

and the alternative signal distribution is

$$f_s(\boldsymbol{x}_1, \ldots, \boldsymbol{x}_K | \boldsymbol{\theta}_{01}, \ldots, \boldsymbol{\theta}_{0K}; \boldsymbol{\theta}_1, \ldots, \boldsymbol{\theta}_K; \varepsilon_1, \ldots, \varepsilon_K) = \prod_{k=1}^{K} [(1 - \varepsilon_k) f_{0k}(\boldsymbol{x}_k | \boldsymbol{\theta}_{0k}) + \varepsilon_k f_k(\boldsymbol{x}_k | \boldsymbol{\theta}_k)].$$

The likelihood ratio is then

$$\ln \lambda = \sum_{k=1}^{K} \left\{ \ln f_{0k}(\boldsymbol{x}_k | \widehat{\boldsymbol{\theta}}_{0k}) - \ln \left[(1 - \hat{\varepsilon}_k) f_{0k}(\boldsymbol{x}_k | \widehat{\boldsymbol{\theta}'}_{0k}) + \hat{\varepsilon}_k f_k(\boldsymbol{x}_k | \widehat{\boldsymbol{\theta}}_k) \right] \right\}.$$

10.6.3 *Tests Based on the Signal Strength*

Instead of using the LR statistic it is often preferable to use a parameter of H_s as test statistic. In the simple example of relation (10.17) the test statistic $t = \alpha_3$ would be a sensible choice. When we want to estimate the significance of a line in a background distribution, instead of the likelihood ratio the number of events which we associate to the line (or the parameter α_3 in example 10.21) is a reasonable test statistic. Compared to the LR statistic it has the advantage to represent a physical parameter but usually the corresponding test is less powerful.

Example 10.12. Example 10.11 continued

Using the fitted fraction of resonance events as test statistic, the p-value for H_0 is $p_f = 2.2 \cdot 10^{-4}$, slightly less stringent than that obtained from the LR. Often physicists compare the number of observed events directly to the prediction from H_0. In our example we have 243 events within two standard deviations around the fitted energy of the resonance compared to the expectation of 200 from a uniform distribution. The probability to observe ≥ 243 for a Poisson distribution with mean 200 is $p_p = 7.3 \cdot 10^{-4}$. This number cannot be compared directly with p_{LR} and p_f because the latter two values include the *look-else-where effect*, i.e. that the simulated resonance may be located at an arbitrary energy. A lower number for p_p is obtained if the background is estimated from the side bands, but then the computation becomes more involved because the error on the expectation has to be included. Primitive methods are only useful for a first crude estimate.

We learn from this example that the LR statistic provides the most powerful test among the considered alternatives. It does not only take into account the excess of events of a signal but also its expected shape. For this reason p_{LR} is smaller than p_f.

Often the significance of a signal s is stated in units of standard deviations σ:

$$s = \frac{N_s}{\sqrt{N_0 + \delta_0^2}} .$$

Here N_s is the number of events associated to the signal, N_0 is the number of events in the signal region expected from H_0 and δ_0 its uncertainty. In the Gaussian approximation it can be transformed into a p-value via (10.14). Unless N_0 is very large and δ_0 is very well known, this p-value has to be considered as a lower limit or a rough guess.

Chapter 11

Statistical Learning

11.1 Introduction

In the process of its mental evolution a child learns to classify objects, persons, animals, and plants. This process partially proceeds through explanations by parents and teachers (supervised learning), but partially also by cognition of the similarities of different objects (unsupervised learning). But the process of learning – of children and adults – is not restricted to the development of the ability merely to classify but it includes also the realization of relations between similar objects, which leads to ordering and quantifying physical quantities, like size, time, temperature, etc.. This is relatively easy, when the laws of nature governing a specific relation have been discovered. If this is not the case, we have to rely on approximations, like inter- or extrapolations.

Also computers, when appropriately programmed, can perform learning processes in a similar way, though to a rather modest degree. The achievements of the so-called artificial intelligence are still rather moderate in most areas, however a substantial progress has been achieved in the fields of supervised learning and classification and there computers profit from their ability to handle a large amount of data in a short time and to provide precise quantitative solutions to well defined specific questions. The techniques and programs that allow computers to learn and to classify are summarized in the literature under the term *machine learning*.

Let us now specify the type of problems which we discuss in this chapter: For an input vector $\boldsymbol{x}$ we want to find an output $\hat{\boldsymbol{y}}$. The input is also called *predictor*, the output *response*. Usually, each input consists of several components (*attributes*, properties), and is written therefore in boldface letters. Normally, it is a metric (quantifiable) quantity but it could also be a categorical quantity like a color or a particle type. The output can also contain several components or consists of a single real or discrete (*Yes* or *No*) variable. Like a human being, a computer program learns from past experience. The teaching process, called training, uses a training sample $\{(\boldsymbol{x}_1, \boldsymbol{y}_1), (\boldsymbol{x}_2, \boldsymbol{y}_2) \dots (\boldsymbol{x}_N, \boldsymbol{y}_N)\}$, where for each input vector $\boldsymbol{x}_i$ the response $\boldsymbol{y}_i$ is known. When we ask for the response to an arbitrary continuous input $\boldsymbol{x}$, usually its estimate $\hat{\boldsymbol{y}}(\boldsymbol{x})$ will be more accurate when the distance to the nearest input vector of the training sample is small than when it is far away. Consequently, the training sample should be as large as possible or affordable. The region of interest should be covered with input vectors homogeneously, and we should be aware that the accuracy of the estimate decreases at the boundary of this region.

Learning which exceeds simple memorizing relies on the existence of more or less simple relations between input and response: Similar input corresponds to similar response. In our approach this translates into the requirement that the responses are similar for input vectors which are close. We can not learn much from erratic distributions.

Example 11.1. Simple empirical relations

The resistance R of a wire is used for a measurement of the temperature T. In the teaching process which here is called calibration, a sample of corresponding values R_i, T_i is acquired. In the application we want to find for a given input R an estimate of T. Usually a simple interpolation will solve this problem.

For more complicated relations, approximations with polynomials, higher spline functions or orthogonal functions are useful.

Example 11.2. Search for common properties

A certain class of molecules has a positive medical effect. The structure, physical and chemical properties $\boldsymbol{x}$ of these molecules are known. In order to find out which combination of the properties is relevant, the distribution of all attributes of the molecules which represent the training objects is investigated. A linear method for the solution of this task is the *principal component analysis*.

Example 11.3. Multi-class classification, pattern recognition

Hand-written letters or figures have to be recognized. Again a sample for which the relation between the written pixels and the letters is known, is used to train the program. Also this problem can be treated by *decision trees, artificial neural networks*, and by *kernel methods*. Here the attributes are the pixel coordinates.

As we have observed also previously, multivariate applications suffer from the *curse of dimensionality*. There are two reasons: i) With increasing number d of dimensions, the distance between the input vectors increases and ii) the surface effects are enhanced. When a fixed number of points is uniformly distributed over a hyper-cube of dimension d, the mean distance between the points is proportional to $\sqrt{d}$. The higher the dimension, the emptier is the space. At the same time the region where estimates become less accurate due to surface effects increases. The fraction of the volume taken by a hyper-sphere inscribed into a hyper-cube is only 5.2% for $d = 5$, and the fraction of the volume within a distance to the surface less than 10% of the edge length increases like $1 - 0.8^d$, this means from 20% for $d = 1$ to 67% for $d = 5$.

Example 11.4. Curse of dimensionality

A training sample of 1000 five-dimensional inputs is uniformly distributed over a hyper-cube of edge length a. To estimate the function value at the center of the region we take all sample elements within a distance of $a/4$ from the center. These are on average $1000 \times 0.052 \times 0.5^5 = 1.6$ elements only, while in one dimension 500 elements would contribute.

In the following, we will first discuss the approximation of measurements afflicted with errors by analytic functions and the interpolation by smoothing techniques. Next we introduce the factor analysis, including the so-called principal component analysis. The last section deals with classification methods, based on artificial neural networks, kernel algorithms, and decision trees. In recent years we observed a fast progress in this field due to new developments, i.e. *support vector machines, boosting,* and the availability of powerful general computer algorithms. This book can only introduce these methods, without claiming completeness. A nice review of the whole field is given in [16].

11.2 Smoothing of Measurements and Approximation by Analytic Functions

We start with two simple examples, which illustrate applications:

i) In a sequence of measurements the gas amplification of an ionization chamber as a function of the applied voltage has been determined. We would like to describe the dependence in form of a smooth curve.

ii) With optical probes it is possible to scan a surface profile point-wise. The objects may be workpieces, tools, or human bodies. The measurements can be used by milling machines or cutting devices to produce replicates or clothes. To steer these machines, a complete surface profile of the objects is needed. The discrete points have to be approximated by a continuous function. When the surface is sufficiently smooth, this may be achieved by means of a spline approximation.

Generally, we are given a number N of measurements y_i with uncertainties δ_i at locations x_i of the independent variables, but are interested in the values of the dependent or response variable y at different values of x, that is, we search for a function $f(x)$ which approximates the measurements, improves their precision and inter- and extrapolates in x. The simplest way to achieve this is to smooth the polygon connecting the data points.

More efficient is the approximation of the measurement by a parameter dependent analytic function $f(x, \boldsymbol{\theta})$. We then determine the parameters by a least square fit, i.e. minimize the sum over the squared and normalized residuals $\sum \left[(y_i - f(x_i, \boldsymbol{\theta})\right]^2 / \delta_i^2$ with respect to $\boldsymbol{\theta}$. The approximation should be compatible with the measurements within their statistical errors but the number of free parameters should be as small as possible. The accuracy of the measurements has a decisive influence on the number of free parameters which we permit in the fit. For large errors we allow also for large deviations of the approximation from the measurements. As a criterion for the number of free parameters, we use statistical tests like the χ^2-test. The value of χ^2 should then be compatible with the number of constraints, i.e. the number of measured points minus the number of fitted parameters. Too low a number of parameters leads to a bias of the predictions, while too many parameters reduce the accuracy, since we profit less from constraints.

Both approaches rely on the presumption that the true function is simple and smooth. Experience tells us that these conditions are justified in most cases.

The approximation of measurements which all have the same uncertainty by analytic functions is called regression analysis. Linear regression had been described in Chapter 6.8.1. In this section we treat the general non-linear case with arbitrary errors.

In principle, the independent variable may also be multi-dimensional. Since then the treatment is essentially the same as in the one-dimensional situation, we will mainly discuss the latter.

11.2.1 *Smoothing Methods*

We use the measured points in the neighborhood of $\boldsymbol{x}$ to get an estimate of the value of $\boldsymbol{y}(\boldsymbol{x})$. We denote the uncertainties of the output vectors of the training sample by δ_j for the component j of $\boldsymbol{y}$. When the points of the training sample have large errors, we average over a larger region than in the case of small errors. The better accuracy of the average for a larger region has to be paid for by a larger bias, due to the possibility of larger fluctuations of the true function in this region. Weighting methods work properly if the function is approximately linear. Difficulties arise in regions with lot of structure and at the boundaries of the region if there the function is not approximately constant.

11.2.1.1 *k-Nearest Neighbors*

The simplest method for a function approximation is similar to the density estimation which we treat in Chapter 9 and which uses the nearest neighbors in the training sample. We define a distance $d_i = |\boldsymbol{x} - \boldsymbol{x}_i|$ and sort the elements of the training sample in the order of their distances $d_i < d_{i+1}$. We choose a number K of nearest neighbors and average over the corresponding output vectors:

$$\hat{\boldsymbol{y}}(\boldsymbol{x}) = \frac{1}{K}\sum_{i=1}^{K} \boldsymbol{y}_i .$$

This relation holds for constant errors. Otherwise for the component j of $\boldsymbol{y}$ the corresponding weighted mean

$$\hat{y}_j(\boldsymbol{x}) = \frac{\sum_{i=1}^{K} y_{ij}/\delta_{ij}^2}{\sum_{i=1}^{K} 1/\delta_{ij}^2}$$

has to be used. The choice of K depends on the density of points in the training sample and the expected variation of the true function $\boldsymbol{y}(\boldsymbol{x})$.

If all individual points in the projection j have mean square errors δ_j^2, the error of the prediction δy_j is given by

$$(\delta y_j)^2 = \frac{\delta_j^2}{K} + \langle y_j(\boldsymbol{x}) - \hat{y}_j(\boldsymbol{x})\rangle^2 . \tag{11.1}$$

The first term is the statistical fluctuation of the mean value. The second term is the bias which is equal to the systematic shift squared, and which is usually difficult to evaluate. There is the usual trade-off between the two error components: with increasing K the statistical term decreases, but the bias increases by an amount depending on the size of the fluctuations of the true function within the averaging region.

11.2.1.2 *k-Nearest Neighbors with Linear Approximation*

The simple average suffers from the drawback that at the boundary of the variable space the measurements contributing to the average are distributed asymmetrically with respect to the point of interest $\boldsymbol{x}$. If, for instance, the function falls strongly toward the left-hand boundary of a one-dimensional space, averaging over points which are predominantly located at the right hand side of x leads to too large a result. (See also the example at the end of this section). This problem can be avoided by fitting a linear function through the K neighboring points instead of using the mean value of $\boldsymbol{y}$.

11.2.1.3 *Gaussian Kernels*

To take all k-nearest neighbors into account with the same weight independent of their distance to $\boldsymbol{x}$ is certainly not optimal. Furthermore, its output function is piecewise constant (or linear) and thus discontinuous. Better should be a weighting procedure, where the weights become smaller with increasing distances. An often used weighting or kernel function[1] is the Gaussian. The sum is now taken over all N training inputs:

$$\hat{\boldsymbol{y}}(\boldsymbol{x}) = \frac{\sum_{i=1}^{N} \boldsymbol{y}_i \mathrm{e}^{-\alpha|\boldsymbol{x}-\boldsymbol{x}_i|^2}}{\sum_{i=1}^{N} \mathrm{e}^{-\alpha|\boldsymbol{x}-\boldsymbol{x}_i|^2}} .$$

The constant α determines the range of the correlation. Therefore the width $s = 1/\sqrt{2\alpha}$ of the Gaussian has to be adjusted to the density of

[1] The denotation *kernel* will be justified later, when we introduce classification methods.

the points and to the curvature of the function. If computing time has to be economized, the sum may of course be truncated and restricted to the neighborhood of $\boldsymbol{x}$, for instance to the distance $2s$. According to (11.1) the mean squared error becomes[2]:

$$(\delta y_j)^2 = \delta_j^2 \frac{\sum e^{-2\alpha|\boldsymbol{x}-\boldsymbol{x}_i|^2}}{\left[\sum e^{-\alpha|\boldsymbol{x}-\boldsymbol{x}_i|^2}\right]^2} + \langle y_j(\boldsymbol{x}) - \hat{y}_j(\boldsymbol{x})\rangle^2 .$$

11.2.2 *Approximation by Orthogonal Functions*

Complete orthogonal function systems offer three attractive features: (i) The fitted function coefficients are uncorrelated, (ii) The function systems are complete and thus able to approximate any well behaved, i.e. square integrable, function, (iii) They are naturally ordered with increasing oscillation frequency.[3] The function system to be used depends on the specific problem, i.e. on the domain of the variable and the asymptotic behavior of the function. Since the standard orthogonal functions are well known to physicists, we will be very brief and omit all mathematical details, they can be looked-up in mathematical handbooks.

Complete normalized orthogonal function systems $\{u_i(x)\}$ defined on the finite or infinite interval $[a, b]$ fulfil the orthogonality and the completeness relations. To simplify the notation, we introduce the inner product (g, h)

$$(g, h) \equiv \int_a^b g^*(x)h(x)\mathrm{d}x$$

and have

$$(u_i, u_j) = \delta_{ij},$$

$$\sum_i u_i^*(x)u_i(x') = \delta(x - x').$$

For instance, the functions of the well known Fourier system for the interval $[a, b] = [-L/2, L/2]$ are $u_n(x) = \frac{1}{\sqrt{L}} \exp(\mathrm{i}2\pi nx/L)$.

Every square integrable function can be represented by the series

$$f(x) = \sum_{i=0}^{\infty} a_i u_i(x), \text{ with } a_i = (u_i, f)$$

---

[2]This relation has to be modified if not all errors are equal.

[3]We use the term *frequency* also for spatial dimensions.

Table 11.1 Characteristics of orthogonal polynomials.

Polynomial	Domain	Weight function
Legendre, $P_i(x)$	$[-1,+1]$	$w(x) = 1$
Hermite, $H_i(x)$	$(-\infty,+\infty)$	$w(x) = \exp(-x^2)$
Laguerre, $L_i(x)$	$[0,\infty)$	$w(x) = \exp(-x)$

in the sense that the squared difference converges to zero with increasing number of terms[4]:

$$\lim_{K\to\infty}\left[f(x) - \sum_{i=0}^{K} a_i u_i(x)\right]^2 = 0. \tag{11.2}$$

The coefficients a_i become small for large i, if $f(x)$ is smooth as compared to the $u_i(x)$, which oscillate faster and faster for increasing i. Truncation of the series therefore causes some smoothing of the function.

The approximation of measurements by orthogonal functions works quite well for very smooth data. When the measurements show strong short range variations, sharp peaks or valleys, then a large number of functions is required to describe the data. Neglecting individually insignificant contributions may lead to a poor approximation. Typically, their truncation may produce spurious oscillations ("ringing") in regions near to the peaks, where the true function is already flat.

For large data sets with equidistant points and equal errors the *Fast Fourier Transform*, FFT, plays an important role, especially for data smoothing and image processing. Besides the trigonometric functions, other orthogonal systems are useful, some of which are displayed in Table 11.1. The orthogonal functions are proportional to polynomials $p_i(x)$ of degree i multiplied by the square root of a weight function $w(x)$, $u_i(x) = p_i(x)\sqrt{w(x)}$. Specifying the domain $[a,b]$ and w, and requiring orthogonality for $u_{i,j}$,

$$(u_i, u_j) = c_i \delta_{ij},$$

fixes the polynomials up to the somewhat conventional normalization factors $\sqrt{c_i}$.

The most familiar orthogonal functions are the trigonometric functions used in the Fourier series mentioned above. From electrodynamics and

[4]At eventual discontinuities, $f(x)$ should be taken as $[f(x+0) + f(x-0)]/2$.

quantum mechanics we are also familiar with Legendre polynomials and spherical harmonics. These functions are useful for data depending on variables defined on the circle or on the sphere, e.g. angular distributions. For example, the distribution of the intensity of the microwave background radiation which contains information about the curvature of the space, the baryon density and the amount of dark matter in the universe, is usually described as a function of the solid angle by a superposition of spherical harmonics. In particle physics the angular distributions of scattered or produced particles can be described by Legendre polynomials or spherical harmonics. Functions extending to $\pm\infty$ are often approximated by the eigenfunctions of the harmonic oscillator consisting of Hermite polynomials multiplied by the exponential $\exp(-x^2/2)$ and functions bounded to $x \geq 0$ by Laguerre polynomials multiplied by $e^{-x/2}$.

In order to approximate a given measurement by one of the orthogonal function systems, one usually has to shift and scale the independent variable x.

11.2.2.1 *Polynomial Approximation*

The simplest function approximation is achieved with a simple polynomial $f(x) = \sum a_k x^k$ or more generally by $f(x) = \sum a_k u_k$ where u_k is a polynomial of order k. Given data y_ν with uncertainties δ_ν at locations x_ν we minimize

$$\chi^2 = \sum_{\nu=1}^{N} \frac{1}{\delta_\nu^2} \left[y_\nu - \sum_{k=0}^{K} a_k u_k(x_\nu) \right]^2 , \tag{11.3}$$

in order to determine the coefficients a_k. To constrain the coefficients, their number $K+1$ has to be smaller than the number N of measurements. All polynomial systems of the same order describe the data equally well but differ in the degree to which the coefficients are correlated. The power of the polynomial is increased until it is compatible within statistics with the data. The decision is based on a χ^2-criterion.

The purpose of this section is to show how we can select polynomials with uncorrelated coefficients. In principle, these polynomials and their coefficients can be computed through diagonalization of the error matrix but they can also be obtained directly with the Gram–Schmidt method. This method has the additional advantage that the polynomials and their coefficients are given by simple algebraic relations.

For a given sample of measured points $y_\nu = f(x_\nu)$ with errors δ_ν, we fix the weights in the usual way

$$w_\nu = w(x_\nu) = \frac{1}{\delta_\nu^2} / \sum_j \frac{1}{\delta_j^2},$$

and now define the inner product of two functions $g(x), h(x)$ by

$$(g, h) = \sum_\nu w_\nu g(x_\nu) h(x_\nu)$$

with the requirement

$$(u_i, u_j) = \delta_{ij}.$$

Minimizing χ^2 is equivalent to minimizing

$$X^2 = \sum_{\nu=1}^{N} w_\nu \left[y_\nu - \sum_{k=0}^{K} a_k u_k(x_\nu) \right]^2 .$$

For $K = N - 1$ the square bracket at the minimum of X^2 is zero,

$$y_\nu - \sum_{k=0}^{N-1} a_k u_k(x_\nu) = 0$$

for all ν, and forming the inner product with u_j we get

$$(y, u_j) = a_j \; . \tag{11.4}$$

This relation produces the coefficients also in the interesting case $K < N-1$.

To construct the orthogonal polynomials, we set $v_0 = 1$,

$$u_i = \frac{v_i}{\sqrt{(v_i, v_i)}}, \tag{11.5}$$

$$v_{i+1} = x^{i+1} - \sum_{j=0}^{i} (u_j, x^{i+1}) u_j. \tag{11.6}$$

The first two terms in the corresponding expansion, $a_0 u_0$ and $a_1 u_1$, are easily calculated. From (11.5), (11.6), (11.4) and the following definition of the moments of the weighted sample

$$\overline{x} = \sum_\nu w_\nu x_\nu, \; s_x^2 = \sum_\nu w_\nu (x_\nu - \overline{x})^2, \; s_{xy} = \sum_\nu w_\nu (x_\nu y_\nu - \overline{x}\overline{y})$$

we find the coefficients and functions which fix the polynomial expansion of y:

$$y = \overline{y} + \frac{s_{xy}}{s_x^2}(x - \overline{x}).$$

We recover the well known result for the best fit by a straight line in the form with independent coefficients: This is of course no surprise, as the functions that are minimized are identical, namely χ^2 in both cases, see Example 6.24 in Chapter 6. The calculation of higher order terms is straight forward but tedious. The uncertainties δa_i of the coefficients are all equal independent of i and given by the simple relation

$$(\delta a_i)^2 = 1/\sum_{\nu=1}^{N} \frac{1}{\delta_\nu^2}.$$

The derivation of this formula is given in the Appendix A.12 together with formulas for the polynomials in the special case where all measurements have equal errors and are uniformly distributed in x.

11.2.2.2 *The Gram–Charlier Series*

The following example for the application of Hermite functions, strictly speaking, does not concern the approximation of measurements by a function but the approximation of an empirical p.d.f. (see Section 12.1.1 in the following Chapter). We discuss it here since it is mathematically closely related to the subject of this section.

The Gram–Charlier series is used to approximate empirical distributions which do not differ very much from the normal distribution. It expresses the quotient of an empirical p.d.f. $f(x)$ to the standard normal distribution $\mathcal{N}(x|0,1)$ as an expansion in the slightly modified Hermite polynomials $\tilde{H}_i(x)$ in the form

$$f(x) = \mathcal{N}(x) \sum_{i=0}^{\infty} a_i \tilde{H}_i(x). \tag{11.7}$$

Here, $\mathcal{N}(x) \equiv (2\pi)^{-1/2} \exp(-x^2/2)$, the standard normal distribution, differs somewhat from the weight function $\exp(-x^2)$ used in the definition of the Hermite polynomials $H(x)$ given above in Table 11.1. The two definitions of the polynomials are related by

$$\tilde{H}_i(x) = \frac{1}{\sqrt{2^i}} H_i\left(\frac{x}{\sqrt{2}}\right).$$

The orthogonality relation of the modified polynomials is

$$(\tilde{H}_i, \tilde{H}_j) = \int_{-\infty}^{+\infty} \mathcal{N}(x) \tilde{H}_i(x) \tilde{H}_j(x) \mathrm{d}x = i!\, \delta_{ij}, \tag{11.8}$$

and their explicit form can be obtained by the simple recursion relation:

$$\tilde{H}_{i+1} = x\tilde{H}_i - i\tilde{H}_{i-1}.$$

With $\tilde{H}_0 = 1$, $\tilde{H}_1 = x$ we get

$$\begin{aligned}\tilde{H}_2 &= x^2 - 1, \\ \tilde{H}_3 &= x^3 - 3x, \\ \tilde{H}_4 &= x^4 - 6x^2 + 3,\end{aligned}$$

and so on.

When we multiply both sides of (11.7) with $\tilde{H}_j(x)$ and integrate, we find, according to (11.8), the coefficients a_i from

$$a_i = \frac{1}{i!}\int f(x)\tilde{H}_i(x)\mathrm{d}x.$$

These integrals can be expressed as combinations of moments of $f(x)$, which are to be approximated by the sample moments of the experimental distribution. First, the sample mean and the sample variance are used to shift and scale the experimental distribution such that the transformed mean and variance equal 0 and 1, respectively. Then $a_{1,2} = 0$, and the empirical skewness and excess of the normalized sample $\gamma_{1,2}$ as defined in Section 3.2 are proportional to the parameters $a_{3,4}$. The approximation to this order is

$$f(x) \approx \mathcal{N}(x)(1 + \frac{1}{3!}\gamma_1\tilde{H}_3(x) + \frac{1}{4!}\gamma_2\tilde{H}_4(x)).$$

As mentioned, this approximation is well suited to describe distributions which are close to normal distributions. This is realized, for instance, when the variate is a sum of independent variates such that the central limit theorem applies. It is advisable to check the convergence of the corresponding Gram–Charlier series and not to truncate the series too early [3].

11.2.3 *Wavelets*

The trigonometric functions used in the Fourier series are discrete in the frequency domain, but extend from minus infinity to plus infinity in the spatial domain and thus are not very well suited to describe strongly localized function variations. To handle this kind of problems, the wavelet system has been invented. Wavelets are able to describe pulse signals and spikes like those generated in electrocardiograms, nuclear magnetic resonance (NMR) records or seismic records, in data transmission, and for the coding of images and hand-written text. For data reduction and storage they have become an indispensable tool.

The simplest orthogonal system with the desired properties are the *Haar wavelets* shown in Fig. 11.1. The lowest row shows three wavelets which are

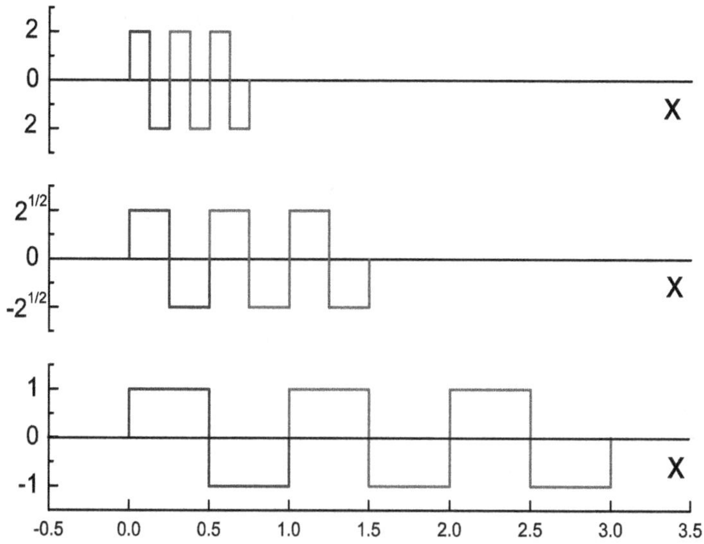

Fig. 11.1 Nine orthogonal, normalized wavelets with three different frequencies.

orthogonal, because they have no overlap. The next higher row contains again three wavelets with one half the length of those below. They are orthogonal to each other and to the wavelets in the lower row. In the same way the higher frequency wavelets in the following row are constructed. We label them with two indexes j, k indicating length and position. We define a *mother function* $\psi(x)$, the bottom left wavelet function of Fig. 11.1.

$$\psi(x) = \begin{cases} 1, & \text{if } 0 \leq x < \frac{1}{2} \\ -1, & \text{if } \frac{1}{2} \leq x < 1 \\ 0, & \text{else} \end{cases}$$

and set $W_{00} = \psi(x)$. The remaining wavelets are then obtained by translations and dilatations in discrete steps from the mother function $\psi(x)$:

$$W_{jk}(x) = 2^{j/2}\psi(2^j x - k).$$

The factor $2^{j/2}$ provides the normalization in the orthonormality relation.[5]

$$\int_{-\infty}^{+\infty} W_{ik}^*(x) W_{jl}(x) \mathrm{d}x = \delta_{ij}\delta_{kl}. \tag{11.9}$$

[5]The Haar wavelets are real, but some types of wavelets are complex.

It is evident that wavelets are much better suited to fit local structures than the sine and cosine functions of the Fourier expansion, since the wavelet expansion coefficients c_{jk} contain information on frequency *and* location of a signal.

The simple Haar wavelets shown in Fig. 11.1 which we have introduced to demonstrate the principal properties of wavelets are rarely used in applications as functions with infinitely sharp edges are usually absent in a realistic phenomenon. More common are the smoother wavelets

$$\psi(x) = \frac{1}{\sqrt{2\pi}\sigma^3} \mathrm{e}^{-x^2/(2\sigma^2)} (1 - \frac{x^2}{\sigma^2}) \text{ (Mexican Hat)}, \qquad (11.10)$$

$$\psi(x) = (\mathrm{e}^{\mathrm{i}x} - c)\mathrm{e}^{-x^2/(2\sigma^2)} \text{ (Morlet-Wavelet)}, \qquad (11.11)$$

and many others. The first function, the Mexican hat, is the second derivative of the Gaussian function, Fig. 11.2. The second, the Morlet function, is a complex monochromatic wave, modulated by a Gaussian. The constant $c = \exp(-\sigma^2/2)$ in the Morlet function can usually be neglected by choosing a wide lowest order function, $\sigma >\sim 5$. In both functions σ defines the width of the window.

The mother function ψ has to fulfill apart from the trivial normalization property 11.9, also the relation

$$\int \psi(x)\mathrm{d}x = 0 .$$

Any square integrable function $f(x)$ fulfilling $\int f(x)\mathrm{d}x = 0$ can be expanded in the discrete wavelet series,

$$f(x) = \sum_{j,k} c_{jk} W_{jk}(x) .$$

As usual, in order to regularize the function $f(x)$, the expansion is truncated when the coefficients become insignificant with increasing j, corresponding to small details or large frequencies.

The calculation of the coefficients c_{jk} is in principle analogous to the calculation of Fourier coefficients by convolution of f with the wavelets.[6] For given measurements a least square fit can be applied. The success of the wavelet applications was promoted by the appearance of fast numerical algorithms, like the multi-scale analysis. They work on a function f which need not integrate to zero, sampled at equidistant points, similarly to the fast Fourier transform (FFT).

An elementary introduction to the wavelet analysis is found in [100]. Programs are available in program libraries and in the internet.

[6]The wavelets (11.10), (11.11) are not orthogonal. Thus the coefficients are correlated.

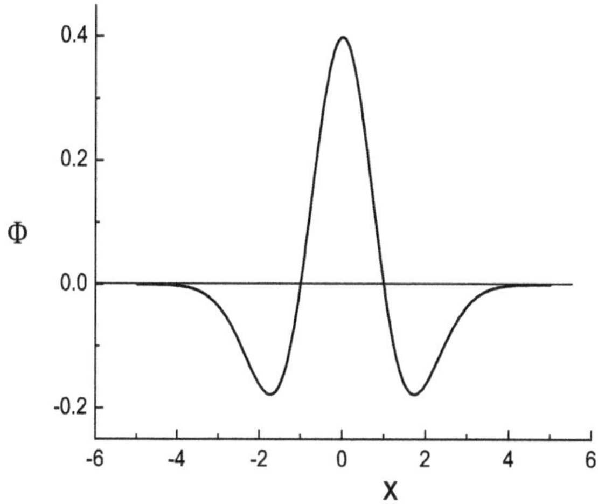

Fig. 11.2 Mexican hat wavelet.

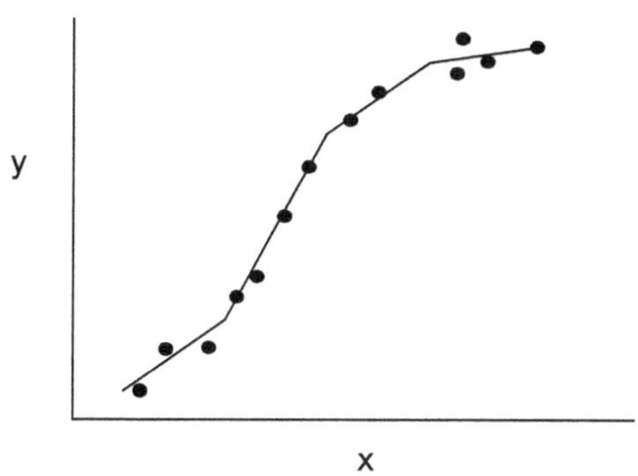

Fig. 11.3 Linear spline approximation.

11.2.4 *Spline Approximation*

The mathematical and numerical treatment of polynomials is especially simple and effective. Therefore, they are often chosen for the approximation of experimental data. A disadvantage of polynomials is however that they tend to infinity for large absolute values of the independent variable. This difficulty is resolved by using piecewise polynomial functions, the splines.

According to the degree of the polynomials used, we distinguish between linear, quadratic, cubic etc. splines.

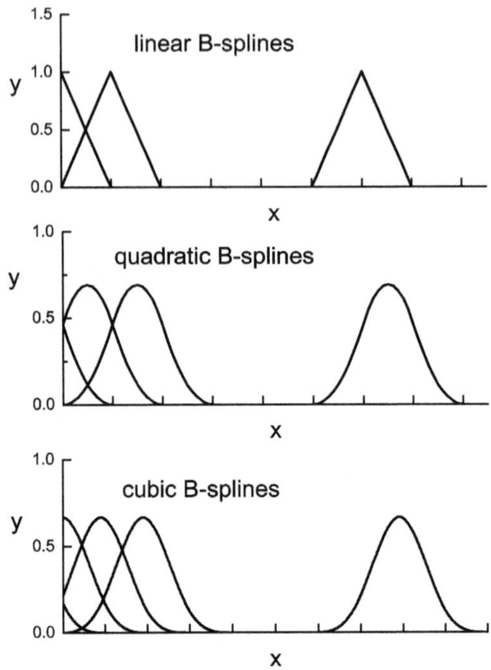

Fig. 11.4 Linear, quadratic and cubic B-splines.

The simplest spline approximation is the linear one, consisting of a polygon. The steps in the independent variable x between the knots are constant (Fig. 11.3). The lower the chosen number of knots and the spline order are, the larger will be on average the deviations of the points from the fitted curve. A sensible choice should take into account the mean squared dispersion of the points, i.e. the χ^2-sum should be of the order of the number of degrees of freedom. When the response values y are exact and equidistant, the points are simply connected by a polygon.

A smoother approximation with no kinks is obtained with quadratic splines. A curve with continuous derivatives up to order n is produced with splines of degree $\geq n + 1$. Since a curve with continuous second derivatives looks smooth to the human eye, splines of degree higher than cubic are rarely used.

Spline approximations are widely used in technical disciplines. They have also been successfully applied to the deconvolution problem [12, 67] (Chapter 9). Instead of adapting a histogram to the true distribution, the amplitudes of spline functions can be fitted. This has the advantage that we obtain a continuous function which incorporates the desired degree of regularization.

For the numerical computations the so called *B-splines* (basis splines) are especially useful. Linear, quadratic and cubic B-splines are shown in Fig. 11.4. The superposition of B-splines fulfills the continuity conditions at the knots. The superposition of the triangular linear B-splines produces a polygon, that of quadratic and cubic B-splines a curve with continuous slope and curvature, respectively.

A B-spline of given degree is determined by the step width b and the position x_0 of its center. Their explicit mathematical expressions are given in Appendix A.13.

The function is approximated by

$$\hat{f}(x) = \sum_{k=0}^{K} a_k B_k(x). \tag{11.12}$$

The amplitudes a_k can be obtained from a least squares fit. For values of the response function y_i and errors δy_i at the input points $x_i, i = 1, \ldots, N$, we minimize

$$\chi^2 = \sum_{i=1}^{N} \frac{\left[y_i - \sum_{k=0}^{K} a_k B_k(x_i)\right]^2}{(\delta y_i)^2}. \tag{11.13}$$

Of course, the number N of input values has to be at least equal to the number K of splines. Otherwise the number of degrees of freedom would become negative and the approximation under-determined.

11.2.4.1 *Spline Approximation in Higher Dimensions*

In principle, the spline approximation can be generalized to higher dimensions. However, there the difficulty is that a grid of intervals (knots) destroys the rotation symmetry. It is again advantageous to work with B-splines. Their definition becomes more complicated: In two dimensions we have instead of triangular functions pyramids and for quadratic splines also mixed terms $\propto x_1 x_2$ have to be taken into account. In higher dimensions the number of mixed terms explodes, another example of the curse of dimensionality.

11.2.5 *Approximation by a Combination of Simple Functions*

There is no general recipe for function approximation. An experienced scientist would try, first of all, to find functions which describe the asymptotic behavior and the rough distribution of the data, and then add further functions to describe the details. This approach is more tedious than using programs from libraries but will usually produce results superior to those of the general methods described above.

Besides polynomials, $a_0 + a_1x + a_2x^2 + \cdots$, rational functions can be used, i.e. quotients of two polynomials (Padé approximation), the exponential function $e^{\alpha x}$, the logarithm ${}^b\log x$, the Gaussian e^{-ax^2}, and combinations like $x^a e^{-bx}$. In many cases a simple polynomial will do. The results usually improve when the original data are transformed by translation and dilatation $x \to a(x+b)$ to a normalized form.

11.2.6 *Example*

In order to compare different methods, we have generated function values y_i according to the function xe^{-x} with superimposed Gaussian fluctuations at equidistant x-values. The measurements generated in this way are smoothed, respectively fitted by different functions. The results are shown in Fig. 10.7. All eight panels show the original function and the measured points connected by a polygon.

In the upper two panels smoothing has been performed by weighting. Typical for both methods are that structures are washed-out and that strong deviations occur at the borders. The Gaussian weighting in the left-hand panel performs much better than the method of nearest neighbors on the right-hand side which also shows spurious short range fluctuations which are typical for this method.

As expected, also the linear spline approximation is not satisfactory but the edges are reproduced better than with the weighting methods. Both quadratic and cubic splines with 10 free parameters describe the measurement points adequately, but the cubic splines show some unwanted oscillations. The structure of the spline intervals is clearly seen. Reducing the number of free parameters to 5 suppresses the spurious fluctuations but then the spline functions cannot follow any more the steep rise at small x. There is only a marginal difference between the quadratic and the cubic spline approximations.

The approximation by a simple polynomial of fourth order, i.e. with only 5 free parameters, works excellently. By the way, it differs substantially from the Taylor expansion of the true function. The polynomial can adapt itself much better to regions of different curvature than the splines with their fixed step width.

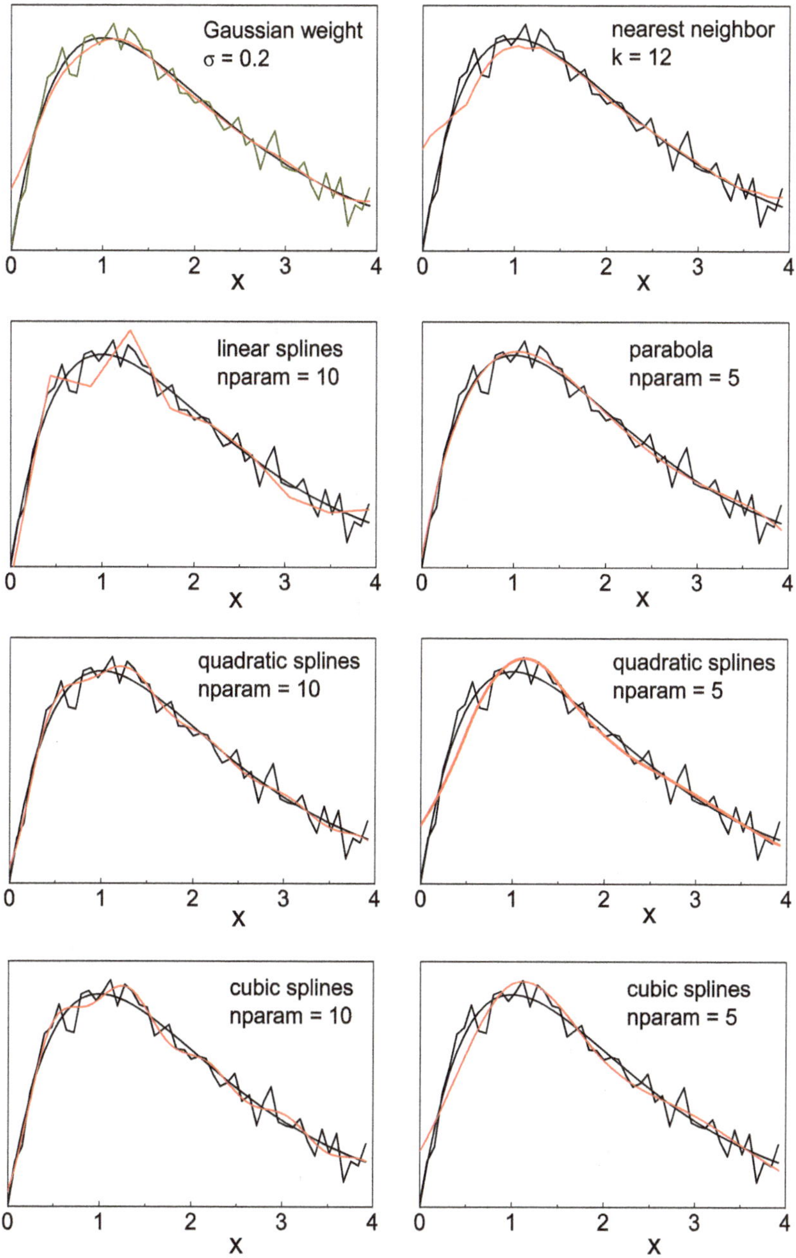

Fig. 11.5 Smoothing and function approximation. The measurements are connected by a polygon. The original function (black) is compared to the smoothed function (red).

To summarize: The physicist will usually prefer to construct a clever parametrization with simple analytic functions to describe his data and avoid the more general standard methods available in program libraries. Those are useful to get a fast overview and for the parametrization of a large amount of data.

As we have already mentioned, the approximation of measurements by the standard set of orthogonal functions works quite well for very smooth functions where sharp peaks and valleys are absent. Peaks and bumps are better described with wavelets than with the conventional orthogonal functions. Smoothing results of measurements with the primitive kernel methods which we have discussed are usually unsatisfactory. A better performance is obtained with kernels with variable width and corrections for a possible boundary bias. The reader is referred to the literature [101]. Spline approximations are useful when the user has no idea about the shape of the function and when the measurements are able to constrain the function sufficiently to suppress fake oscillations.

11.3 Linear Factor Analysis and Principal Components

Factor analysis and *principal component analysis* (PCA) both reduce a multi-dimensional variate space to lower dimensions. In the literature there is no clear distinction between the two techniques.

Often several features of an object are correlated or redundant, and we want to express them by a few uncorrelated components with the hope to gain deeper insight into latent relations. One would like to reduce the number of features to as low a number of components, called factors, as possible.

Let us imagine that for 20 cuboids we have determined 6 geometrical and physical quantities: volume, surface, basis area, sum of edge lengths, mass, and principal moments of inertia. We submit these data which may be represented by a 6-dimensional vector to a colleague without further information. He will look for similarities and correlations, and he might guess that these data can be derived for each cuboid from only 4 parameters, namely length, width, height, and density. The search for these basis parameters, the components or factors is called *factor analysis* [102, 103].

A general solution for this problem cannot be given, without an ansatz for the functional relationship between the feature matrix, in our example build from the 20 six-dimensional data vectors and its errors. Our example indicates though, in which direction we might search for a solution of this problem. Each body is represented by a point in a six-dimensional feature space. The points are, however, restricted to a four-dimensional subspace, the component space. The problem is to find this subspace. This is relatively simple if it is linear.

In general, and in our example, the subspace is not linear, but a linear approximation might be justified if the cuboids are very similar such that the components depend approximately linearly on the deviations of the input vectors from a center of gravity. Certainly in the general situation it is reasonable to look first for a linear relationship between features and parameters. Then the subspace is a linear vector space and easy to identify. In the special situation where only one component exists, all points lie approximately on a straight line, deviations being due to measurement errors and non-linearity. To identify the multi-dimensional plane, we have to investigate the correlation matrix. Its transformation into diagonal form delivers the *principal components* – linear combinations of the feature vectors in the direction of the principal axes. The principal components are the eigenvectors of the correlation matrix ordered according to decreasing eigenvalues. When we ignore the principal components with small eigenvalues, the remaining components form the planar subspace.

Factor analysis or PCA has been developed in psychology, but it is widely used also in other descriptive fields, and there are numerous applications in chemistry and biology. Its moderate computing requirements which are at the expense of the restriction to linear relations, are certainly one of the historical reasons for its popularity. We sketch it below, because it is still in use, and because it helps to get a quick idea of hidden structures in multi-dimensional data. When no dominant components are found, it may help to disprove expected relations between different observations.

A typical application is the search for factors explaining similar properties between different objects: Different chemical compounds may act similarly, e.g. decrease the surface tension of water. The compounds may differ in various features, as molecular size and weight, electrical dipole moment, and others. We want to know which parameter or combination of parameters is relevant for the interesting property. Another application is the search for decisive factors for a similar curing effect of different drugs. The knowledge of the principal factors helps to find new drugs with the same positive effect.

In physics factor analysis does not play a central role, mainly because its results are often difficult to interpret and, as we will see below, not unambiguous. It is not easy, therefore, to find examples from our discipline.

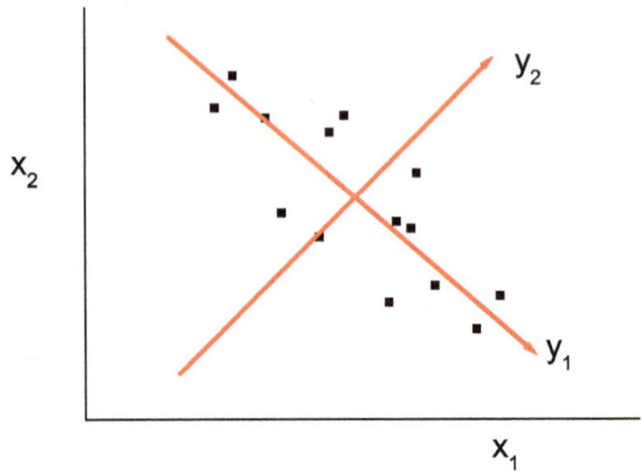

Fig. 11.6 Scatter diagram of two attributes of 11 measured objects.

Here we illustrate the method with an artificially constructed example taken from astronomy.

Example 11.5. Principal component analysis

Galaxies show the well known red-shift of the spectrum which is due to their escape velocity. Besides the measurement value or feature *red-shift* x_1 we know the *brightness* x_2 of the galaxies. To be independent of scales and mean values, we transform these quantities in such a way that sample mean and variance are zero, respectively unity. To demonstrate the concept, we have invented some date which are shown in Fig. 11.6. The two coordinates are strongly correlated. The correlation is eliminated in a rotated coordinate system where the objects have coordinates y_1 and y_2 which are linear combinations of red-shift and brightness in the directions of the principal axes of the correlation matrix. Now we consider as important those directions, where the observed objects show the largest differences. In our case this is the direction of y_1, while y_2 has apparently only a minor influence on both features. We may conclude that red-shift and brightness have mainly one and the same cause which determines the value of y_1. In our example, we know that this is the distance, both brightness and red shift depend on it. Since, apparently, the distance determines y_1, we can use it, after a suitable calibration, as a measure for the distance.

We will now put these ideas into concrete terms.

The input data for the factor analysis are given in the form of a matrix X of N rows and P columns. The element x_{np} is the measured value of the feature p of the object n, thus X is a rectangular matrix. In a first step we determine the correlations between the P input attributes. By a simple transformation, we obtain uncorrelated linear combinations of the features. The hope is that there are few dominant combinations and that the others can be neglected. Then the data can be described by a small number of $Q < P$ linear combinations, the principal components.

We first transform the data X_{np} into standardized form where the sample mean and variance are zero, respectively unity. We get the normalized variables[7]

$$x_{np} = \frac{X_{np} - \overline{X}_p}{\delta_p}$$

with

$$\overline{X}_p = \frac{1}{N}\sum_{n=1}^{N} X_{np},$$

$$\delta_p^2 = \frac{1}{N-1}\sum_{n=1}^{N}(X_{np} - \overline{X}_p)^2.$$

The quantity x_{np} is the normalized deviation of the measurement value of type p for the object n from the average over all objects for this measurement.

In the same way as in Chapter 4 we construct the correlation matrix for our sample by averaging the $P \times P$ products of the components $x_{n1} \dots x_{nP}$ over all N objects:

$$\mathsf{R} = \frac{1}{N-1}\mathsf{X}^T\mathsf{X},$$

$$R_{pq} = \frac{1}{N-1}\sum_{n=1}^{N} x_{np}x_{nq}.$$

It is a symmetric positive definite $P \times P$ matrix. Due to the normalization the diagonal elements are equal to unity.

Then this matrix is brought into diagonal form by an orthogonal transformation corresponding to a rotation in the P-dimensional feature space.

$$\mathsf{R} \to \mathsf{V}^T\mathsf{R}\mathsf{V} = \mathrm{diag}(\lambda_1 \dots \lambda_P).$$

[7]The normalization (division by δ_p) is not always required.

The uncorrelated feature vectors in the rotated space $\boldsymbol{y}_n = \{y_{n1}, \ldots, y_{nP}\}$ are given by

$$\boldsymbol{y}_n = \mathsf{V}^T \boldsymbol{x}_n, \quad \boldsymbol{x}_n = \mathsf{V}\boldsymbol{y}_n .$$

To obtain eigenvalues and -vectors we solve the linear equation system

$$(\mathsf{R} - \lambda_p \mathsf{I})\boldsymbol{v}_p = 0, \qquad (11.14)$$

where λ_p is the eigenvalue belonging to the eigenvector $\boldsymbol{v}_p$:

$$\mathsf{R}\boldsymbol{v}_p = \lambda_p \boldsymbol{v}_p .$$

The P eigenvalues are found as the solutions of the characteristic equation

$$\det(\mathsf{R} - \lambda \mathsf{I}) = 0 .$$

In the simple case described above of only two features, this is a quadratic equation

$$\begin{vmatrix} R_{11} - \lambda & R_{12} \\ R_{21} & R_{22} - \lambda \end{vmatrix} = 0 ,$$

that fixes the two eigenvalues. The eigenvectors are calculated from (11.14) after substituting the respective eigenvalue. As they are fixed only up to an arbitrary factor, they are usually normalized. The rotation matrix V is constructed by taking the eigenvectors $\boldsymbol{v}_p$ as its columns: $v_{qp} = (\boldsymbol{v}_p)_q$.

Since the eigenvalues are the diagonal elements in the rotated, diagonal correlation matrix, they correspond to the variances of the data distribution with respect to the principal axes. A small eigenvalue means that the projection of the data on this axis has a narrow distribution. The respective component is then, presumably, only of small influence on the data, and may perhaps be ignored in a model of the data. Large eigenvalues belong to the important principal components.

Factors f_{np} are obtained by standardization of the transformed variables y_{np} by division by the square root of the eigenvalues λ_p:

$$f_{np} = \frac{y_{np}}{\sqrt{\lambda_p}} .$$

By construction, these factors represent variates with zero mean and unit variance. In most cases they are assumed to be normally distributed. Their relation to the original data x_{np} is given by a linear (not orthogonal) transformation with a matrix A, the elements of which are called factor *loadings*. Its definition is

$$\boldsymbol{x}_n = \mathsf{A}\boldsymbol{f}_n, \text{ or } \mathsf{X}^T = \mathsf{A}\,\mathsf{F}^T . \qquad (11.15)$$

Its components show, how strongly the input data are influenced by certain factors.

In the classical factor analysis, the idea is to reduce the number of factors such that the description of the data is still satisfactory within tolerable deviations ε:

$$\begin{aligned} x_1 &= a_{11} f_1 + \cdots + a_{1Q} f_Q + \varepsilon_1 \\ x_2 &= a_{21} f_1 + \cdots + a_{2Q} f_Q + \varepsilon_2 \\ \vdots & \quad \vdots \\ x_P &= a_{P1} f_1 + \cdots + a_{PQ} f_Q + \varepsilon_P \end{aligned}$$

with $Q < P$, where the "factors" (latent random variables) $f_1, \ldots, f_Q$ are considered as uncorrelated and distributed according to $N(0,1)$, plus uncorrelated zero-mean Gaussian variables ε_p, with variances σ_p^2, representing the residual statistical fluctuations not described by the linear combinations. As a first guess, Q is taken as the index of the smallest eigenvalue λ_Q which is considered to be still significant. In the ideal case $Q = 1$ only one decisive factor would be the dominant element able to describe the data.

Generally, the aim is to estimate the loadings a_{pq}, the eigenvalues λ_p, and the variances σ_p^2 from the sampling data, in order to reduce the number of relevant quantities responsible for their description.

The same results that have been obtained above by the traditional method by solving the eigenvalue problem for the correlation matrix[8] can be obtained directly by using the *singular value decomposition* (SVD) of the matrix X (remember that it has N rows and P columns):

$$\mathsf{X} = \mathsf{U}\,\mathsf{D}\,\mathsf{V}^T,$$

where U and V are orthogonal. U is not a square matrix, nevertheless $\mathsf{U}^T\mathsf{U} = \mathsf{I}$, where the unit matrix I has dimension P. D is a diagonal matrix with elements $\sqrt{\lambda_p}$, ordered according to decreasing values, and called here singular values. The decomposition (11.15) is obtained by setting $\mathsf{F} = \mathsf{U}$ and $\mathsf{A} = \mathsf{V}\,\mathsf{D}$.

The decomposition (11.15) is not unique: If we multiply both F and A with a rotation matrix R from the right we get an equivalent decomposition:

$$\mathsf{X} = \tilde{\mathsf{F}}\,\tilde{\mathsf{A}}^T = \mathsf{F}\,\mathsf{R}(\mathsf{A}\,\mathsf{R})^T = \mathsf{F}\,\mathsf{R}\,\mathsf{R}^T\mathsf{A}^T = \mathsf{U}\,\mathsf{D}\,\mathsf{V}^T, \qquad (11.16)$$

which is the same as (11.15), with factors and loadings being rotated.

There exist program packages which perform the numerical calculation of principal components and factors.

[8] Physicists may find the method familiar from the discussion of the inertial momentum tensor and many similar problems.

11.4 Classification

We have come across classification of events already when we have treated goodness-of-fit. There the problem was either to accept or to reject a hypothesis without a clear alternative. Now we consider a situation where we have to associate an observation to one out of several distinct event classes.

The assignment of an object according to some quality, to a class or category is described by a so-called categorical variable. In most cases it makes sense, to give as a result instead of a discrete classification a continuous variable as a measure for the correctness of the classification.

Classification is indispensable in data analysis in many fields. Examples in particle physics are the identification of particles from shower profiles or from Cerenkov ring images, beauty, top or Higgs particles from kinematics and secondaries and the separation of rare interactions from frequent ones. In astronomy the classification of galaxies and other stellar objects is of interest. But classification is also a precondition for decisions in many scientific fields and in every day life.

Classification of an event is simple, if the distributions of the test variables are known for the different classes. In this case the *discriminant analysis* is applied.

More involved is the situation where the class of an event has to be decided from the similarity to other events whose class is known. The most important methods for this kind of problems are the *artificial neural nets*, *kernel* or *weighting* methods, and *decision trees*. In the last years, remarkable progress in these fields could be realized with the development of *support vector machines,*[9] *boosted decision trees*, and *random forests* classifiers.

We start with two examples:

A patient suffers from certain symptoms: stomach-ache, diarrhoea, temperature, head-ache. The doctor has to give a diagnosis. He will consider further factors, as age, sex, earlier diseases, blood values possibility of infection, duration of the illness, etc.. The diagnosis is based on the experience and education of the doctor. A computer program that could support the doctor in this matter should be able to learn from past cases and to compare new inputs in a sensible way with the stored data.

[9]The classifying computer programs are called machines. The adjustment of their parameters is called *machine learning*.

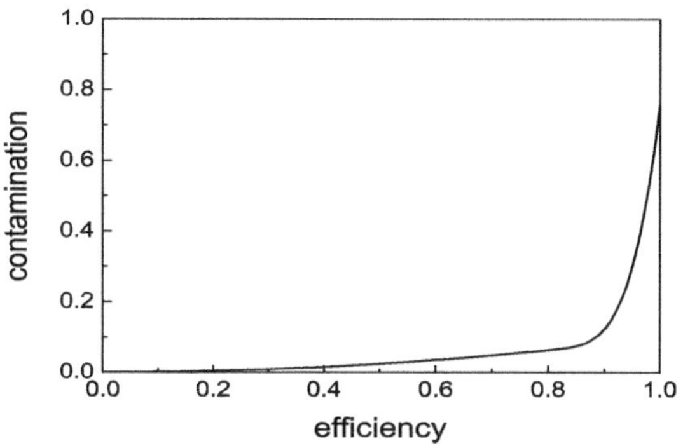

Fig. 11.7 Fraction of wrongly assigned events as a function of the efficiency.

The interactions of electrons and hadrons in calorimeter detectors of particle physics differ in a many parameters. Calorimeters consist of a large number of detector elements, for which the signal heights are evaluated and recorded. The system should learn from test measurements with known particle beams to classify electrons and hadrons.

The concrete realization of the goal to classify new data proceeds in several steps. First a training sample has to be established. Each element consists of a set of input variables and a classification output. The data may have been obtained in the past from documented cases, from previous experiments or from a Monte Carlo simulation. Parameters of a computer program (the machine) are adjusted to reproduce the results of a fraction of the training sample. The remaining fraction is used to validate the adjusted machine. In this way it is avoided that the parameters of the machine are specific for a sample that is possibly not representative.

When we optimize a given method, it is not only the percentage of right decisions which is of interest, but we will also consider the consequences of the various kinds of errors. In statistics this is taken into account by a *loss function* which has to be defined by the user. In the standard situation where we want to select a certain class of events, we have to consider *efficiency* and *contamination.*[10] The larger the efficiency, the larger is also the relative contamination by wrongly assigned events. A typical curve showing this relation if plotted in Fig. 11.7. The user will select the value of his cut variable on the bases of this curve.

[10]One minus the contamination is called purity.

11.4.1 *The Discriminant Analysis*

The classical discriminant analysis as developed by Fisher is a special case of the classification method that we introduce in the following. We follow our discussions of Chapter 6, Section 6.3.

If we know the p.d.f.s $f_1(\boldsymbol{x})$ and $f_2(\boldsymbol{x})$ for two classes of events it is easy to assign an observation $\boldsymbol{x}$ to one of the two classes in such a way that the error rate is minimal (case 1):

$$\begin{aligned} \boldsymbol{x} &\rightarrow \text{ class 1, if } f_1(\boldsymbol{x}) > f_2(\boldsymbol{x}), \\ \boldsymbol{x} &\rightarrow \text{ class 2, if } f_1(\boldsymbol{x}) < f_2(\boldsymbol{x}). \end{aligned}$$

Normally we will get a different number of wrong assignments for the two classes: observations originating from the broader distribution will be miss-assigned more often, see Fig. 11.8) than those of the narrower distribution. In most cases it will matter whether an input from class 1 or from class 2 is wrongly assigned. An optimal classification is then reached using an appropriately adjusted likelihood ratio:

$$\begin{aligned} \boldsymbol{x} &\rightarrow \text{ class 1, if } f_1(\boldsymbol{x})/f_2(\boldsymbol{x}) > c, \\ \boldsymbol{x} &\rightarrow \text{ class 2, if } f_1(\boldsymbol{x})/f_2(\boldsymbol{x}) < c. \end{aligned}$$

If we want to have the same error rates (case 2), we must choose the constant c such that the integrals over the densities in the selected regions are equal:

$$\int_{f_1/f_2>c} f_1(\boldsymbol{x})\mathrm{d}\boldsymbol{x} = \int_{f_1/f_2<c} f_2(\boldsymbol{x})\mathrm{d}\boldsymbol{x}. \tag{11.17}$$

This assignment has again a minimal error rate, but now under the constraint (11.17). We illustrate the two possibilities in Fig. 11.8 for univariate functions.

For normal distributions we can formulate the condition for the classification explicitly: For case 2 we choose that class for which the observation $\boldsymbol{x}$ has the smallest distance to the mean measured in standard deviations. This condition can then be written as a function of the exponents. With the usual notations we get

$$\begin{aligned} (\boldsymbol{x}-\boldsymbol{\mu}_1)^T\mathsf{V}_1(\boldsymbol{x}-\boldsymbol{\mu}_1) - (\boldsymbol{x}-\boldsymbol{\mu}_2)^T\mathsf{V}_2(\boldsymbol{x}-\boldsymbol{\mu}_2) < 0 &\rightarrow \text{ class 1}, \\ (\boldsymbol{x}-\boldsymbol{\mu}_1)^T\mathsf{V}_1(\boldsymbol{x}-\boldsymbol{\mu}_1) - (\boldsymbol{x}-\boldsymbol{\mu}_2)^T\mathsf{V}_2(\boldsymbol{x}-\boldsymbol{\mu}_2) > 0 &\rightarrow \text{ class 2}. \end{aligned}$$

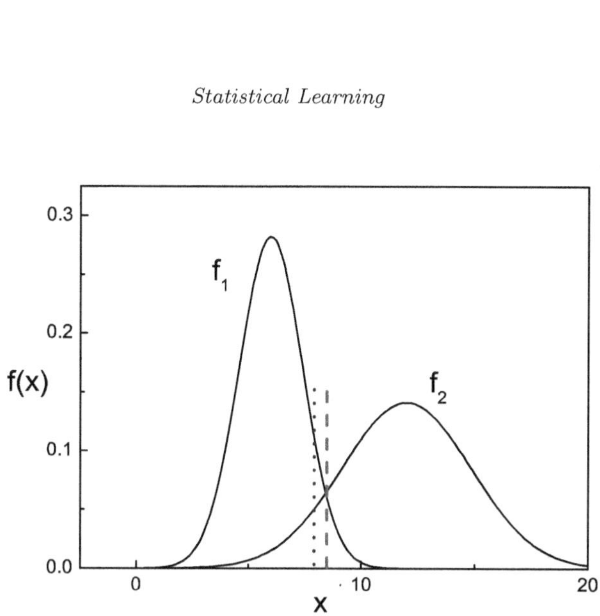

Fig. 11.8 Separation of two classes. The dashed line separates the events such that the error rate is minimal, the dotted line such that the wrongly assigned events are the same in both classes.

This condition can easily be generalized to more than two classes; the assignment according to the standardized distances will then, however, no longer lead to equal error rates for all classes.

The classical discriminant analysis sets $\mathsf{V}_1 = \mathsf{V}_2$. The left-hand side in the above relations becomes a linear combination of the x_p. The quadratic terms cancel. Equating it to zero defines a hyperplane which separates the two classes. The sign of this linear combination thus determines the class membership. Note that the separating hyperplane is cutting the line connecting the distribution centers under a right angle only for spherical symmetric distributions.

If the distributions are only known empirically from representative samples, we approximate them by continuous distributions, usually by a normal distribution, and fix their parameters to reproduce the empirical moments. In situations where the empirical distributions strongly overlap, for instance when a narrow distribution is located at the center of a broad one, the simple discriminant analysis does no longer work. The classification methods introduced in the following sections have been developed for this and other more complicated situations and where the only source of information on the population is a training sample. The various approaches are all based on the continuity assumption that observations with similar attributes lead to similar outputs.

11.4.2 *Artificial Neural Networks*

11.4.2.1 *Introduction*

The application of *artificial neural networks*, ANN, has seen a remarkable boom in the last decades, parallel to the exploding computing capacities. From its many variants, in science the most popular are the relatively simple forward ANNs with back-propagation, to which we will restrict our discussion. The interested reader should consult the broad specialized literature on this subject, where fascinating self organizing networks are described which certainly will play a role also in science, in the more distant future. It could e.g. be imagined that a self-organizing ANN would be able to classify a data set of events produced at an accelerator without human intervention and thus would be able to discover new reactions and particles. A comprehensive introduction to the field of ANN is given in [104].

ANNs are used in many fields for a broad variety of problems. Examples are pattern recognition, e.g. for hand-written letters or figures, or the forecast of stock prices. They are successful in situations where the relations between many parameters are too complex for an analytic treatment. In particle physics they have been used, among other applications, to distinguish electron from hadron cascades and to identify reactions with heavy quarks.

A forward ANNs with back-propagation approximates an algebraic function which transforms the input vector $\boldsymbol{x}$ into the response vector $\boldsymbol{y}$,

$$\boldsymbol{y} = f(\boldsymbol{x}|w) .$$

Here w symbolizes a large set of parameters, depending on the application, their number can vary between about ten to many thousands. In a training procedure the parameters are adjusted to produce the correct answer. What is correct has to be learned from a data sample of pairs $\boldsymbol{x}'$, $\boldsymbol{y}'$, where both the input and the answer are known. For each pair the result $\boldsymbol{y} = f(\boldsymbol{x}'|w)$ is compared to the target result $\boldsymbol{y}'$ and the parameters of the network are updated in such a way that the agreement is improved. We talk about supervised learning.

With the large number of parameters it is evident that the solution is not always unique. Networks with different parameters can perform the same function within the desired accuracy.

For the fitting of the large number of parameters minimizing programs like simplex (see Appendix A.10) are not suited. The gradient descent method is applied.

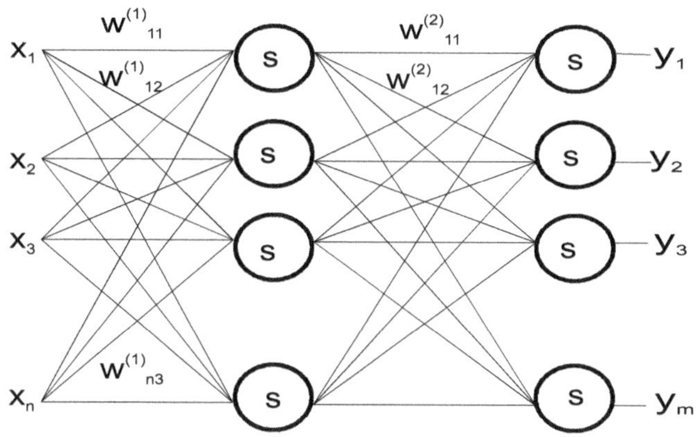

Fig. 11.9 Backpropagation. At each knot the sigmoid function of the sum of the weighted inputs is computed.

11.4.2.2 *Network Structure*

We consider a network consisting of two layers of knots (neurons), see Fig. 11.9. Each component x_k of the n-component input vector $\boldsymbol{x}$ is transmitted to all knots, labeled $i = 1, \ldots, m$, of the first layer. Each individual data line $k \to i$ is ascribed a weight $W_{ik}^{(1)}$. In each unit the weighted sum $u_i = \sum_k W_{ik}^{(1)} x_k$ of the data components connected to it is calculated. Each knot symbolizes a non-linear so-called activation function $x_i' = s(u_i)$, which is identical for all units. The first layer produces a new data vector $\boldsymbol{x}'$. The second layer, with m' knots, acts analogously on the outputs of the first one. We call the corresponding $m \times m'$ weight matrix $\mathsf{W}^{(2)}$. It produces the output vector $\boldsymbol{y}$. The first layer is called hidden layer, since its output is not observed directly.

Early studies indicated that a second hidden layers would not improve appreciably the performance of the net. In more recent investigations with increased computing power it has been shown that adding further layers can increase the performance.

The ANN with two layers executes the following functions:

$$x_j' = s\left(\sum_k W_{jk}^{(1)} x_k\right),$$

$$y_i = s\left(\sum_j W_{ij}^{(2)} x_j'\right).$$

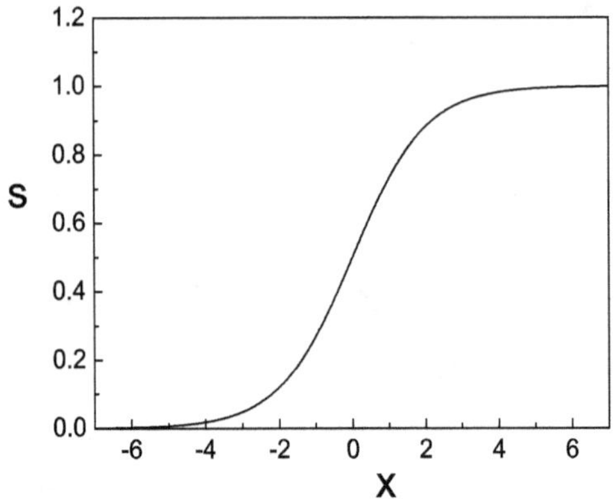

Fig. 11.10 Sigmoid function.

This leads to the final result:

$$y_i = s\left\{\sum_j W_{ij}^{(2)} s\left(\sum_k W_{jk}^{(1)} x_k\right)\right\}. \tag{11.18}$$

Sometimes it is appropriate to shift the input of each unit in the first layer by a constant amount (bias). This is easily realized by specifying an artificial additional input component $x_0 \equiv 1$.

The number of weights (the parameters to be fitted) is, when we include the component x_0, $(n+1) \times m + mm'$.

11.4.2.3 *The Activation Function*

The activation function $s(x)$ has to be non-linear, in order to achieve that the superposition (11.18) is able to approximate widely arbitrary functions. It is plausible that it should be more sensitive to variations of the arguments near zero than for very large absolute values. The input bias x_0 helps to shift input parameters which have a large mean value into the sensitive region. The activation function is usually standardized to vary between zero and one. The most popular activation function is the sigmoid function

$$s(u) = \frac{1}{\mathrm{e}^{-u} + 1},$$

which is similar to the Fermi function. It is shown in Fig. 11.10.

11.4.2.4 *The Training Process*

In the training phase the weights are adapted after each new input object. The output vector of the network $\boldsymbol{y}$ is compared with the target vector $\boldsymbol{y}_t$. The loss function E:

$$E = (\boldsymbol{y} - \boldsymbol{y}_t)^2, \tag{11.19}$$

measures for each training object the deviation of the response from the expected one.

To reduce the error E we walk downwards in the weight space. This means, we change each weight component by ΔW, proportional to the sensitivity $\partial E/\partial W$ of E to changes of W:

$$\begin{aligned}\Delta W &= -\frac{1}{2}\alpha\frac{\partial E}{\partial W}\\ &= -\alpha(\boldsymbol{y} - \boldsymbol{y}_t)\cdot\frac{\partial \boldsymbol{y}}{\partial W}.\end{aligned}$$

The proportionality constant α, the learning rate, determines the step width.

We now have to find the derivatives. Let us start with s:

$$\frac{\mathrm{d}s}{\mathrm{d}u} = \frac{\mathrm{e}^{-u}}{(\mathrm{e}^{-u}+1)^2} = s(1-s). \tag{11.20}$$

From (11.18) and (11.20) we compute the derivatives with respect to the weight components of the first and the second layer,

$$\frac{\partial y_i}{\partial W^{(2)}_{ij}} = y_i(1-y_i)x'_j,$$

and

$$\frac{\partial y_i}{\partial W^{(1)}_{jk}} = y_i(1-y_i)x'_j W^{(2)}_{ij}(1-x'_j)x_k.$$

It is seen that the derivatives depend on the same quantities which have already been calculated for the determination of $\boldsymbol{y}$ (the forward run through the net). Now we run backwards, change first the matrix $\mathsf{W}^{(2)}$ and then with already computed quantities also $\mathsf{W}^{(1)}$. This is the reason why this process is called *back propagation.* The weights are changed in the following way:

$$\begin{aligned}W^{(1)}_{jk} &\to W^{(1)}_{jk} - \alpha(\boldsymbol{y} - \boldsymbol{y}_t)\sum_i y_i(1-y_i)x'_j W^{(2)}_{ij}(1-x'_j)x_k,\\ W^{(2)}_{ij} &\to W^{(2)}_{ij} - \alpha(\boldsymbol{y} - \boldsymbol{y}_t)y_i(1-y_i)x'_j.\end{aligned}$$

11.4.2.5 *Test and Validation*

The gradient descending minimum search has not necessarily reached the minimum after processing the training sample a single time, especially when the available sample is small. Then the sample should be used several times (e.g. 10 or 100 times). On the other hand it may happen for too small a training sample that the network performs correctly for the training data, but produces wrong results for new data. The network has, so to say, learned the training data by heart. Similar to other minimizing concepts, the network interpolates and extrapolates the training data. When the number of fitted parameters (here the weights) become of the same order as the number of constraints from the training data, the network will occasionally, after sufficient training time, describe the training data exactly but fail for new input data. This effect is called *over-fitting* and is common to all fitting schemes when too many parameters are adjusted.

It is therefore indispensable to validate the network function after the optimization, with data not used in the training phase or to perform a cross validation. If in the training phase simulated data are used, it is easy to generate new data for testing. If only experimental data are available with no possibility to enlarge the sample size, usually a certain fraction of the data is reserved for testing. If the validation result is not satisfactory, one should try to solve the problem by reducing the number of network parameters or the number of repetitions of the training runs with the same data set.

The neural network generates from the input data the response through the fairly complicated function (11.18). It is impossible by an internal analysis of this function to gain some understanding of the relation between input and resulting response. Nevertheless, it is not necessary to regard the ANN as a "black box". We have the possibility to display graphically correlations between input quantities and the result, and all functional relations. In this way we gain some insight into possible connections. If, for instance, a physicist would have the idea to train a network with an experimental data sample to predict for a certain gas the volume as a function of the pressure and the temperature, he would be able to reproduce, with a certain accuracy, the results of the van-der-Waals equation. He could display the relations between the three quantities graphically. Of course the analytic form of the equation and its interpretation cannot be delivered by the network.

Often a study of the optimized weights makes it possible to simplify the net. Very small weights can be set to zero, i.e. the corresponding connections between knots are cut. We can check whether switching off certain neurons has a sizable influence on the response. If this is not the case, these neurons can be eliminated. Of course, the modified network has to be trained again.

11.4.2.6 *Practical Hints for the Application*

Computer Programs for ANNs with back-propagation are relatively simple and available at many places but the effort to write an ANN program is also not very large. The number of input vector components n and the number of knots m and m' are parameters to be chosen by the user, thus the program is universal, only the loss function has to be adapted to the specific problem.

- The number of units in each layer should more or less match the number of input components. Some experts plead for a higher number. The user should try to find the optimal number.
- The sigmoid function has values only between zero and unity. Therefore the output or the target value has to be appropriately scaled by the user.
- The raw input components are usually correlated. The network is more efficient if the user orthogonalizes them. Then often some of the new components have negligible effect on the output and can be discarded.
- The weights have to be initialized at the beginning of the training phase. This can be done by a random number generator or they can be set to fixed values.
- The loss function E (11.19) has be adjusted to the problem to be solved.
- The learning rate α should be chosen relatively high at the beginning of a training phase, e.g. $\alpha = 10$. In the course of fitting it should be reduced to avoid oscillations.
- The convergence of minimizing process is slow if the gradient is small. If this is the case, and the fit is still bad, it is recommended to increase the learning constant for a certain number of iterations.
- Networks often have several local minima. It is recommended to train the network with different start values of the weights.

- Other possibilities for the improvement of the convergence and the elimination of local minima can be found in the substantial literature. An ANN program package that proceeds automatically along many of the proposed steps is described in [105].

11.4.2.7 *Example: Čerenkov circles*

Charged, relativistic particles traversing a thin layer of a material can emit photons due to the Čerenkov effect. Under special experimental conditions the photons are detected as points that are located on a circle. Of interest are radius and center of this circle, since they provide information on direction and velocity of the emitting particle. The number of photons and the coordinates where they hit the detector fluctuate statistically and are disturbed by spurious noise signals. It has turned out that ANNs can reconstruct the parameters of interest from the available coordinates with good efficiency and accuracy.

We study this problem by a Monte Carlo simulation. In a simplified model, we assume that exactly 5 photons are emitted by a particle and that the coordinate pairs are located on a circle and registered. The center, the radii, and the hit coordinates are simulated. The input vector of the network thus consists of 10 components, the 5 coordinate pairs. The output is a single value, the radius R. The loss function is $(R - R_{true})^2$, where the true value R_{true} is known from the simulation.

The relative accuracy of the reconstruction as a function of the iteration step is shown in Fig. 11.11. Different sequences of the learning rate have been tried. Typically, the process is running by steps, where after a flat phase follows a rather abrupt improvement. The number of iterations required to reach the minimum is quite large.

11.4.3 *Weighting Methods*

For the decision whether to assign an observation at the location $\boldsymbol{x}$ to a certain class, an obvious option is, to do this according to the classification of neighboring objects of a training sample. One possibility is to consider a certain region around $\boldsymbol{x}$ and to take a "majority vote" of the training objects inside this region to decide about the class membership of the input. The region to be considered here can be chosen in different ways; it can be a fixed volume around $\boldsymbol{x}$, or a variable volume defined by requiring that it contains a fixed number of observations, or an infinite volume, introducing weights for the training objects which decrease with their distance from $\boldsymbol{x}$.

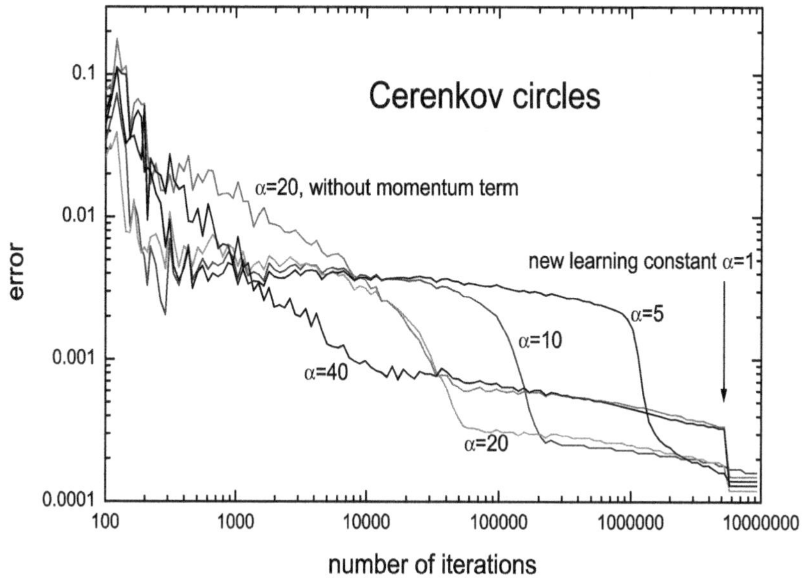

Fig. 11.11 Reconstruction of radii of circles through 5 points by means of an ANN with different sequences of learning constants a.

In any case we need a metric to define the distance. The choice of a metric in multi-dimensional applications is often a rather intricate problem, especially if some of the input components are physically of very different nature. A way-out seems to be to normalize the different quantities to equal variance and to eliminate global correlations by a linear variable transformation. This corresponds to the transformation to principal components discussed above (see Section 11.3) with subsequent scaling of the principal components. An alternative but equivalent possibility is to use a direction dependent weighting. The same result is achieved when we apply the Mahalanobis metric, which we have introduced in Section 10.4.8.

For a large training sample the calculation of all distances is expensive in computing time. A drastic reduction of the number of distances to be calculated is in many cases possible by the so-called *support vector machines* which we will discuss below. Those are not machines, but programs which reduce the training sample to a few, but decisive inputs, without impairing the results.

11.4.3.1 *k-Nearest Neighbors*

We choose a number K which of course will depend on the size of the training sample and the overlap of the classes. For an input $\boldsymbol{x}$ we determine the K nearest neighbors and the numbers k_1, $k_2 = K - k_1$, of observations that belong to class I and II, respectively. For a ratio k_1/k_2 greater than α, we assign the new observation to class I, in the opposite case to class II:

$$\begin{aligned} k_1/k_2 > \alpha &\Longrightarrow \text{ class I}, \\ k_1/k_2 < \alpha &\Longrightarrow \text{ class II}. \end{aligned}$$

The choice of α depends on the loss function. When the loss function treats all classes alike, then α will be unity and we get a simple majority vote. To find the optimal value of K we minimize the average of the loss function computed for all observations of the training sample.

11.4.3.2 *Distance Dependent Weighting*

Instead of treating all training vector inputs $\boldsymbol{x}'$ within a given region in the same way, one should attribute a larger weight to those located nearer to the input $\boldsymbol{x}$. A sensible choice is again a Gaussian kernel,

$$K(\boldsymbol{x}, \boldsymbol{x}') \sim \exp\left(-\frac{(\boldsymbol{x}-\boldsymbol{x}')^2}{2s^2}\right).$$

With this choice we obtain for the class β the weight w_β,

$$w_\beta = \sum_i K(\boldsymbol{x}, \boldsymbol{x}_{\beta i}), \tag{11.21}$$

where $\boldsymbol{x}_{\beta i}$ are the locations of the training vectors of the class β.

If there are only two classes, writing the training sample as

$$\{\boldsymbol{x}_1, y_1 \ldots \boldsymbol{x}_N, y_N\}$$

with the response vector $y_i = \pm 1$, the classification of a new input $\boldsymbol{x}$ is done according to the value ± 1 of the classifier $\hat{y}(\boldsymbol{x})$, given by

$$\hat{y}(\boldsymbol{x}) = \operatorname{sign}\left(\sum_{y_i=+1} K(\boldsymbol{x}, \boldsymbol{x}_i) - \sum_{y_i=-1} K(\boldsymbol{x}, \boldsymbol{x}_i)\right) = \operatorname{sign}\left(\sum_i y_i K(\boldsymbol{x}, \boldsymbol{x}_i)\right). \tag{11.22}$$

For a direction dependent density of the training sample, we can use a direction dependent kernel, eventually in the Mahalanobis form mentioned above:

$$K(\boldsymbol{x}, \boldsymbol{x}') \sim \exp\left[-\frac{1}{2}(\boldsymbol{x}-\boldsymbol{x}')^T \mathsf{C}(\boldsymbol{x}-\boldsymbol{x}')\right].$$

with the weight matrix C. When we first normalize the sample, this complication is not necessary. The parameter s of the matrix C, which determines the width of the kernel function, again is optimized by minimizing the loss for the training sample.

11.4.3.3 *Support Vector Machines*

Support vector machines (SVMs) produce similar results as ordinary distance depending weighting methods, but they require less memory for the storage of learning data and the classification is extremely fast. Therefore, they are especially useful in on-line applications.

The assignment of elements $\boldsymbol{x}$ to a class is usually confined to simply connected regions. Very often, in a two case classification, there are only two regions separated by a hypersurface. For short range kernels it is obvious then that for the classification of observations, the knowledge of the position is essential only of those input vectors of the training sample that are located in the vicinity of the hypersurface. These input vectors are called *support vectors* [106]. SVMs are programs which try to determine them, respectively their weights, in an optimal way, setting the weights of all other inputs vectors to zero.

In the one-dimensional case with non-overlapping classes it is sufficient to know those inputs of each class which are located nearest to the dividing limit between the classes. Sums like (11.21) are then running over one element only. This, of course, makes the calculation extremely fast.

In higher dimensional spaces with overlapping classes and for more than two classes the problem to determine support vectors is of course more complicated. But also in these circumstances the number of relevant training inputs can be reduced drastically. The success of SVMs is based on the so-called *kernel trick*, by which non-linear problems in the input space are treated as linear problems in some higher-dimensional space by well known optimization algorithms. For the corresponding algorithms and proofs we refer to the literature, e.g. [16, 107, 108]. A short introduction is given in Appendix A.14.

11.4.3.4 *Example and Discussion*

In Fig. 11.12 are shown in the top panel two overlapping training samples of 500 inputs each. The loss function is the number of wrong assignments independent of the respective class. Since the distributions are quite similar in both coordinates we do not change the metric. We use a Gaussian kernel.

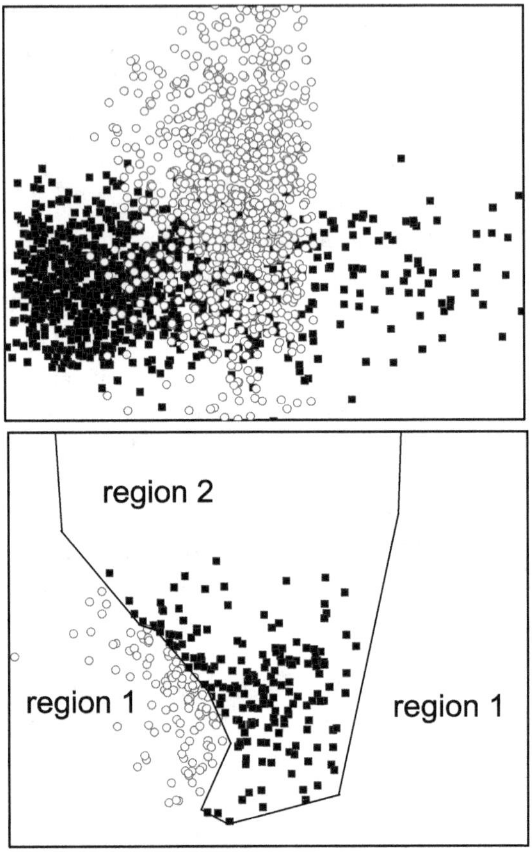

Fig. 11.12 Separation of two classes. Top: learning sample, bottom:: wrongly assigned events of a test sample.

The optimization of the parameter s by means of the training sample shows only a small change of the error rate for a change of s by a factor four. The lower panel displays the result of the classification for a test sample of the same size (500 inputs per class). Only the wrong assignments are shown.

We realize that wrongly assigned training observations occur in two separate, non overlapping regions which can be separated by a curve or a polygon chain as indicated in the figure. Obviously all new observations would be assigned to the class corresponding to the region in which they are located. If we would have used instead of the distance-depending weighting the k-nearest neighbors method, the result would have been almost identical. In spite of the opposite expectation, this more primitive method is more expensive in both the programming and in the calculation, when compared to the weighting with a distance dependent kernel.

Since for the classification only the separation curve between the classes is required, it must be sufficient to know the class assignment for those training observations which lie near this curve. They would define the support vectors of a SVM. Thus the number of inputs needed for the assignment of new observations would be drastically reduced. However, for a number of assignments below about 10^6 the effort to determine support vectors usually does not pay. The SVMs are useful for large event numbers in applications where computing time is relevant.

11.4.4 *Decision Trees*

11.4.4.1 *Simple Trees*

We study the simple case, the two class classification, i.e. the assignment of N events (also called input vectors) to one of two classes I and II. An event is characterized by M features $x_1, x_2, \ldots, x_M$.

As shown in Fig. 11.13 a tree consists of nods, branches, and leaves. At a nod a decision is taken which of the two branches to follow. After the last nod figure leaves that decide on the classification.

In a first training step we consider the first feature of all N events of the training sample. We determine a separation value x_{c1} which optimally divides the two classes and obtain a division of the training sample into two parts A and B. Each of these parts is now treated separately. We take the subspace A, look at the second feature (y in Fig. 11.13) and divide the sample of A in the same way as before the full space, again into two parts. Analogously we treat the subspace B. Now we can switch to the next feature or return to the first feature, if there are only two features, and perform further splittings. The sequence of divisions leads to smaller and smaller subspaces, each of them assigned to a certain class. The growing of the tree is stopped by a pruning rule. The final partitions are the leaves.

In Fig. 11.13 we show schematically the subdivision into subspaces and the corresponding decision tree for a training sample of 32 elements with only two features. The training sample which determines the decisions is indicated. At the end of the tree (here at the bottom) the decision about the class membership is taken. The classes blue, red, blue, red are assigned

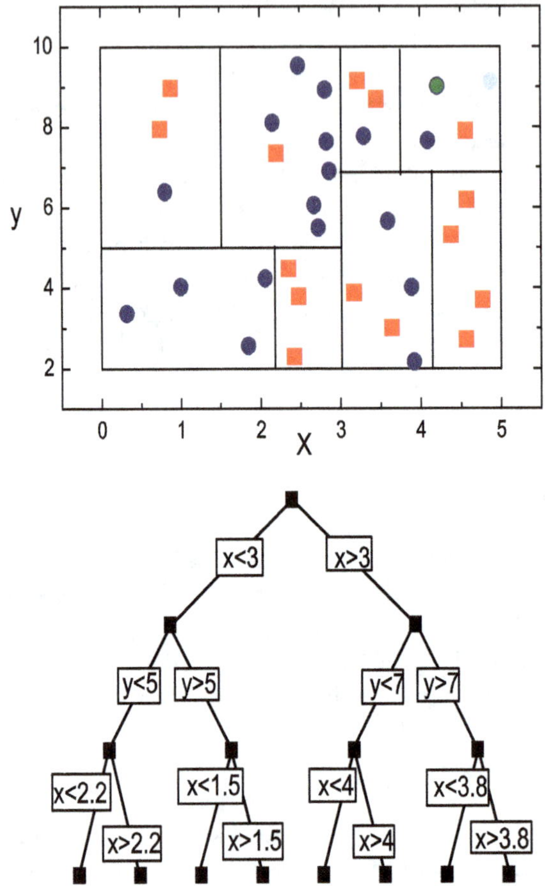

Fig. 11.13 Decision tree (bottom) corresponding to the classification shown above.

to the four leaves from left to right as shown in the upper graph of the figure.

It is not obvious, how one should optimize the sequence of partitions and the position of cuts, and also not, under which circumstances the procedure should be stopped.

For the optimization of splits we must again define a loss function which will depend on the given problem. A simple possibility in the case of two classes is to maximize the distance to the closest events of the two subspaces.

We used this possibility in the example of Fig.11.13. To some extend the exact position of the splitting line is still arbitrary, the loss function changes its value only when it hits the nearest input. Often the importance of efficiency and purity is different for the two classes. Then we would chose an asymmetric loss function.

Very popular is the following, slightly more complicated criterion: We define the purity P_I of class I

$$P_I = \frac{N_I}{N_I + N_{II}}. \tag{11.23}$$

N_I, N_{II} are the number of events that belong to class I and class II. For an optimal classification P_1 would be either 1 or 0. The quantity

$$G = P_I(1 - P_I) + P_{II}(1 - P_{II}) \tag{11.24}$$
$$= 1 - P_I^2 - P_{II}^2 \tag{11.25}$$

the *Gini-index*, should be as small as possible. For each separation of a parent node E with Gini index G_E into two children nodes A, B with G_A, respectively G_B, we minimize the sum $G_A + G_B$.

The difference

$$D = G_E - G_A - G_B$$

is taken as stopping or pruning parameter. The quantity D measures the increase in purity, it is large for a parent node with large G and two children nodes with small G. When D becomes less than a certain critical value D_c the branch will not be split further and ends at a leave. The leave is assigned to the class which has the majority in it.

Besides the Gini index, also other measures for the purity or impurity are used [16]. An interesting quantity is entropy $S = -P_I \ln P_I - P_{II} \ln P_{II}$, a well known measure of disorder, i.e. of impurity.

The purity parameter, e.g. G, is also used to organize the splitting sequence. We choose always that input vector component in which the splitting produces the most significant separation.

Unfortunately, the classification by decision trees is usually not perfect. The discontinuity at the boundaries and the fixed splitting sequence impair the accuracy. On the other hand, decision trees are simple, transparent and the corresponding computer programs are extremely fast.

11.4.4.2 *Boosted Decision Trees*

Boosting [109] is based on a simple idea: By a weighted superposition of many moderately effective classifiers it should be possible to reach a fairly precise assignment. Instead of only one decision tree, many different trees are grown. Each time, before the development of a new tree is started, wrongly assigned training inputs are *boosted* to higher weights in order to lower their probability of being wrongly classified in the following tree. The final class assignment is then done by averaging the decisions from all trees. Obviously, the computing effort for these *boosted decision trees* is increased, but the precision is significantly enhanced. The results of boosted decision trees are sometimes as good as those of ANNs. Their algorithm is very well suited for parallel processing. There are first applications in particle physics [110].

Before the first tree is constructed, all training events have the weight 1. In the following run each input gets a weight w_i, determined by a certain boosting algorithm (see below) which depends on the particular method. The definition of the node impurity P for calculating the loss function

$$P = \frac{\sum_I w_i}{\sum_I w_i + \sum_{II} w_i},$$

where the sums $\sum_I$, $\sum_{II}$ run over all events in class I or II, respectively. The weights are boosted and the next run starts. Typically $M \approx 1000$ trees are generated in this way.

If we indicate the decision of a tree m for the input $\boldsymbol{x}_i$ by $T_m(\boldsymbol{x}_i) = 1$ (for class I) and $= -1$ (for class II), the final result will be given by the sign of the weighted sum over the results from all trees,

$$T_M(\boldsymbol{x}_i) = \text{sign}\left(\sum_{m=1}^{M} \alpha_m T_m(\boldsymbol{x}_i)\right).$$

We proceed in the following way: To the first tree we assign a weight $\alpha_1 = 1$. The weights of the wrongly assigned input vectors are increased. The weight[11] α_2 of the second tree $T_2(\boldsymbol{x})$ is chosen such that the overall loss from all input vectors of the training sample is minimal for the combination $[\alpha_1 T_1(\boldsymbol{x}) + \alpha_2 T_2(\boldsymbol{x})] \,/\, [\alpha_1 + \alpha_2]$. We continue in the same way and add further trees. For tree i the weight α_i is optimized such that the existing trees are complemented in an optimal way. How this is done depends of course on the loss function.

A well tested recipe for the choice of weights is AdaBoost [109]. The training algorithm proceeds as follows:

[11] We have two kinds of weight, weights of input vectors (w_i) and weights of trees (α_m).

- The i-th input $\boldsymbol{x}_i$ gets the weight $w_i = 1$ and the value $y_i = 1$, $(= -1)$, if it belongs to class I, (II).
- $T_m(\boldsymbol{x}_i) = 1$ $(= -1)$, if the input ends in a leave belonging to class I (II). $S_m(\boldsymbol{x}_i) = (1 - y_i T_m(\boldsymbol{x}_i))/2 = 1$ $(= 0)$, if the assignment is wrong (right).
- The fraction of the weighted wrong assignments ε_m is used to change the weights for the next iteration:

$$\varepsilon_m = \sum_i w_i S_m(\boldsymbol{x}_i) / \sum_i w_i,$$

$$\alpha_m = \ln \frac{1 - \varepsilon_m}{\varepsilon_m},$$

$$w_i \to w_i \mathrm{e}^{\alpha_m S_m}.$$

Weights of correctly assigned training inputs thus remain unchanged. For example, for $\varepsilon_m = 0.1$, wrongly assigned inputs will be boosted by a factor $0.9/0.1 = 9$. Note that $\alpha_m > 0$ if $\varepsilon < 0.5$; this is required because otherwise the replacement $T_m(x_i) \to -T_m(x_i)$ would produce a better decision tree.

- The response for a new input which is to be classified is

$$T_M(\boldsymbol{x}_i) = \mathrm{sign}\left(\sum_{m=1}^{M} \alpha_m T_m(\boldsymbol{x}_i)\right).$$

For $\varepsilon_m = 0.1$ the weight of the tree is $\alpha_m = \ln 9 \approx 2.20$. For certain applications it may be useful to reduce the weight factors α_m somewhat, for instance $\alpha_m = 0.5 \ln\left((1 - \varepsilon_m)/\varepsilon_m\right)$ [110].

11.4.5 *Bagging and Random Forest*

11.4.5.1 *Bagging*

The concept of bagging was first introduced by Breiman [111]. He has shown that the performance of unstable classifiers can be improved considerably by training many classifiers with bootstrap replicates and then using a majority vote of those: From a training sample containing N input vectors, N vectors are drawn at random with replacement. Some vectors will be contained several times. This bootstrap[12] sample is used to train a classifier. Many, 100 or 1000 classifiers are produced in this way. New inputs are run through all trees and each tree "votes" for a certain classification. The classification receiving the majority of votes is chosen. In a

[12]We will discuss bootstrap methods in the following chapter.

study of real data [111] a reduction of error rates by bagging between 20% and 47% was found. There the bagging concept had been applied to simple decision trees, however, the bagging concept is quite general and can be adopted also to other classifiers.

11.4.5.2 *Random Forest*

Another new development [112] which includes the bootstrap idea, is the extension of the decision tree concept to the *random forest* classifier.

Many trees are generated from bootstrap samples of the training sample, but now part of the input vector components are suppressed. A tree is constructed in the following way: First m out of the M components or attributes of the input vectors are selected at random. The tree is grown in a m-dimensional subspace of the full input vector space. It is not obvious how m is to be chosen, but the author proposes $m \ll M$ and says that the results show little dependence on m. With large m the individual trees are powerful but strongly correlated. The value of m is the same for all trees.

From the N truncated bootstrap vectors, N_b are separated, put into a *bag* and reserved for testing. A fraction $f = N_b/N \approx 1/3$ is proposed. The remaining ones are used to generate the tree. For each split that attribute out of the m available attributes is chosen which gives the smallest number of wrong classifications. Each leave contains only elements of a single class. There is no pruning.

Following the bagging concept, the classification of new input vectors is obtained by the majority vote of all trees.

The out-of-the-bag (oob) data are used to estimate the error rate. To this end, each oob-vector of the k-th sample is run through the k-th tree and classified. The fraction of wrong classifications from all oob vectors is the error rate. (For T trees there are in total $T \times N_b$ oob vectors.) The oob data can also be used to optimize the constant m.

The random forest classifier has received quite some interest. The concept is simple and seems to be similarly powerful as that of other classifiers. It is especially well suited for large data sets in high dimensions.

11.4.6 *Comparison of the Methods*

We have discussed various methods for classification. Each of them has its advantages and its drawbacks. It depends on the special problem, which one is the most suitable.

The *discriminant analysis* offers itself for one- or two dimensional continuous distributions (preferably normal or other unimodal distributions). It is useful for event selection in simple situations.

Kernel methods are relatively easy to apply. They work well if the division line between classes is sufficiently smooth and transitions between different classes are continuous. Categorical variables cannot be treated. The variant with support vectors reduces computing time and the memory space for the storage of the training sample. In standard cases with not too extensive statistics one should avoid this additional complication. Kernel methods can perform event selection in more complicated environments than is possible with the primitive discriminant analysis. For the better performance the possibility of interpreting the results is diminished, however.

Artificial neural networks are, due to the enormous number of free parameters, able to solve any problem in an optimal way. They suffer from the disadvantage that the user usually has to intervene to guide the minimizing process to a correct minimum. The user has to check and improve the result by changing the network structure, the learning constant and the start values of the weights. New program packages are able to partially take over these tasks. ANN are able to separate classes in very involved situations and extract very rare events from large samples.

Decision trees are a very attractive alternative to ANN. One should use boosted decision trees, random forest or apply bagging though, since those discriminate much better than simple trees. The advantage of simple trees is that they are very transparent and that they can be displayed graphically. Like ANN, decision trees can, with some modifications, also be applied to categorical variables.

At present, there is lack of theoretical framework and experimental information on some of the new developments. We would like to know to what extent the different classifiers are equivalent and which classifier should be selected in a given situation. There will certainly be answers to these questions in the near future.

Chapter 12

Auxiliary Methods

12.1 Probability Density Estimation

12.1.1 *Introduction*

In the subsection *function approximation* we have considered measurements $\boldsymbol{y}$ at fixed locations $\boldsymbol{x}$ where $\boldsymbol{y}$ due to statistical fluctuations deviates from an unknown function. Now we start from a sample $\{\boldsymbol{x}_1, \ldots, \boldsymbol{x}_N\}$ which follow an unknown statistical distribution which we want to approximate. We have to estimate the density $\hat{f}(\boldsymbol{x})$ at the location $\boldsymbol{x}$ from the frequency of observations $\boldsymbol{x}_i$ in the vicinity of $\boldsymbol{x}$. The corresponding technique, *probability density estimation* (PDE), is strongly correlated with function approximation. Both problems are often treated together under the title *smoothing methods*. In this section we discuss only non-parametric approaches; a parametric method, where parameters are adjusted to approximate Gaussian like distributions has been described in Section 11.2.2. We will essentially present results and omit the derivations. For details the reader has to consult the specialized literature.

PDE serves mainly to visualize an empirical frequency distribution. Visualization of data is an important tool of scientific research. It can lead to new discoveries and often constitutes the basis of experimental decisions. PDE also helps to classify data and sometimes the density which has been estimated from some ancillary measurement is used in subsequent Monte Carlo simulations of experiments. However, to solve certain problems like the estimation of moments and other characteristic properties of a distribution, it is preferable to deal directly with the sample instead of performing a PDE. This path is followed by the bootstrap method which we will discuss in a subsequent section. When we have some knowledge about the shape of a distribution, then PDE can improve the precision of the bootstrap estimates. For instance there may exist good reasons to assume that the distribution has only one maximum and/or it may be known that the random variable is restricted to a certain region with known boundary.

The PDE $\hat{f}(\boldsymbol{x})$ of the true density $f(\boldsymbol{x})$ is obtained by a smoothing procedure applied to the discrete experimental distribution of observations. This means, that some kind of averaging is done which introduces a bias which is especially large if the distribution $f(\boldsymbol{x})$ varies strongly in the vicinity of $\boldsymbol{x}$.

The simplest and most common way to measure the quality of the PDE is to evaluate the integrated square error (ISE) L_2

$$L_2 = \int_{-\infty}^{\infty} \left[\hat{f}(\boldsymbol{x}) - f(\boldsymbol{x})\right]^2 \mathrm{d}x$$

and its expectation value $E(L_2)$, the *mean integrated square error (MISE)*.[1] The mean quadratic difference $E(\left[\hat{f}(\boldsymbol{x}) - f(\boldsymbol{x})\right]^2)$ has two components, according to the usual decomposition:

$$E\left(\left[\hat{f}(\boldsymbol{x}) - f(\boldsymbol{x})\right]^2\right) = \mathrm{var}\left(\hat{f}(\boldsymbol{x})\right) + \mathrm{bias}^2\left(\hat{f}(\boldsymbol{x})\right).$$

The first term, the variance, caused by statistical fluctuations, decreases with increasing smoothing and the second term, the bias squared, decreases with decreasing smoothing. The challenge is to find the optimal balance between these two contributions.

We will give a short introduction to PDE mainly restricted to one-dimensional distributions. The generalization of the simpler methods to multi-dimensional distributions is straight forward but for the more sophisticated ones this is more involved. A rather complete and comprehensive overview can be found in the books by J.S. Simonoff [113], A.W. Bowman and A. Azzalini [101], D. W. Scott [114] and W. Härdle *et al.* [116]. Several results presented below have been taken from a summary by D. W. Scott and S. R. Sain [82]. We follow [82]. Detailed derivations are omitted.

12.1.2 *Fixed Interval Methods*

12.1.2.1 *Histogram Approximation*

The simplest and most popular method of density estimation is *histogramming* the approximation of the true function $f(x)$ by a histogram with fixed

[1]The estimate $\hat{f}(\boldsymbol{x})$ is a function of the set $\{\boldsymbol{x}_1, \dots, \boldsymbol{x}_N\}$ of random variables and thus also a random variable.

bin width. For the number ν_k of N events falling into bin B_k and bin width h the estimated density is

$$\hat{f}(x) = \frac{\nu_k}{Nh} \text{ for } x \in B_k .$$

It is easy to construct, transparent, does not contain hidden parameters which often are included in other more sophisticated methods and indicates quite well which distributions are compatible with the data. However it has, as we have repeatedly stated, the disadvantage of the rather arbitrary binning and its discontinuity. Fine binning provides a good resolution of structures and low bias but has to be paid for by large fluctuations. Histograms with wide bins have the advantage of small statistical errors but are biased. A sensible choice for the bin width h is derived from the requirement that the mean squared integrated error should be as small as possible. The mean integrated square error, $MISE$, for a histogram is [82]:

$$MISE = \frac{1}{Nh} + \frac{1}{12}h^2 \int f'(x)^2 \mathrm{d}x + O(1/N + h^4). \tag{12.1}$$

The approximation is optimal for a constant function and deteriorates with increasing slope $f'(x)$. The integral $\int f'(x)^2 \mathrm{d}x = R(f')$ is called *roughness*. For a normal density with variance σ^2 it is $R = (4\sqrt{\pi}\sigma^3)^{-1}$. Neglecting the small terms ($h \to 0$) we can derive [82] the optimal bin width h^* and the corresponding *asymptotic mean integrated square error AMISE*:

$$\begin{aligned} h^* &\approx \left[\frac{6}{N\int f'(x)^2 \mathrm{d}x}\right]^{1/3} \approx 3.5\sigma N^{-1/3}, \qquad (12.2)\\ AMISE &\approx \left[\frac{9\int f'(x)^2 \mathrm{d}x}{16N^2}\right]^{1/3} \approx 0.43 N^{-2/3}/\sigma . \end{aligned}$$

The second part of relation (12.2) holds for a Gaussian p.d.f. with variance σ^2 and is a reasonable approximation for other distributions with the same variance. Even though the derivative f' and the bandwidth[2] σ are not precisely known, they can be estimated from the data. As expected, the optimal bin width is proportional to the bandwidth, whereas its $N^{-1/3}$ dependence on the sample size N is less obvious.

In d dimensions similar relations hold. Of course the N-dependence has to be modified. For d-dimensional cubical bins the optimal bin width scales with $N^{-1/(d+2)}$ and the mean square error scales with $N^{-2/(d+2)}$.

---

[2]Contrary to what is understood usually under bandwidth, in PDE this term is used to describe the typical width of structures. For a Gaussian it equals the standard deviation.

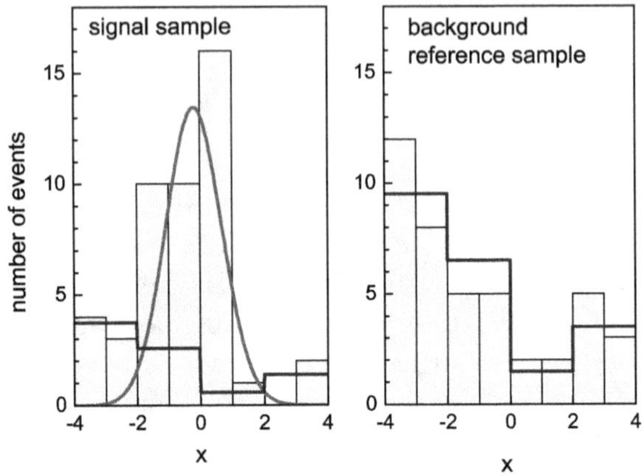

Fig. 12.1 Experimental signal with some background (left) and background reference sample with two times longer exposure time (right). The fitted signal and background functions are indicated.

Example 12.1. PDE of a background distribution and signal fit

We analyze a signal sample containing a Gaussian signal $\mathcal{N}(x|\mu,\sigma)$ with unknown location and scale parameters μ, σ containing some unknown background. In addition, we dispose of a reference sample containing only background. From the data taking times and fluxes we know that the background in the signal sample should nominally be half ($r = 0.5$) of that in the reference sample. In Fig. 12.1 we show the two experimental distributions. From the shape of the experimental background distribution we estimate the slope $y' = 0.05$ of its p.d.f. $h(x)$, and, using relation 12.2, find a bin width of 2 units. The heights $\beta_1, \beta_2, \beta_3, \beta_4$ of the 4 equally wide bins of the histogram distribution are left as free parameters in the fit. Because of the normalization, we have $\beta_4 = 1 - \beta_1 - \beta_2 - \beta_3$. Further parameters are the expected rate of background events in the reference sample ρ and the fraction ϕ of true signal events in the signal sample. These 7 parameters are to be determined in a likelihood fit. The log-likelihood function $\ln L = \ln L_1 + \ln L_2 + \ln L_3 + \ln L_4$ comprises 4 terms, with:

1. L_1, the likelihood of the n_s events in the signal sample (superposition of signal and background distribution):

$$\ln L_1(\mu,\sigma,\phi,\beta_1,\beta_2,\beta_3) = \sum_{i=1}^{n_s} \ln\left[\phi\mathcal{N}(x_i|\mu,\sigma) + (1-\phi)h(x_i|\beta_1,\beta_2,\beta_3)\right].$$

2. L_2, the likelihood of the n_r events of the reference sample (background distribution):

$$\ln L_2(\beta_1, \beta_2, \beta_3) = \sum_{i=1}^{n_r} \ln h(x_i|\beta_1, \beta_2, \beta_3).$$

3. L_3, the likelihood to observe n_r reference events where ρ are expected (Poisson distribution):

$$\ln L_3 = -\rho + n_r \ln \rho - \ln n_r!.$$

4. L_4, the likelihood to get $n_s(1-\phi)$ background events[3] in the signal sample where $r\rho$ are expected (Poisson distribution):

$$\ln L_4 = -r\rho + n_s(1-\phi)\ln(r\rho) - \ln\{[n_s(1-\phi)]!\}.$$

The results of the fit are indicated in the Fig. 12.1 which is a histogram of the observed events.The MLE of the interesting parameters are $\mu = -0.18 \pm 0.32$, $\sigma = 0.85 \pm 0.22$, $\phi = 0.60^{+0.05}_{-0.09}$, the correlation coefficients are of the order of 0.3. We abstain from showing the full error matrix. The samples have been generated with the nominal parameter values $\mu_0 = 0$, $\sigma_0 = 1$, $\phi = 0.6$. To check the influence of the background parametrization, we repeat the fit with only two bins. The results change very little to $\mu = -0.12 \pm 0.34$, $\sigma = 0.85 \pm 0.22$, $\phi = 0.57^{+0.07}_{-0.11}$. When we represent the background p.d.f. by a polygon (see next chapter) instead of a histogram, the result again remains stable. We then get $\mu = -0.20 \pm 0.30$, $\sigma = 0.82 \pm 0.21$, $\phi = 0.60^{+0.05}_{-0.09}$. The method which we have applied in the present example is more precise than that of Section 7.4 but depends to a certain degree on the presumed shape of the background distribution.

12.1.2.2 *Linear and Higher Order Parabolic Approximation*

In Chapter 9 we had adjusted spline functions to unfolded distributions. Similarly, we can use them to approximate the probability density. We will consider here only the linear approximation by a polygon but it is obvious that the method can be extended to higher order parabolic functions. The discontinuity corresponding to the steps between bins is avoided when we transform the histogram into a polygon. We just have to connect the points corresponding to the histogram functions at the center of the bins. It can be shown that this reduces the $MISE$ considerably, especially for large samples. The optimum bin width in the one-dimensional case now depends on the average second derivative f'' of the p.d.f. and is much wider than for a histogram and the error is smaller [82] than in the corresponding histogram case:

$$h^* \approx 1.6 \left[\frac{1}{N \int f''(x)^2 \mathrm{d}x} \right]^{1/5} ,$$

$$MISE^* \approx 0.5 \left[\frac{\int f''(x)^2 \mathrm{d}x}{N^4} \right]^{1/5} .$$

In d dimensions the optimal bin width for polygon bins scales with $N^{-1/(d+4)}$ and the mean square error scales with $N^{-4/(d+4)}$.

12.1.3 *Fixed Number and Fixed Volume Methods*

To estimate the density at a point $\boldsymbol{x}$ an obvious procedure is to divide the number k of observations in the neighborhood of $\boldsymbol{x}$ by the volume V which they occupy, $\hat{f}(\boldsymbol{x}) = k/V$. Either we can fix k and compute the corresponding volume $V(\boldsymbol{x})$ or we can choose V and count the number of observations contained in that volume. The quadratic uncertainty is $\sigma^2 = k + bias^2$, hence the former emphasizes fixed statistical uncertainty and the latter rather aims at small variations of the bias.

The k-nearest neighbors method avoids large fluctuations in regions where the density is low. We obtain a constant statistical error if we estimate the density from the spherical volume V taken by the k nearest neighbors of point $\boldsymbol{x}$:

$$\hat{f}(\boldsymbol{x}) = \frac{k}{V_k(\boldsymbol{x})} . \tag{12.3}$$

As many other PDE methods, the k-nearest neighbors method is problematic in regions with large curvature of f and at boundaries of $\boldsymbol{x}$.

Instead of fixing the number of observations k in relation (12.3) we can fix the volume V and determine k. Strong variations of the bias in the k-nearest neighbors method are somewhat reduced but both methods suffer from the same deficiencies, the boundary bias and a loss of precision due to the sharp cut-off due to either fixing k or V. Furthermore it is not guaranteed that the estimated density is normalized to one. Hence a renormalization has to be performed.

The main advantage of fixed number and fixed volume methods is their simplicity.

12.1.4 *Kernel Methods*

We now generalize the fixed volume method and replace (12.3) by

$$\hat{f}(\boldsymbol{x}) = \frac{1}{NV}\sum K(\boldsymbol{x} - \boldsymbol{x}_i)$$

where the kernel K is equal to 1 if $\boldsymbol{x}_i$ is inside the sphere of volume V centered at $\boldsymbol{x}$ and 0 otherwise. Obviously, smooth kernel functions are more attractive than the uniform kernel of the fixed volume method. An obvious candidate for the kernel function $K(u)$ in the one-dimensional case is the Gaussian $\propto \exp(-u^2/2h^2)$. A very popular candidate is also the parabolically shaped Epanechnikov kernel $\propto (1 - c^2u^2)$, for $|cu| \leq 1$, and else zero. Here c is a scaling constant to be adjusted to the bandwidth of f. Under very general conditions the Epanechnikov kernel minimizes the asymptotic mean integrated square error $AMISE$ obtained in the limit where the effective binwidth tends to zero, but other kernels perform nearly as well. The $AMISE$ of the Gaussian kernel is only 5% larger and that of the uniform kernel by 8% [113].

The optimal bandwidth of the kernel function obviously depends on the true density. For example for a Gaussian true density $f(x)$ with variance σ^2 the optimal bandwidth h of a Gaussian kernel is $h_G \approx 1.06\sigma N^{-1/5}$ [113] and the corresponding constant c of the Epanechnikov kernel is $c \approx 2.2/(2h_G)$. In practice, we will have to replace the Gaussian σ in the relation for h_0 by some estimate depending on the structure of the observed data. $AMISE$ of the kernel PDE is converging at the rate $N^{-4/5}$ while this rate was only $N^{-2/3}$ for the histogram.

12.1.5 *Problems and Discussion*

The simple PDE methods sketched above suffer from several problems, some of which are unavoidable:

1. The boundary bias: When the variable x is bounded, say $x < a$, then $\hat{f}(x)$ is biased downwards unless $f(a) = 0$ in case the averaging process includes the region $x > a$ where we have no data. When the averaging is restricted to the region $x < a$, the bias is positive (negative) for a distribution decreasing (increasing) towards the boundary. In both cases the size of the bias can be estimated and corrected for, using so-called boundary kernels.
2. Many smoothing methods do not guarantee normalization of the estimated probability density. While this effect can easily be corrected by a renormalization $\hat{f}$, it indicates a problem of the method.

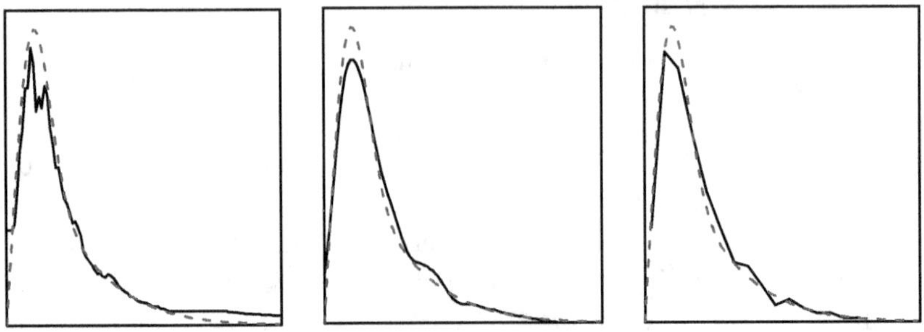

Fig. 12.2 Estimated probability density. Left hand: k-nearest neighbors, center: Gaussian Kernel, right hand: Polygon.

3. Fixed bandwidth methods over-smooth in regions where the density is high and tend to produce fake bumps in regions where the density is low. Variable bandwidth kernels are able to avoid this effect partially. Their bandwidth is chosen inversely proportional to the square root of the density, $h(x_i) = h_0 f(x_i)^{-1/2}$. Since the true density is not known, f must be replaced by a first estimate obtained for instance with a fixed bandwidth kernel.
4. Kernel smoothing corresponds to a convolution of the discrete data distribution with a smearing function and thus unavoidably tends to flatten peaks and to fill-up valleys. This is especially pronounced at locations where the distribution has sharp structures, that is where the second derivative f'' is large. Convolution and thus also PDE implies a loss of information contained in the original data. This defect may be acceptable if we gain sufficiently due to knowledge about f that we put into the smoothing program. In the simplest case this is only the fact that the distribution is continuous and differentiable but in some situations also the asymptotic behavior of f may be given, or we may know that it is unimodal. Then we will try to implement this information into the smoothing method.

Some of the remedies for the difficulties mentioned above use estimates of f and its derivatives. Thus iterative procedures seem to be the solution. However, the iteration process usually does not converge and thus has to be supervised and stopped before artifacts appear.

In Fig. 12.2 three simple smoothing methods are compared. A sample of 1000 events has been generated from the function shown as a dashed curve in the figure. A k-nearest neighbors approximation of the p.d.f. of a sample is shown in the left hand graph of Fig. 12.2. The value of k chosen was 100 which is too small to produce enough smoothing but too large to follow the distribution at the left hand border. The result of the PDE with a Gaussian kernel with fixed width is presented in the central graph and a polygon approximation is shown in the right hand graph. All three graphs show the typical defects of simple smoothing methods, broadening of the peak and fake structures in the region where the statistics is low.

Alternatively to the standard smoothing methods, complementary approaches often produce better results than the former. The typical smoothing problems can partially be avoided when the p.d.f. is parametrized and adjusted to the data sample in a likelihood fit. A simple parametrization is the superposition of normal distributions,

$$f(\boldsymbol{x}) = \sum \alpha_i \mathcal{N}(\boldsymbol{x}; \boldsymbol{\mu}_i, \Sigma_i),$$

with the free parameters, weights α_i, mean values $\boldsymbol{\mu}_i$ and covariance matrices Σ_i.

If information about the shape of the distribution is available, more specific parametrizations which describe the asymptotic behavior can be applied. Distributions which resemble a Gaussian should be approximated by the Gram-Charlier series (see last paragraph of Section 11.2.2). If the data sample is sufficiently large and the distribution is unimodal with known asymptotic behavior the construction of the p.d.f. from the moments as described in [115] is quite efficient.

Physicists use PDE mainly for the visualization of the data. Here, in one dimension, histogramming is the standard method. When the estimated distribution is used to simulate an experiment, frequency polygons are to be preferred. Whenever a useful parametrization is at hand, then PDE should be replaced by an adjustment of the corresponding parameters in a likelihood fit. Only in rare situations it pays to construct complicated kernels. For a quick qualitative illustration of a distribution off-the-shelf programs may do. For most quantitative evaluations of moments and parameters of the unknown distribution we recommend to use the bootstrap method which is discussed in the following section.

12.2 Resampling Techniques

12.2.1 *Introduction*

In the previous section we have discussed a method to construct a distribution approximately starting from a sample drawn from it. The constructed distribution allows us, to calculate certain parameters, like moments or quantiles. In most cases, however, it is preferable to determine the wanted quantities directly from the sample. A trivial example for this approach is the estimation of the mean value and the variance from a series of measurements as we have discussed in Chapter 4. We had used the sample mean $\overline{x} = \sum x_i/N$ and the empirical variance $s^2 = \left[\sum(x_i - \overline{x})^2\right]/(N-1)$. In a similar way we can also determine higher moments, correlations, quantiles and other statistical parameters. However, the analytical derivation of the corresponding expressions is often not as simple as that of the mean value or the variance. Here the bootstrap concept helps, which to a large extend has been developed by Efron. A comprehensive presentation of the method is given in Efron's book [77], which has served as bases for this section.

The name *bootstrap* goes back to the famous book of Erich Raspe in which he narrates the adventures of the lying Baron von Münchhausen. Münchhausen had pretended to have saved himself out of a swamp by pulling himself up with his own bootstraps.[4] In statistics, the expression bootstrap is used because from a small sample the quasi complete distribution is generated. There is not quite as much lying as in Münchhausen's stories.

The bootstrap concept is based upon a simple idea: The sample itself replaces the unknown distribution. The sample *is* the distribution from which we draw individual observations. In fact, the bootstrap idea is also used, when we associate the errors to an observed Poisson distributed number n. We do not know the correct mean value μ of the distribution and replace it by the observation n.

As already mentioned, the bootstrap method permits us, apart from error estimation, to compute p-values for significance tests and the error rate in classifications. It relies, as will be shown in subsequent examples, on the combination of randomly selected observations.

A subvariant of the bootstrap technique is called *jackknife* which is mainly used to estimate biases from subsets of the data.

In Chapter 10 where we had evaluated the distribution of the energy test statistic in two-sample problems, we have used another resampling

[4] In the original version he is pulling himself with his hair.

technique. We had reshuffled the elements of two partitions applying *random permutations*. Whereas in the bootstrap method, elements are drawn with replacement, while permutations generate samples where every element occurs only a single time.

The reason for not using all possible permutations is simply that their number is in most cases excessively large and that a finite random sample provides sufficiently precise results.

Remark: We may ask with some right, whether resampling makes sense since randomly choosing elements from a sample does not seem to be as efficient as a systematic evaluation of the compete sample. Indeed, it should always be optimal to evaluate the interesting parameter directly using all elements of the sample with the same weight, either analytically or numerically, but, as in Monte Carlo simulations, the big advantage of parameter estimation by randomly selecting elements relies on the simplicity of the approach. Nowadays, lack of computing capacity is not a problem and in the limit of an infinite number of drawn combinations of observations the complete available information is exhausted.

12.2.2 *Bootstrap and Simple Examples*

We sort the N observations of a given data sample $\{x_1, x_2, \ldots, x_N\}$ according to their value, $x_i \leq x_{i+1}$, and associate to each of them the probability $1/N$. We call this discrete distribution $P_0(x_i) = 1/N$ the *sample distribution.* We obtain a bootstrap sample $\{x_1^*, x_2^*, \ldots, x_M^*\}$ by generating M observations following P_0. Bootstrap observations are marked with a star "*". A bootstrap sample may contain the same observation several times. The evaluation of interesting quantities follows closely that which is used in Monte Carlo simulations. The p.d.f. used for event generation in Monte Carlo procedures is replaced by the sample distribution.

Example 12.2. Bootstrap evaluation of the accuracy of the estimated mean value of a distribution

We dispose already of an efficient method to estimate the variance. Here we present an alternative approach in order to introduce and illustrate the bootstrap method. Given be the sample of $N = 10$ observations $\{0.80, 0.85, 0.30, 0.09, 0.75, 2.31, 0.12, 0.35, 1.11, 0.65\}$. The estimate of the mean value is obviously $\widehat{\mu} = \overline{x} = \sum x_i/N = 0.74$. We have also derived in Section 3.2

a formula to estimate the uncertainty, $\delta_\mu = \sum(x_i - \overline{x})/N/\sqrt{N-1} = 0.21$. We treat the sample as representative of the distribution and produce an empirical distribution of the mean value: We draw from the data sample sequentially N observations (*drawing with replacement*) and get for instance the bootstrap sample {0.75, 0.35, 0.85, 0.35, 0.11, 2.31, 0.80, 0.80, 0.75, 1.11}. We compute the sample mean and repeat this procedure B times and obtain in this way B mean values μ_k^*. The number of bootstrap replicates should be large compared to N, for example B typically equal to 100 or 1000 for $N = 10$. From the distribution of the values μ_b, $b = 1, \dots, B$ we can compute again the mean value $\hat{\mu}^*$ and its uncertainty δ_μ:

$$\hat{\mu}^* = \frac{1}{B}\sum \mu_b^*,$$

$$\delta_\mu^{*2} = \frac{1}{B}\sum(\mu_b^* - \hat{\mu}^*)^2.$$

Figure 12.3 shows the sample distribution corresponding to the 10 observations and the bootstrap distribution of the mean values. The bootstrap estimates $\hat{\mu}^* = 0.74$, $\delta_\mu^* = 0.19$, agree reasonably well with the directly obtained values. The larger value of δ_μ compared to δ_μ^* is due to the bias correction in its evaluation. The bootstrap values are closer to the maximum likelihood estimates. The distribution of Fig. 12.3 contains further information. We realize that the distribution is asymmetric, the reason being that the sample was drawn from an exponential distribution. We could, for example, also derive the skewness or the frequency that the mean value exceeds 1.0 from the bootstrap distribution.

While we know the exact solution for the estimation of the mean value and mean squared error of an arbitrary function $u(x)$, however, it is difficult to compute the same quantities for more complicated functions like the median or for correlations, while it is easy with the bootstrap method.

Example 12.3. Acceptance of weighted events

We resume an example from Section 4.2.4 where we had presented an analytic solution. Now we propose a simpler solution: We dispose of a sample of N Monte Carlo generated events with weights w_i, $i = 1, \dots, N$, where we know for each of them whether it is accepted, $\varepsilon_i = 1$, or not,

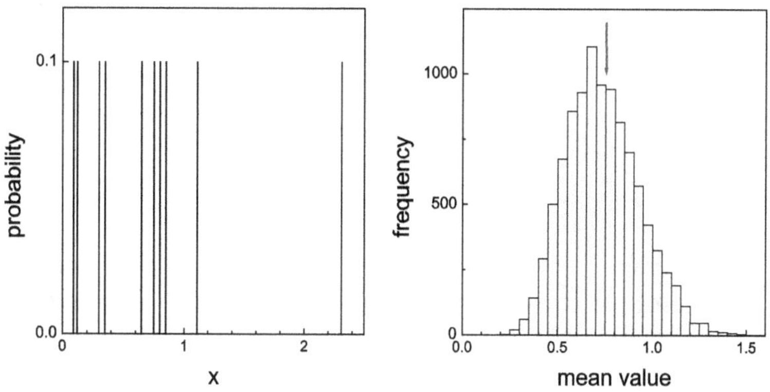

Fig. 12.3 Sample distribution (left) and distribution of bootstrap sample mean values (right).

$\varepsilon_i = 0$. The mean acceptance is $\varepsilon = \sum w_i\varepsilon_i / \sum w_i$. Now we draw from the sample randomly B new bootstrap samples $\{(w_1^*, \varepsilon_1^*), \ldots, (w_N^*, \varepsilon_N^*)\}$ and compute in each case ε^*. The empirical variance $\sigma^2$of the distribution of ε^* is the bootstrap estimate of the error squared, $\delta_\varepsilon^2 = \sigma^2$, of the acceptance ε.

12.2.3 *Precision of the Error Estimate*

Usually we are not interested in the uncertainty σ_δ of the error estimate δ. This is a higher order effect, yet we want to know how many bootstrap samples are required to avoid additional error contributions related to the method.

The standard deviation has two components, σ_t, which depends on the shape of the true distribution and the sample size N, and, σ_B, which depends on the number B of bootstrap replicates. Since the two causes are independent and of purely statistical nature, we can put

$$\sigma_\delta^2 = \sigma_t^2 + \sigma_B^2 ,$$
$$\frac{\sigma_\delta^2}{\delta^2} = \frac{c_1}{N} + \frac{\dot{c}_2}{B} .$$

We can only influence the second term, N being given. Obviously it is sufficient to choose the number B of bootstrap replicates large compared to the number N of experimental observations. For a normal distribution the two constants c_1, c_2 are both equal to 1/2. (A derivation is given in [77].) For distributions with long tails, i.e. large excess γ_2, they are larger. (Remember: $\gamma_2 = 0$ for the normal distribution.) The value of c_2 in the general case is given by [77]:

$$c_2 = \frac{\gamma_2 + 2}{4} .$$

An estimate for γ_2 can be derived from the empirical fourth moment of the sample. Since error estimates are rather crude anyway, we are satisfied with the choice $B \gg N$.

12.2.4 *Precision of Classifiers*

Classifiers like decision trees and ANNs usually subdivide the learning sample in two parts, one part is used to train the classifier and a smaller part is reserved to test the classifier. The precision can be enhanced considerably by using bootstrap samples for both training and testing.

12.2.5 *Random Permutations*

In Chapter 10 we have treated the two-sample problem: "Do two experimental distributions belong to the same population?" In one of the tests, the energy test, we had used permutations of the observations to determine the distribution of the test quantity. The same method can be applied to an arbitrary test quantity which is able to discriminate between samples.

Example 12.4. Two-sample test with a decision tree

Let us assume that we want to test whether the two samples $\{\boldsymbol{x}_1, \dots, \boldsymbol{x}_{N_1}\}$, $\{\boldsymbol{x}_1, \dots, \boldsymbol{x}_{N_2}\}$ of sizes N_1 and N_2 belong to the same population. This is our null hypothesis. Instead of using one of the established two-sample methods we may train a decision tree to separate the two samples. As a test quantity serves the number of wrong classifications $\tilde{N}$ which of course is smaller than $(N_1 + N_2)/2$, half the size of the total sample. Now we combine the two samples, draw from the combined sample two new random samples of sizes N_1 and N_2, train again a decision tree to identify for each element the sample index and count the number of wrong classifications. We repeat this procedure many, say 1000, times and obtain this way the distribution of the test statistic under the null hypothesis.

The fraction of cases where the random selection yields a smaller number of wrong classifications than the original samples is equal to the p-value of the null hypothesis.

Instead of a decision tree we can use any other classifier, for instance a ANN. The corresponding tests are potentially very powerful but also quite involved. Even with nowadays computer facilities, training of some 1000 decision trees or artificial neural nets is quite an effort.

12.2.6 *Jackknife and Bias Correction*

Jackknife has been invented in the 50ties of the last century by Maurice Quenouille and John Tukey. The name *jackknife* had been chosen to indicate the simplicity of the tool.

Jackknife is mainly used for bias removal.[5]

Estimates derived from a sample of N observations $x_1, ..., x_N$ are frequently biased. The bias of a consistent estimator vanishes in the limit $N \to \infty$. Let us assume that he bias decreases proportional to $1/N$. This assumption holds in the majority of cases. To infer the size of the bias $b = \hat{t}_N - t$ of the estimate $\hat{t}_N$ of the true parameter t, we estimate $\hat{t}_{N-1}$ for a sample of $N-1$ events and use the $1/N$ relation to compute the bias. For the expected values $\mathrm{E}(\hat{t}_N)$ and $\mathrm{E}(\hat{t}_{N-1})$ we get

$$\frac{\mathrm{E}(\hat{t}_N) - t}{\mathrm{E}(\hat{t}_{N-1}) - t} = \frac{N-1}{N}$$

and obtain

$$\begin{aligned}
t &= N\mathrm{E}(\hat{t}_N) - (N-1)\mathrm{E}(\hat{t}_{N-1}), \\
\mathrm{E}(b) &= t - \mathrm{E}(\hat{t}_N) = (N-1)\left[\mathrm{E}(\hat{t}_N) - \mathrm{E}(\hat{t}_{N-1})\right], \\
\hat{b} &= (N-1)(\hat{t}_N - \hat{t}_{N-1}).
\end{aligned}$$

To estimate the actual bias, we have to replace the expected values by the observed values. To determine $\hat{t}_{N-1}$, we exclude observation x_i from the sample and compute the estimate $\hat{t}_i$ from the corresponding subsample

[5] Remember, bias corrections should be applied to MLEs only in exceptional situations

that contains $N-1$ observations and average over the results for all values of i:

$$\hat{t}_{N-1} = \frac{1}{N}\sum_{i=1}^{N}\hat{t}_i.$$

The remaining bias after the jackknife correction is zero, of order $1/N^2$, or smaller.

Example 12.5. Jackknife bias correction

When we estimate the variance σ^2 of a distribution from sample $x_1, ..., x_N$ of size N, using the formula

$$\delta_N^2 = \sum_{i=1}^{N}(x_i - \overline{x})^2/N$$

then δ_N^2 is biased, its expected value is smaller than σ^2. We remove one observation at a time and compute each time the mean squared error $\delta_{N-1,i}^2$ and average the results:

$$\delta_{N-1}^2 = \frac{1}{N}\sum_{i=1}^{N}\delta_{N-1,i}^2.$$

Thus an improved estimate is $\delta_c^2 = N\delta_N^2 - (N-1)\delta_{N-1}^2$.

Inserting the known expectation values (see Chapter 3),

$$E(\delta_N^2) = \sigma^2\frac{N-1}{N},$$

we confirm the bias corrected result $E(\delta_c^2) = \sigma^2$.

Appendix

A.1 Large Number Theorems

A.1.1 *Chebyshev Inequality and Law of Large Numbers*

For a probability density $f(x)$ with expected value μ, finite variance σ and arbitrary given positive δ, the following inequality, known as *Chebyshev inequality*, is valid:

$$P\{|x-\mu| \geq \delta\} \leq \frac{\sigma^2}{\delta^2}. \tag{A.1}$$

This very general theorem says that a given, fixed deviation from the expected value becomes less probable when the variance becomes smaller. It is also valid for discrete distributions.

To prove the inequality, we use the definition

$$P_I \equiv P\{x \in I\} = \int_I f(x)\mathrm{d}x,$$

where the domain of integration I is given by $1 \leq |x-\mu|/\delta$. The assertion follows from the following inequalities for the integrals:

$$\int_I f(x)\mathrm{d}x \leq \int_I \left(\frac{x-\mu}{\delta}\right)^2 f(x)\mathrm{d}x \leq \int_{-\infty}^{\infty} \left(\frac{x-\mu}{\delta}\right)^2 f(x)\mathrm{d}x = \sigma^2/\delta^2.$$

Applying (A.1) to the arithmetic mean $\overline{x}$ from N independent identical distributed random variables $x_1, \ldots, x_N$ results in one of the so-called laws of large numbers:

$$P\{|\overline{x} - \langle x\rangle| \geq \delta\} \leq \mathrm{var}(x)/(N\delta^2), \tag{A.2}$$

with the relations $\langle\overline{x}\rangle = \langle x\rangle$, $\mathrm{var}(\overline{x}) = \mathrm{var}(x)/N$ obtained in Sects. 3.2.2 and 3.2.3. The right-hand side disappears for $N \to \infty$, thus in this limit the *probability* to observe an arithmetic mean value outside an arbitrary interval centered at the expected value approaches zero. We talk about stochastic convergence or convergence in probability, here of the arithmetic mean against the expected value.

We now apply (A.2) to the indicator function $\mathcal{I}_I(x) = 1$ for $x \in I$, else 0. The sample mean $\overline{\mathcal{I}_I} = \sum \mathcal{I}_I(x_i)/N$ is the observed relative frequency of events $x \in I$ in the sample. The expected value and the variance are

$$\begin{aligned} \langle \mathcal{I}_I \rangle &= \int \mathcal{I}_I(x) f(x) \mathrm{d}x \\ &= P_I , \\ \mathrm{var}(\mathcal{I}_I) &= \int \mathcal{I}_I^2(x) f(x) \mathrm{d}x - \langle \mathcal{I}_I \rangle^2 \\ &= P_I(1 - P_I) \leq 1/4 , \end{aligned}$$

where, as above, P_I is the probability $P\{x \in I\}$ to find x in the set I. When we insert these results into (A.2), we obtain

$$P\{|\overline{\mathcal{I}_I} - P_I| \geq \delta\} \leq 1/(4N\delta^2) . \tag{A.3}$$

The relative frequency of events of a certain type in a sample converges with increasing N stochastically to the probability to observe an event of that type.[6]

A.1.2 *Central Limit Theorem*

The central limit theorem states that the distribution of the sample mean $\overline{x}$,

$$\overline{x} = \frac{1}{N} \sum_{i=1}^{N} x_i ,$$

of N i.i.d. variables x_i with finite variance σ^2 in the limit $N \to \infty$ will approach a normal distribution with variance σ^2/N independent of the form of the distribution $f(x)$. The following proof assumes that its characteristic function exists.

To simplify the notation, we transform the variable to $y = (x - \mu)/(\sqrt{N}\sigma)$, where μ is the mean of x. The characteristic function of a p.d.f. with mean zero and variance $1/N$ and thus also the p.d.f. of y is of the form

$$\phi(t) = 1 - \frac{t^2}{2N} + c\frac{t^3}{N^{3/2}} + \cdots$$

[6]This theorem was derived by the Dutch-Swiss mathematician Jakob I. Bernoulli (1654–1705).

The characteristic function of the sum $z = \sum_{i=1}^{N} y_i$ is given by the product

$$\phi_z = \left[1 - \frac{t^2}{2N} + c\frac{t^3}{N^{3/2}} + \cdots\right]^N$$

which in the limit $N \to \infty$, where only the first two terms survive, approaches the characteristic function of the standard normal distribution $\mathrm{N}(0,1)$:

$$\lim_{N\to\infty} \phi_z = \lim_{N\to\infty} \left[1 - \frac{t^2}{2N}\right]^N = e^{-t^2/2}.$$

It can be shown that the convergence of characteristic functions implies the convergence of the distributions. The distribution of $\overline{x}$ for large N is then approximately

$$f(\overline{x}) \approx \frac{\sqrt{N}}{\sqrt{2\pi}\sigma} \exp\left[-\frac{N(\overline{x}-\mu)^2}{2\sigma^2}\right].$$

Remark: The law of large numbers and the central limit theorem can be generalized to sums of independent but not identically distributed variates. The convergence is relatively fast when the variances of all variates are of similar size.

A.2 Consistency, Bias and Efficiency of Estimators

The following estimator properties are essential in frequentist statistics.

Throughout this chapter we assume that samples consist of N i.i.d. variables x_i, the true parameter value is θ_0, estimates are $\hat{\theta}$ or t.

A.2.1 *Consistency*

We expect from an useful estimator that it becomes more accurate with increasing size of the sample, i.e. larger deviations from the true value should become more and more improbable.

A sequence of estimators t_N of a parameter θ is called consistent, if their p.d.f.s for $N \to \infty$ are shrinking towards a central value equal to the true parameter value θ_0, or, expressing it mathematically, if

$$\lim_{N\to\infty} P\{|t_N - \theta_0| > \varepsilon\} = 0 \tag{A.4}$$

is valid for arbitrary ε. A sufficient condition for consistency which can be easier checked than (A.4), is the combination of the two requirements

$$\lim_{N\to\infty} \langle t_N \rangle = \theta_0, \quad \lim_{N\to\infty} \mathrm{var}(t_N) = 0,$$

where of course the existence of mean value and variance for the estimator t_N has to be assumed.

For instance, as implied by the law of large numbers, the sample moments

$$t_m = \frac{1}{N}\sum_{i=1}^{N} x_i^m$$

are consistent estimators for the respective m-th moments μ_m of $f(x)$ if this moments exist.

A.2.2 *Bias of Estimates*

The bias of an estimate has been already introduced in Section 6.9.3: An estimate t_N for θ is unbiased, if already for finite N (eventually $N > N_0$) and all parameter values considered, the estimator satisfies the condition

$$\langle t_N \rangle = \theta .$$

The bias of an estimate is defined as:

$$b = \langle t_N \rangle - \theta .$$

Obviously, consistent estimators are asymptotically unbiased:

$$\lim_{N\to\infty} b(N) = 0 .$$

The bias of a consistent estimator can be removed without affecting the consistency by multiplying the estimate with a factor like $(N+a)/(N+b)$ which approaches unity for $N \to \infty$.

A.2.3 *Efficiency*

An important characteristics is of course the accuracy of the statistical estimate. A useful measure for accuracy is the mean square deviation $\langle (t - \theta_0)^2 \rangle$ of the estimate from the true value of the parameter. According to (3.11) it is related to the variance of the estimator and the bias by

$$\left\langle (t-\theta_0)^2 \right\rangle = \mathrm{var}(t) + b^2 . \tag{A.5}$$

Definition: An estimator t is (asymptotically) efficient for the parameter θ , if for all permitted parameter values it fulfills the following conditions for $N \to \infty$:

(1) $\sqrt{N}(t-\theta)$ approaches a normal distribution of constant width and mean equal to zero.
(2) $\mathrm{var}(t) \le \mathrm{var}(t')$ for any other estimator t' which satisfies condition 1.

In other words, an efficient estimate is asymptotically normally distributed and has minimal variance. According to condition 1, its variance decreases with $1/N$. An efficient estimator therefore reaches the same accuracy as a competing one with a smaller sample size N, and is therefore superior. Not in all situations an efficient estimator exists.

Example A.1. Efficiency of different estimates of the location parameter of a Gaussian [117]

Let us compare three methods to estimate the expected value μ of a Gaussian $\mathcal{N}(x|\mu,\sigma)$ with given width σ from a sample $\{x_i\}$, $i = 1,\ldots,N$. For large N we obtain for var(t):

Method 1: sample mean σ^2/N

Method 2: sample median $\sigma^2/N \cdot \pi/2$

Method 3: $(x_{min} + x_{max})/2$ $\sigma^2/N \cdot N\pi^2/(12\ln N)$

Obviously methods 2 and 3 are not efficient. Especially the third method, taking the mean of the two extremal values found in the sample, performs badly here. For other distributions, different results will be found. For the rather exotic two-sided exponential distribution (an exponential distribution of the absolute value of the variate, also called Laplace distribution) method 2 would be efficient and equal to the MLE. For a uniform distribution the estimator of method 3 would be efficient and also equal to the MLE.

While it is of interest to find the estimator which provides the smallest variance, it is not obvious how we could prove this property, since a comparison with all thinkable methods is of course impossible. Here a useful tool is the Cramer–Rao inequality. It provides a lower bound of the variance of an estimator. If we reach this minimum, we can be sure that the optimal accuracy is obtained.

The *Cramer–Rao inequality* states:

$$\mathrm{var}(t) \geq \frac{[1 + (\mathrm{d}b/\mathrm{d}\theta)]^2}{N\langle(\partial \ln f/\partial\theta)^2\rangle} . \tag{A.6}$$

The denominator of the right-hand side is also called, after R. A. Fisher, the information[7] about the parameter θ from a sample of size N of i.i.d. variates.

[7]This is a special use of the word *information* as a technical term.

To prove this inequality, we define the random variable $y = \sum y_i$ with

$$y_i = \frac{\partial \ln f_i}{\partial \theta}, \quad f_i \equiv f(x_i|\theta). \tag{A.7}$$

It has the expected value

$$\begin{aligned}
\langle y_i \rangle &= \int \frac{1}{f_i} \frac{\partial f_i}{\partial \theta} f_i \, \mathrm{d}x_i \\
&= \int \frac{\partial f_i}{\partial \theta} \, \mathrm{d}x_i \\
&= \frac{\partial}{\partial \theta} \int f_i \, \mathrm{d}x_i \\
&= \frac{\partial}{\partial \theta} 1 = 0.
\end{aligned} \tag{A.8}$$

Because of the independence of the y_i we have $\langle y_i y_j \rangle = \langle y_i \rangle \langle y_j \rangle = 0$ and

$$\mathrm{var}(y) = N \langle y_i^2 \rangle = N \left\langle \left(\frac{\partial \ln f}{\partial \theta} \right)^2 \right\rangle . \tag{A.9}$$

Using the definition $L = \prod f_i$, we find for $\mathrm{cov}(ty) = \langle (t - \langle t \rangle)(y - \langle y \rangle) \rangle$:

$$\begin{aligned}
\mathrm{cov}(ty) &= \int t \frac{\partial \ln L}{\partial \theta} L \, \mathrm{d}x_1 \cdots \mathrm{d}x_N \\
&= \int t \frac{\partial}{\partial \theta} L \, \mathrm{d}x_1 \cdots \mathrm{d}x_N \\
&= \frac{\partial}{\partial \theta} \langle t \rangle \\
&= 1 + \frac{\mathrm{d}b}{\mathrm{d}\theta}.
\end{aligned} \tag{A.10}$$

From the Cauchy–Schwarz inequality

$$[\mathrm{cov}(ty)]^2 \le \mathrm{var}(t) \mathrm{var}(y)$$

and (A.9), (A.10) follows (A.6).

The equality sign in (A.6) is valid if and only if the two factors t, y in the covariance are proportional to each other. In this case t is called a Minimum Variance Bound (MVB) estimator. It can be shown to be also minimal sufficient.

In most of the literature efficiency is defined by the stronger condition: An estimator is called efficient, if it is bias-free and if it satisfies the MVB.

A.3 Properties of the Maximum Likelihood Estimator

A.3.1 *Consistency*

The maximum likelihood estimator (MLE) is consistent under mild assumptions.

To prove this property, we consider the expected value of

$$\ln L(\theta|\boldsymbol{x}) = \sum_{i=1}^{N} \ln f(x_i|\theta) \tag{A.11}$$

which is to be calculated by integration over the variables[8] $\boldsymbol{x}$ using the true p.d.f. (with the true parameter θ_0). First we prove the inequality

$$\langle \ln L(\theta|\boldsymbol{x}) \rangle < \langle \ln L(\theta_0|\boldsymbol{x}) \rangle , \tag{A.12}$$

for $\theta \neq \theta_0$: Since the logarithm is a strongly concave function, there is always $\langle \ln(\ldots) \rangle < \ln \langle (\ldots) \rangle$, hence

$$\left\langle \ln \frac{L(\theta|\boldsymbol{x})}{L(\theta_0|\boldsymbol{x})} \right\rangle < \ln \left\langle \frac{L(\theta|\boldsymbol{x})}{L(\theta_0|\boldsymbol{x})} \right\rangle = \ln \int \frac{L(\theta|\boldsymbol{x})}{L(\theta_0|\boldsymbol{x})} L(\theta_0|\boldsymbol{x}) \mathrm{d}\boldsymbol{x} = \ln 1 = 0 .$$

In the last step we used

$$\int L(\theta|\boldsymbol{x}) \mathrm{d}\boldsymbol{x} = \int \prod f(x_i|\theta) \mathrm{d}x_1 \cdots \mathrm{d}x_N = 1 .$$

Since $\ln L(\theta|\boldsymbol{x})/N = \sum \ln f(x_i|\theta)/N$ is an arithmetic sample mean which, according to the law of large numbers (A.2), converges stochastically to the expected value for $N \to \infty$, we have also (in the sense of stochastic convergence)

$$\ln L(\theta|\boldsymbol{x})/N \to \langle \ln f(x|\theta) \rangle = \sum \langle \ln f(x_i|\theta) \rangle /N = \langle \ln L(\theta|\boldsymbol{x}) \rangle /N ,$$

and from (A.12)

$$\lim_{N\to\infty} P\{\ln L(\theta|\boldsymbol{x}) < \ln L(\theta_0|\boldsymbol{x})\} = 1 ,\ \theta \neq \theta_0 . \tag{A.13}$$

On the other hand, the MLE $\hat{\theta}$ is defined by the extremal condition

$$\ln L(\hat{\theta}|\boldsymbol{x}) \geq \ln L(\theta_0|\boldsymbol{x}) .$$

A contradiction to (A.13) can be avoided only, if also

$$\lim_{N\to\infty} P\{|\hat{\theta} - \theta_0| < \varepsilon\} = 1$$

is valid. This means consistency of the MLE.

[8] We keep the form of the argument list of L, although now $\boldsymbol{x}$ is not considered as fixed to the experimentally sampled values, but as a random vector with given p.d.f..

A.3.2 *Efficiency*

Since the MLE is consistent, it is unbiased asymptotically for $N \to \infty$. Under certain assumptions in addition to the usually required regularity[9] the MLE is also efficient asymptotically.

Proof:

With the notations of the last paragraph with $L = \prod f_i$ and using (A.8), the expected value and variance of $y = \sum y_i = \partial \ln L / \partial \theta$ are given by the following expressions:

$$\langle y \rangle = \int \frac{\partial \ln L}{\partial \theta} L \, d\boldsymbol{x} = 0, \tag{A.14}$$

$$\sigma_y^2 = \mathrm{var}(y) = \left\langle \left(\frac{\partial \ln L}{\partial \theta} \right)^2 \right\rangle = -\left\langle \frac{\partial^2}{\partial \theta^2} \ln L \right\rangle . \tag{A.15}$$

The last relation follows after further differentiation of (A.14) and from the relation

$$\int \frac{\partial^2 \ln L}{\partial \theta^2} L \, d\boldsymbol{x} = -\int \frac{\partial \ln L}{\partial \theta} \frac{\partial L}{\partial \theta} d\boldsymbol{x} = -\int \frac{\partial \ln L}{\partial \theta} \frac{\partial \ln L}{\partial \theta} L \, d\boldsymbol{x} .$$

From the Taylor expansion of $\partial \ln L / \partial \theta|_{\theta = \hat{\theta}}$ which is zero by definition and with (A.15) we find

$$\begin{aligned} 0 = \frac{\partial \ln L}{\partial \theta}|_{\theta=\hat{\theta}} &\approx \frac{\partial \ln L}{\partial \theta}|_{\theta=\theta_0} + (\hat{\theta} - \theta_0) \frac{\partial^2 \ln L}{\partial \theta^2}|_{\theta=\theta_0} \\ &\approx y - (\hat{\theta} - \theta_0)\sigma_y^2 , \end{aligned} \tag{A.16}$$

where the consistency of the MLE guaranties the validity of this approximation in the sense of stochastic convergence. Following the central limit theorem, y/σ_y being the sum of i.i.d. variables, is asymptotically normally distributed with mean zero and variance unity. The same is then true for $(\hat{\theta} - \theta_0)\sigma_y$, i.e. $\hat{\theta}$ follows asymptotically a normal distribution with mean θ_0 and asymptotically vanishing variance $1/\sigma_y^2 \sim 1/N$, as seen from (A.9).

A.3.3 *Asymptotic Form of the Likelihood Function*

A similar result as derived in the last paragraph for the p.d.f. of the MLE $\hat{\theta}$ can be derived for the likelihood function itself.

[9] The boundaries of the domain of x must not depend on θ and the maximum of L should not be reached at the boundary of the range of θ.

If one considers the Taylor expansion of $y = \partial \ln L / \partial\theta$ around the MLE $\hat{\theta}$, we get with $y(\hat{\theta}) = 0$

$$y(\theta) \approx (\theta - \hat{\theta}) y'(\hat{\theta}). \tag{A.17}$$

As discussed in the last paragraph, we have for $N \to \infty$

$$y'(\hat{\theta}) \to y'(\theta_0) \to \langle y' \rangle = -\sigma_y^2 = const.$$

Thus $y'(\hat{\theta})$ is independent of $\hat{\theta}$ and higher derivatives disappear. After integration of (A.17) over θ we obtain a parabolic form for $\ln L$:

$$\ln L(\theta) = \ln L(\hat{\theta}) - \frac{1}{2}\sigma_y^2(\theta - \hat{\theta})^2,$$

where the width of the parabola decreases with $\sigma_y^{-2} \sim 1/N$ (A.9). Up to the missing normalization, the likelihood function has the same form as the distribution of the MLE with $\hat{\theta} - \theta_0$ replaced by $\theta - \hat{\theta}$.

A.3.4 *Properties of the Maximum Likelihood Estimate for Small Samples*

The criterion of asymptotic efficiency, fulfilled by the MLE for large samples, is usually extended to small samples, where the normal approximation of the sampling distribution does not apply, in the following way: A bias-free estimate t is called a *minimum variance* (MV) estimate if $\mathrm{var}(t) \leq \mathrm{var}(t')$ for any other bias-free estimate t'. If, moreover, the Cramer–Rao inequality (A.6) is fulfilled as an equality, one speaks of a *minimum variance bound* (MVB) estimate, often also called efficient or most efficient, estimate (not to be confused with the asymptotic efficiency which we have considered before in Appendix A.2). The latter, however, exists only for a certain function $\tau(\theta)$ of the parameter θ if it has a one-dimensional sufficient statistic (see 6.6.1). It can be shown [3] that under exactly this condition the MLE for τ will be this MVB estimate, and therefore bias-free for any N. The MLE for any non-linear function of τ will in general be biased, but still optimal in the following sense: if bias-corrected, it becomes an MV estimate, i.e. it will have the smallest variance among all unbiased estimates.

Example A.2. : Efficiency of small sample MLEs

The MLE for the variance σ^2 of a normal distribution with known mean μ,

$$\widehat{\sigma^2} = \frac{1}{N}\sum (x_i - \mu)^2,$$

is unbiased and efficient, reaching the MVB for all N. The MLE for σ is of course

$$\hat{\sigma} = \sqrt{\widehat{\sigma^2}},$$

according to the relation between σ and σ^2. It is biased and thus not efficient in the sense of the above definition. A bias-corrected estimator for σ is (see for instance [118])

$$\hat{\sigma}_{corr} = \sqrt{\frac{N}{2}} \frac{\Gamma\left(\frac{N}{2}\right)}{\Gamma\left(\frac{N+1}{2}\right)} \hat{\sigma}.$$

This estimator can be shown to have the smallest variance of all unbiased estimators, independent of the sample size N.

In the above example a one-dimensional sufficient statistic exists. If this is not the case, the question whether the MLE is optimal for small samples, from a frequentist point of view cannot be answered.

In summary, also for finite N the MLE for a certain parameter achieves the optimal – from the frequentist point of view – properties of an MVB estimator, if the latter does exist. Of course these properties cannot be preserved for other parametrizations, since variance and bias are not invariant properties.

A.4 The Expectation Maximization (EM) Algorithm

The EM method finds iteratively the MLE in situations where the statistical model depends on latent variables. The method goes back to the sixties, has been invented several times and has been made popular by Dempster, Laird and Rubin [68]. A very comprehensive introduction to the expectation maximization (EM) algorithm is given in the Wikipedia article Ref. [125]. The EM algorithm exists in many different variants. We will restrict our discussion to its application to classification problems.

To get an idea of the method, we consider a simple standard example. Let us assume that we have a sample of observations $x_1, ..., x_N$ each drawn from one of M overlapping normal distribution $f_m(x|\mu_m) \sim \mathcal{N}(\mu_m, s)$ with unknown mean values μ_1,..., μ_M and given standard deviation s.

The log-likelihood of the parameters is

$$\ln L(\mu_1, ..., \mu_N) = \sum_{m=1}^{M} \frac{\Sigma_{i=1}^{N} z_{mi} x_i}{\Sigma_{i=1}^{N} z_{mi}} - \mu_m)$$

where the classification variable $z_{mi} = 1$ if x_i belongs to the normal distribution m and $z_{mi} = 0$ otherwise. If we know the classification variables, we get the MLE of the parameter μ_m:

$$\hat{\mu}_m = \frac{\Sigma_{i=1}^{N} z_{mi} x_i}{\Sigma_{i=1}^{N} z_{mi}}.$$

If this is not the case, we can estimate z_{mi} from the observed distribution. In the EM formalism z_{mi} is called *missing* or *latent variable.* We can solve our problem iteratively with two altenating steps, an expectation and a maximization step. We start with a first guess $\mu_m^{(1)}$ of the parameters of interest and estimate the missing data. In the *expectation step* k we compute the probability $g_{mi}^{(k)}$that x_i belongs to subdistribution m. It is proportional to the value of the distribution $f_m(x_i|\mu_m)$ at x_i:

$$g_{mi}^{(k)} = \frac{f_m(x_i|\hat{\mu}_m^{(k)})}{\Sigma_{j=1}^{M} f_j(x_i|\hat{\mu}_j^{(k)})}.$$

The probability g_{mi} is the expected value of the latent variable z_{mi}.The *expected* log-likelihood is

$$Q(\boldsymbol{\mu}, \widehat{\boldsymbol{\mu}}^{(k)}) = \sum_{m=1}^{M} \left(\sum_{i=1}^{N} g_{mi}^{(k)} x_i - \mu_m \right).$$

In the *maximization* step we obtain the MLEs

$$\hat{\mu}_m^{(k+1)} = \Sigma_{i=1}^{N} g_{mi}^{(k)} x_i.$$

which are used in the following expectation step. The iteration converges to the overall MLE.

Let us generalize this procedure. Given be a probability distribution $p(\boldsymbol{x}, \boldsymbol{z}|\boldsymbol{\theta}) = g(\boldsymbol{z}|\boldsymbol{x}\ ,\boldsymbol{\theta})p_1(\boldsymbol{x}|\boldsymbol{\theta})$ depending on a parameter vector $\boldsymbol{\theta}$ and a sample of observations $x_1, ..., x_N$. The distribution g of the latent variables $\boldsymbol{z}$ is a function of $\boldsymbol{\theta}$ and $\boldsymbol{x}$.

- *Expectation step*: For the parameter vector $\boldsymbol{\theta}^{(k)}$ we compute the distribution g of the hidden variables $\boldsymbol{z}$. We form the log-likelihood function $\ln L(\theta|\boldsymbol{x}, \boldsymbol{z})$ which is a random variable as it depends on the random z. Averaging over z, we obtain the expected value $Q(\boldsymbol{\theta}, \widehat{\boldsymbol{\theta}}^{(k)})$ of the log-likelihood:

$$Q(\boldsymbol{\theta}, \widehat{\boldsymbol{\theta}}^{(k)}) = \boldsymbol{E}_{z|x,\theta^{(k)}} [\ln L(\boldsymbol{\theta}|\boldsymbol{x}, \boldsymbol{z})].$$

The conditional expectation means that we average over $\boldsymbol{z}$ given the distribution of $g(\boldsymbol{z})$ obtained for fixed values $\boldsymbol{x}$ and $\widehat{\boldsymbol{\theta}}^{(k)}$:

$$Q(\boldsymbol{\theta}, \widehat{\boldsymbol{\theta}}^{(k)}) = \int_{\mathcal{Z}} \ln L(\boldsymbol{\theta}|\boldsymbol{x}, \boldsymbol{z}) g\left(\boldsymbol{z}|\boldsymbol{x}, \widehat{\boldsymbol{\theta}}^{(k)}\right) dz.$$

If the values of the vector components $\boldsymbol{z}$ are discrete, the integral is replaced by a sum over all J possible values:

$$Q(\boldsymbol{\theta}, \widehat{\boldsymbol{\theta}}^{(k)}) = \sum_{j=1}^{J} \ln L(\boldsymbol{\theta}|\boldsymbol{x}, \boldsymbol{z}_j) g\left(\boldsymbol{z}_j|\boldsymbol{x}, \widehat{\boldsymbol{\theta}}^{(k)}\right).$$

Alternatively, somewhat less efficient, we can insert the expected values of the latent variables:

$$Q(\boldsymbol{\theta}, \widehat{\boldsymbol{\theta}}^{(k)}) = \ln L\left(\boldsymbol{\theta}|\boldsymbol{x}, \boldsymbol{E}(\boldsymbol{z})\right).$$

- *Maximization step*: The MLE $\boldsymbol{\theta}^{(k+1)}$ is computed:

$$\boldsymbol{\theta}^{(k+1)} = \arg\max_{\theta} Q(\theta|\theta^{(k)}).$$

The procedure is started with a first $\boldsymbol{\theta}^{(1)}$ guess of the parameters and iterated. It converges to a minimum of the log-likelihood. To avoid that the iteration is caught by a local minimum, different starting values can be selected. It is especially useful in classification problems in connection with p.d.f.s of the exponential family[10] where the maximization step is relatively simple.

Example A.3. Unfolding a histogram

Experimental data are collected in form of a histogram with N bins. The number of events in bin i be d_i. The experiment suffers from an imperfect resolution and from acceptance losses which we have to correct for. The "true" histogram with M bins contains θ_j events in bin j. Knowing the measurement device we can simulate the experimental effects and determine the matrix A which relates $\boldsymbol{\theta}$ with the expected values of the numbers $\boldsymbol{d}$: $\boldsymbol{E}(d_i) = \Sigma_{j=1}^{M} A_{ij}\theta_j$. The element A_{ij} is the probability to observe an event in bin i which belongs to the bin j in the true histogram. The missing

[10]To the exponential family belong among others the normal, Poisson, exponential, gamma, χ^2-distribution.

information is the number of events d_{ij} in an observed bin i that belong to the true bin j. Hence there are M missing variables per bin. The number d_{ij} is Poisson distribution with mean $A_{ij}\theta_j$. The likelihood depends only on the hidden variables:

$$\ln L(\boldsymbol{\theta}|d_{11}, ..., d_{NM}) = \sum_{j=1}^{M}\sum_{i=1}^{N}[-A_{ij}\theta_j + d_{ij}\ln A_{ij}\theta_j].$$

The following alternating steps are repeated:

- Expectation step: We have

$$\begin{aligned} Q(\boldsymbol{\theta}, \widehat{\boldsymbol{\theta}}_j^{(k)}) &= \boldsymbol{E}_{d_{ik}} \ln L \\ &= \sum_{j=1}^{M}\sum_{i=1}^{N}[-A_{ij}\theta_j + \boldsymbol{E}(d_{ij}^{(k)})\ln A_{ij}\theta_j]. \end{aligned}$$

  The expected value $\boldsymbol{E}(d_{ij}^{(k)})$ conditioned on d_i and $\widehat{\boldsymbol{\theta}}^{(k)}$ is given by d_i times the probability that an event of bin i belongs to true bin j:

$$\boldsymbol{E}(d_{ij}^{(k)}) = d_i \frac{A_{ij}\hat{\theta}_j^{(k)}}{\sum_{j=1}^{M} A_{ij}\hat{\theta}_j^{(k)}}.$$

  We get

$$Q(\boldsymbol{\theta}, \widehat{\boldsymbol{\theta}}_j^{(k)}) = \sum_{j=1}^{M}\sum_{i=1}^{N}[-A_{ij}\theta_j + d_i \frac{A_{ij}\hat{\theta}_j^{(k)}}{\sum_{m=1}^{M} A_{im}\hat{\theta}_m^{(k)}} \ln A_{ij}\theta_j].$$

- Maximization step:

The computation of the maximum of Q is easy, because the components of the parameter vector $\boldsymbol{\theta}$ appear in independent summands.

$$\frac{\partial Q}{\partial \theta_j} = \sum_{i=1}^{N} \left[-A_{ij} + d_i \frac{A_{ij}\hat{\theta}_j^{(k)}}{\sum_{j=1}^{M} A_{ij}\hat{\theta}_j^{(k)}} \frac{1}{\theta_j} \right] = 0,$$

$$\hat{\theta}_j^{(k+1)} \sum_{i=1}^{N} A_{ij} = \sum_{i=1}^{N} d_i \frac{A_{ij}\hat{\theta}_j^{(k)}}{\sum_{j=1}^{M} A_{ij}\hat{\theta}_j^{(k)}},$$

$$\theta_j^{(k+1)} = \sum_{i=1}^{N} d_i \frac{A_{ij}\theta_j^{(k)}}{\sum_{j=1}^{M} A_{ij}\theta_j^{(k)}} / \alpha_j .$$

$\Sigma_{i=1}^{N} A_{ij} = \alpha_j$ is the average acceptance of the events in the true bin j.

A.5 χ^2 Test for Composite Hypotheses

In most cases measurements do not serve to verify a fixed theory but to estimate one or several parameters of it. The method of least squares for parameter estimation has been discussed in Section 6.8. To fit a curve $y = t(x, \boldsymbol{\theta})$ to measured points y_i with Gaussian errors δ_i, $i = 1, \ldots, N$, that are independent of the parameters,

$$\chi^2 = \sum_{i=1}^{N} \frac{(y_i - t_i(x_i, \theta_1, \ldots, \theta_Z))^2}{\delta_i^2}, \tag{A.18}$$

is minimized with respect to the Z free parameters. In what follows it is shown that for linear functions t_i of the parameters, after the adjustment, χ^2 follows a χ^2-distribution of $N - Z$ degrees of freedom.

It is plausible that with increasing number of free parameters the description of the data improves, χ^2 decreases. In the extreme case where the number of parameters is equal to the number of measured points or histogram bins, it becomes zero. It has to be shown, that if Z parameters are adjusted, and the functions t_i depend linearly on the parameters, χ^2 follows a χ^2-distribution of $f = N - Z$ degrees of freedom.

We set $z_i = (y_i - t(x_i, \boldsymbol{\theta}))/\delta_i$ in (A.18) and differentiate χ^2 with respect to the parameter θ_1. The result has to be zero, because χ^2 has been minimized.

$$-\frac{1}{2}\frac{\partial \chi^2}{\partial \theta_1} = \sum_{i=1}^{N} z_i \frac{\partial t_i}{\partial \theta_1}/\delta_i = 0.$$

As the functions t_i are linear function of the parameters and as the errors δ_i are independent of the parameters, the derivative $\partial t_i/\partial \theta_1$ is a constant. We obtain a linear relation

$$\sum_{i=1}^{Z} c_i z_i/\delta_i = 0.$$

The differentiation with respect to the other parameters produces similar relations. The Z relations constrain the N variables z_i. Thus we remain with $f = N - Z$ degrees of freedom. We can use the Z relations to reduce the N variables to only f variables z_i, $i = 1, \dots, f$. Consequently, under the quoted conditions, χ^2 follows a χ^2-distribution of f degrees of freedom and a standard χ^2 test can be applied.

In the case of a fit of a function with Z parameters to a histogram with B bins, we have $f = B - Z$ degrees of freedom. If the histogram is normalized, we add one constraint and remain with $f = B - Z - 1$ degrees of freedom.

A.6 *p*-Values for EDF-Statistics

The formulas reviewed here are taken from the book of D'Agostino and Stephens [123] and generalized to include the case of the two-sample comparison.

Calculation of the Test Statistics

The calculation of the supremum statistics D and of $V = D_+ + D_-$ is simple enough, so we will skip a further discussion.

The quadratic statistics W^2, U^2 and A^2 are calculated after a probability integral transformation (PIT). The PIT transforms the expected theoretical distribution of x into a uniform distribution. The new variate z is found from the relation $z = F(x)$, whereby F is the integral distribution function of x.

With the transformed observations z_i, ordered according to increasing values, we get for W^2, U^2 and A^2:

$$W^2 = \frac{1}{12N} + \sum_{i=1}^{N}(z_i - \frac{2i-1}{2N})^2, \tag{A.19}$$

$$U^2 = \frac{\sum_{i=2}^{N}(z_i - z_{i-1})^2}{\sum_{i=1}^{N} z_i^2},$$

$$A^2 = -N + \sum_{i=1}^{N-1}(z_i - 1)\left(\ln z_i + \ln(1 - z_{N+1-i})\right). \tag{A.20}$$

If we know the distribution function only through a Monte Carlo simulation but not analytically, the z-value for an observation x is approximately $z \approx$ (number of Monte-Carlo observations with $x_{MC} < x$) / (total number of Monte Carlo observations). (Somewhat more accurate is an interpolation). For the comparison with a simulated distribution is N to be taken as the equivalent number of observations

$$\frac{1}{N} = \frac{1}{N_{\text{exp}}} + \frac{1}{N_{MC}}.$$

Here N_{exp} and N_{MC} are the experimental respectively the simulated sample sizes.

Calculation of p-Values

After normalizing the test variables with appropriate powers of N they follow p.d.f.s which are independent of N. The test statistics' D^*, W^{2*}, A^{2*} modified in this way are defined by the following empirical relations

$$D^* = D_{max}(\sqrt{N} + 0.12 + \frac{0.11}{\sqrt{N}}), \tag{A.21}$$

$$W^{2*} = (W^2 - \frac{0.4}{N} + \frac{0.6}{N^2})(1.0 + \frac{1.0}{N}), \tag{A.22}$$

$$A^{2*} = A^2. \tag{A.23}$$

The relation between these modified statistics and the p-values is given in Fig. A.1.

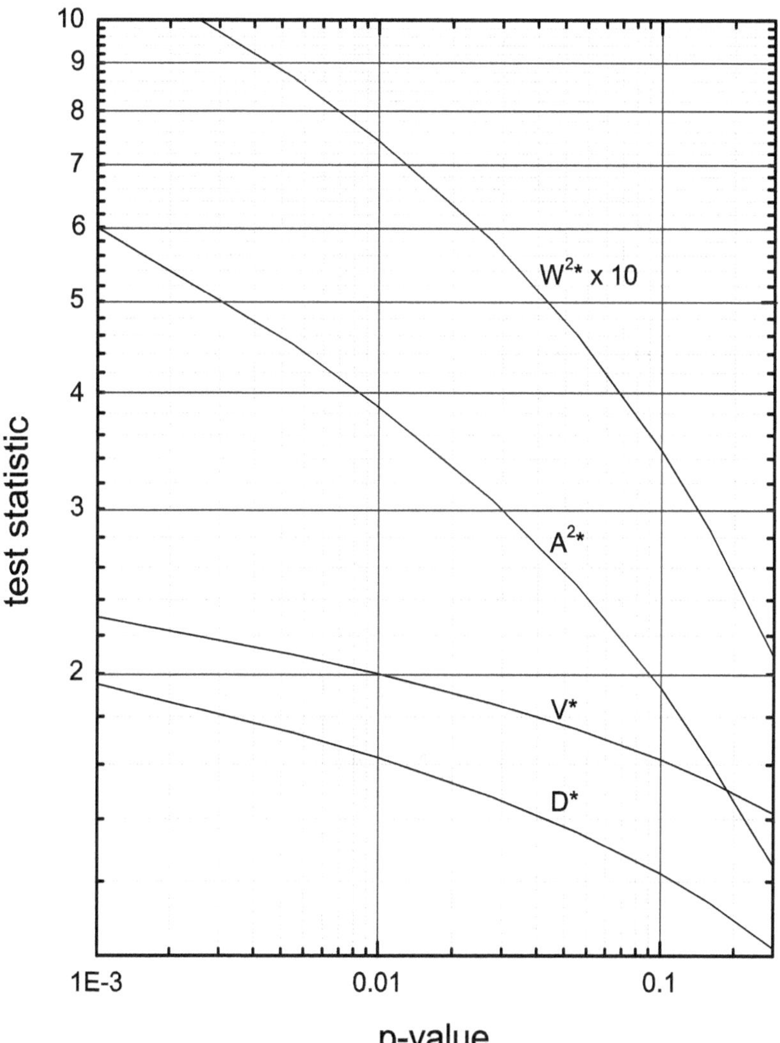

Fig. A.1 p-values for empirical test statistics.

A.7 Fisher–Yates Shuffle

The Fisher–Yates Shuffle produces a random permutation of elements of a sample. It is rather trivial. Given be a set of n numbered elements, an element j is randomly selected from the first $n-1$ elements. The elements n and j are exchanged. Then a new element j' is randomly selected from the first $n-2$ elements of the modified arrangement and exchanged with the element $n-1$. The procedure is continued until the beginning of the queue is reached.

The time needed for the shuffle is of order n.

A.8 Comparison of Histograms Containing Weighted Events, Goodness-of-Fit and Parameter Estimation

In the main text we have treated goodness-of-fit and parameter estimation from a comparison of two histograms in the simple situation where the statistical errors of one of the histograms (generated by Monte Carlo simulation) was negligible compared to the uncertainties of the other histogram. Here we take the errors of both histograms into account and also permit that the histogram bins contain weighted entries.

A.8.1 *Comparison of Two Poisson Numbers with Different Normalization*

We compare $c_n n$ with $c_m m$ where the normalization constants c_n, c_m are known and n, m are Poisson distributed. Only c_n/c_m matters and for example c_n could be set equal to one, but we prefer to keep both constants because then the formulas are more symmetric. The null hypothesis H_0 is that n is drawn from a Poisson distribution with mean λ/c_n and m from a Poisson distribution with mean λ/c_m. To compare the numbers we define the χ^2 expression:

$$\chi^2 = \frac{(c_n n - c_m m)^2}{\delta^2} . \tag{A.24}$$

The denominator δ^2 is the expected variance of the parenthesis in the numerator under the null hypothesis. To compute δ we have to estimate λ. The p.d.f.s of n and m are $\mathcal{P}(n|\lambda/c_n)$ and $\mathcal{P}(m|\lambda/c_m)$ leading to the corresponding log-likelihood of λ

$$\ln L(\lambda) = n \ln \frac{\lambda}{c_n} - \frac{\lambda}{c_n} + m \ln \frac{\lambda}{c_m} - \frac{\lambda}{c_m} + const.$$

with the MLE

$$\hat{\lambda} = \frac{n+m}{1/c_n + 1/c_m} = c_n c_m \frac{n+m}{c_n + c_m} . \tag{A.25}$$

We assume now that n is distributed according to a Poisson distribution with mean $\hat{n} = \hat{\lambda}/c_n$ and m according to a Poisson distribution with, mean $\hat{m} = \hat{\lambda}/c_m$. We get

$$\begin{aligned}\delta^2 &= c_n^2 \hat{n} + c_m^2 \hat{m} \\ &= (c_n + c_m)\hat{\lambda} \\ &= c_n c_m (n + m)\end{aligned}$$

and inserting the result into (A.24), we obtain

$$\chi^2 = \frac{1}{c_n c_m} \frac{(c_n n - c_m m)^2}{n + m}. \tag{A.26}$$

As mentioned, only the relative normalization c_n/c_m is relevant.

A.8.2 *Comparison of Weighted Sums*

When we compare experimental data to a Monte Carlo simulation, the simulated events frequently are weighted. We generalize our result to the situation where both numbers n, m consist of a sums of weights, v_k, respectively w_k, $n = \sum v_k$, $m = \sum w_k$. In Appendix A.9.1 it is shown that the sum of weights for not too small event numbers can be approximated by a scaled Poisson distribution and that this approximation is superior to the approximation with a normal distribution. Now the equivalent numbers of unweighted events (see relation (3.64)) $\tilde{n}$ and $\tilde{m}$,

$$\tilde{n} = \frac{\left[\sum v_k\right]^2}{\sum v_k^2}, \quad \tilde{m} = \frac{\left[\sum w_k\right]^2}{\sum w_k^2}, \tag{A.27}$$

are approximately Poisson distributed. We simply have to replace (A.26) by

$$\chi^2 = \frac{1}{\tilde{c}_n \tilde{c}_m} \frac{(\tilde{c}_n \tilde{n} - \tilde{c}_m \tilde{m})^2}{\tilde{n} + \tilde{m}} \tag{A.28}$$

where now $\tilde{c}_n, \tilde{c}_m$ are the relative normalization constants for the equivalent numbers of events. We summarize in short the relevant relations, assuming that and as before $c_n n$ is supposed to agree with $c_m m$ as before. As discussed in 3.7.3 we find with $\tilde{c}_n \tilde{n} = c_n n$, $\tilde{c}_m \tilde{m} = c_m m$:

$$\tilde{c}_n = c_n \frac{\sum v_k^2}{\sum v_k}, \quad \tilde{c}_m = c_m \frac{\sum w_k^2}{\sum w_k}. \tag{A.29}$$

A.8.3 χ^2 *of Histograms of Weighted Events*

We have to evaluate the expression (A.28) for each bin and sum over all B bins

$$\chi^2 = \sum_{i=1}^{B} \left[\frac{1}{\tilde{c}_n \tilde{c}_m} \frac{(\tilde{c}_n \tilde{n} - \tilde{c}_m \tilde{m})^2}{\tilde{n} + \tilde{m}} \right]_i \tag{A.30}$$

where the prescription indicated by the index i means that all quantities in the bracket have to be evaluated for bin i. In case the entries are not weighted the tilde is obsolete. The constants c_n, c_m in (A.29) usually are overall normalization constants and equal for all bins of the corresponding histogram. If the histograms are normalized with respect to each other, we have $c_n \Sigma n_i = c_m \Sigma m_i$ and we can set $c_n = \Sigma m_i = M$ and $c_m = \Sigma n_i = N$.

A.8.3.1 χ^2-*Goodness-of-Fit Test*

This expression (A.30) can be used for goodness-of-fit tests. In case the normalization constants are given externally, for instance through the luminosity, χ^2 follows approximately a χ^2-distribution of B degrees of freedom. Frequently the histograms are normalized with respect to each other. Then we have one degree of freedom less, i.e. $B - 1$. If P parameters have been adjusted in addition, then we have $B - P - 1$ degrees of freedom.

A.8.3.2 *Likelihood Ratio Test*

In Chapter 10, Section 10.4.3 we have introduced the likelihood ratio test for histograms. For a pair of Poisson numbers n, m the likelihood ratio is the ratio of the maximal likelihood under the condition that the two numbers are drawn from the same distribution to the unconditioned maximum of the likelihood for the observation of n. The corresponding difference of the logarithms is our test statistic V (see likelihood ratio test for histograms)

$$\begin{aligned} V &= n \ln \frac{\lambda}{c_n} - \frac{\lambda}{c_n} - \ln n! + m \ln \frac{\lambda}{c_m} - \frac{\lambda}{c_m} - \ln m! - [n \ln n - n - \ln n!] \\ &= n \ln \frac{\lambda}{c_n} - \frac{\lambda}{c_n} + m \ln \frac{\lambda}{c_m} - \frac{\lambda}{c_m} - \ln m! - n \ln n + n . \end{aligned}$$

We now turn to weighted events and perform the same replacements as above:

$$V = \tilde{n} \ln \frac{\tilde{\lambda}}{\tilde{c}_n} - \frac{\tilde{\lambda}}{\tilde{c}_n} + \tilde{m} \ln \frac{\tilde{\lambda}}{\tilde{c}_m} - \frac{\tilde{\lambda}}{\tilde{c}_m} - \ln \tilde{m}! - \tilde{n} \ln \tilde{n} + \tilde{n} .$$

Here the parameter $\tilde{\lambda}$ is the MLE corresponding to (A.25) for weighted events.

$$\tilde{\lambda} = \tilde{c}_n \tilde{c}_m \frac{\tilde{n} + \tilde{m}}{\tilde{c}_n + \tilde{c}_m} \tag{A.31}$$

The test statistic of the full histogram is the sum of the contributions from all bins.

$$V = \sum_{i=1}^{B} \left[\tilde{n} \ln \frac{\tilde{\lambda}}{\tilde{c}_n} - \frac{\tilde{\lambda}}{\tilde{c}_n} + \tilde{m} \ln \frac{\tilde{\lambda}}{\tilde{c}_m} - \frac{\tilde{\lambda}}{\tilde{c}_m} - \ln \tilde{m}! - \tilde{n} \ln \tilde{n} + \tilde{n} \right]_i .$$

The variables and parameters of this formula are given in relations (A.31), (A.27) and (A.29) and depend on c_n, c_m. As stated above, only the ratio c_n, c_m matters. The ratio is either given or obtained from the normalization $c_n \Sigma n_i = c_m \Sigma m_i$.

The distribution of the test statistic under H_0 for large event number follows approximately a χ^2-distribution of B degrees of freedom if the normalization is given or of $B - 1$ degrees of freedom in the usual case where the histograms are normalized to each other. For small event numbers the distribution of the test statistic has to be obtained by simulation.

A.8.4 *Parameter Estimation*

When we compare experimental data to a parameter dependent Monte Carlo simulation, one of the histograms depends on the parameter, e.g. $m(\theta)$ and the comparison is used to determine the parameter. During the fitting procedure, the parameter is modified and this implies a change of the weights of the Monte Carlo events. The experimentally observed events are not weighted. Then (A.30) simplifies with $\tilde{n}_i = n_i$, $\tilde{c}_n = c_n$, $\tilde{c}_m = c_m w$ and $\tilde{m}_i$ is just the number of unweighted Monte Carlo events in bin i.

$$\chi^2 = \sum_{i=1}^{B} \left[\frac{1}{c_n \tilde{c}_m} \frac{(c_n n - \tilde{c}_m \tilde{m})^2}{(n + \tilde{m})} \right]_i$$

For each minimization step we have to recompute the weights and with (A.27) and (A.29) the LS parameter χ^2. If the relative normalization of the simulated and observed data is not known the ratio c_n / c_m is a free parameter in the fit. As only the ratio matters, we can set for instance $c_m = 1$.

We do not recommend to apply a likelihood fit, because the approximation of the distribution of the sum of weights by a scaled Poisson distribution is not valid for small event numbers where the statistical errors of the simulation are important.

A.9 The Compound Poisson Distribution and Approximations of It

This section is based on Ref. [26].

A.9.1 *Equivalence of Two Definitions of the CPD*

The CPD describes

(i) the sum $x = \sum_{i=1}^{N} k_i w_i$, with a given discrete, positive weight distribution, w_i, $i = 1, 2, .., N$ and Poisson distributed numbers k_i with mean values λ_i,
(ii) the sum $x = \sum_{i=1}^{k} w_i$ of a Poisson distributed number k of independent and identical distributed positive weights w_i.

The equivalence of the two definitions is related to the following identity:

$$\prod_{i=1}^{N} \mathcal{P}_{\lambda_i}(k_i) = \mathcal{P}_\lambda(k) \mathcal{M}^k_{\varepsilon_1,\dots,\varepsilon_N}(k_1, ..., k_N). \tag{A.32}$$

The left hand side describes N independent Poisson processes with mean values λ_i and random variables k_i, and the right hand side corresponds to a single Poisson process with $\lambda = \Sigma\lambda_i$ and the random variable $k = \Sigma k_i$ where the numbers k_i follow a multinomial distribution

$$\mathcal{M}^n_{\varepsilon_1,\dots,\varepsilon_N}(k_1, ..., k_N) = \frac{k!}{\prod_{i=1}^{N} k_i!} \prod_{i=1}^{N} \varepsilon_i^{k_i} .$$

Here k is distributed to the N different classes with probabilities $\varepsilon_i = \lambda_i/\lambda$.

The validity of (A.32) for the binomial case

$$\mathcal{P}_\lambda \mathcal{M}^k_{\lambda_1/\lambda,\lambda_2/\lambda} = \frac{e^{-\lambda}\lambda^k}{k!} \frac{k!}{k_1!k_2!} \frac{\lambda_1^{k_1}\lambda_2^{k_2}}{\lambda^{k_1}\lambda^{k_2}} = \frac{e^{-(\lambda_1+\lambda_2)}\lambda_1^{k_1}\lambda_2^{k_2}}{k_1!k_2!} = \mathcal{P}_{\lambda_1}\mathcal{P}_{\lambda_2} \tag{A.33}$$

can easily be generalized to several Poisson processes. The multinomial distribution describes a random distribution of k events into N classes. If to each class i is attributed a weight w_i, then to the k events are randomly associated weights w_i with probabilities λ_i/λ.

If all probabilities are equal, $\varepsilon_i = 1/N$, the multinomial distribution describes a random selection of the weights w_i out of the N weights with equal probabilities $1/N$. It does not matter whether we describe the distribution of $x = \Sigma w_i$ by independent Poisson distributions or by the product of a

Poisson distribution with a multinomial distribution. To describe a continuous weight distribution $f(w)$, the limit $N \to \infty$ has to be considered. The formulas (3.62), (3.63) remain valid with $\varepsilon N = 1$.

A.9.2 *Approximation by a Scaled Poisson Distribution*

The scaled Poisson distribution (SPD) is fixed by the requirement that the first two moments of the weighted sum have to be reproduced. We define an equivalent mean value $\tilde{\lambda}$,

$$\tilde{\lambda} = \frac{\lambda \mathrm{E}(w)^2}{\mathrm{E}(w^2)}, \tag{A.34}$$

an equivalent random variable $\tilde{k} \sim \mathcal{P}_{\tilde{\lambda}}$ and a scale factor s,

$$s = \frac{\mathrm{E}(w^2)}{\mathrm{E}(w)}, \tag{A.35}$$

such that the expected value $\mathrm{E}(s\tilde{k}) = \mu$ and $\mathrm{var}(s\tilde{k}) = \sigma^2$. The cumulants of the scaled distribution are $\tilde{\kappa}_m = s^m \tilde{\lambda}$.

We compare the cumulants of the two distributions and form the ratios $\kappa_m/\tilde{\kappa}_m$. Per definition the ratios for $m = 1, 2$ agree because the two lowest moments agree.

The skewness and excess for the two distributions are in terms of the moments $\mathrm{E}(w^m)$ of w:

$$\gamma_1 = \frac{\mathrm{E}(w^3)}{\sigma^3} = \frac{\mathrm{E}(w^3)}{\lambda^{1/2}\mathrm{E}(w^2)^{3/2}}, \tag{A.36}$$

$$\gamma_2 = \frac{\mathrm{E}(w^4)}{\sigma^4} = \frac{\mathrm{E}(w^4)}{\lambda \mathrm{E}(w^2)^2} \tag{A.37}$$

$$\tilde{\gamma}_1 = \left[\frac{\mathrm{E}(w^2)}{\lambda \mathrm{E}(w)^2}\right]^{1/2};, \tag{A.38}$$

$$\tilde{\gamma}_2 = \frac{\mathrm{E}(w^2)}{\lambda \mathrm{E}(w)^2}, \tag{A.39}$$

and the ratios are

$$\frac{\gamma_1}{\tilde{\gamma}_1} = \frac{\mathrm{E}(w^3)\mathrm{E}(w)}{\mathrm{E}(w^2)^2} \geq 1, \tag{A.40}$$

$$\frac{\gamma_2}{\tilde{\gamma}_2} = \frac{\mathrm{E}(w^4)\mathrm{E}(w)^2}{\mathrm{E}(w^2)^3} \geq 1. \tag{A.41}$$

To proof these relations, we use Hölders inequality,

$$\sum_i a_i b_i \leq \left(\sum_i a_i^p\right)^{1/p} \left(\sum_i b_i^{p/(p-1)}\right)^{(p-1)/p},$$

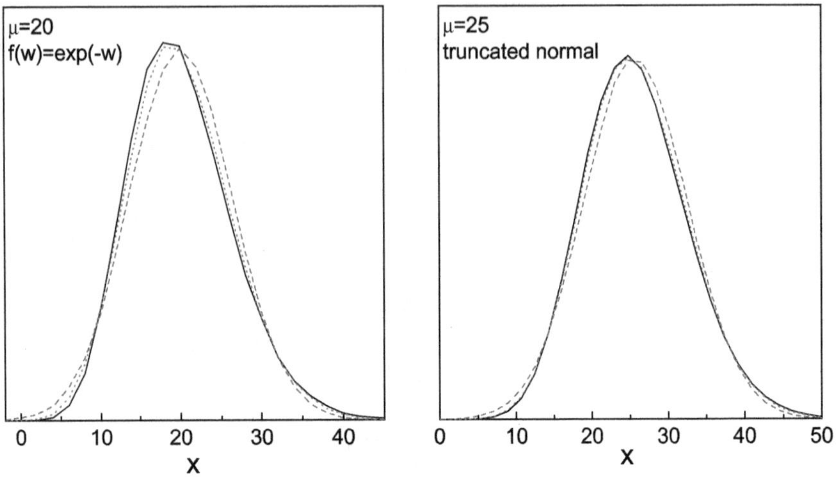

Fig. A.2 comparison of the CPD (black) with the SPD approximation (red) and the normal distribution (blue), see text below.

where a_i, b_i are non-negative and $p > 1$. For $p = 2$ one obtains the Cauchy–Schwartz inequality. Setting $a_i = w_i^{3/2}$, respectively $b_i = w_i^{1/2}$, we get immediately the relation (A.40) for the skewness:

$$\left(\sum_i w_i^2\right)^2 \leq \sum_i w_i^3 \sum_i w_i .$$

In general, with $p = n - 1$ and $a_i = w_i^{n/(n-1)}$, $b_i = w_i^{(n-2)/(n-1)}$, the inequality becomes

$$\left(\sum_i w_i^2\right)^{n-1} \leq \sum_i w_i^n \left(\sum_i w_i\right)^{n-2}$$

which includes (A.41).

The values $\tilde{\gamma}_1, \tilde{\gamma}_2$ of the SPD lie between those of the CPD and the normal distribution. Thus, the SPD is expected to be a much better approximation of the CPD than the normal distribution [26].

Example A.4. Comparison of the CPD with the SPD approximation and the normal distribution

In Fig. A.2 the results of a simulation of CPDs with two different weight distributions is shown (black). The simulated events are collected into histogram bins but the histograms are displayed as line graphs which are easier to read than column graphs. Corresponding SPD distributions are generated with the parameters chosen according to the relations (A.34) and (A.35). They are indicated by dotted lines (red). The approximations by normal distributions are shown as dashed lines (blue). In the left hand graph the weights are exponentially distributed and the weight distribution of the right hand graph is a truncated, re-normalized normal distribution $\mathcal{N}_t(x|1,1) = c\mathcal{N}(x|1,1)$, $x > 0$ with mean and variance equal to 1 where negative values are cut. In this case the approximation by the SPD is hardly distinguishable from the CPD. The exponential weight distribution includes large weights with low frequency where the approximation by the SPD is less good. Still it models the CPD reasonably well. The examples show, that the approximation by the SPD is close to the CPD and superior to the approximation by the normal distribution.

A.10 Extremum Search

If we apply the maximum-likelihood method for parameter estimation, we have to find the maximum of a function in the parameter space. This is, as a rule, not possible without numerical tools. An analogous problem is posed by the method of least squares. Minimum and maximum search are principally not different problems, since we can invert the sign of the function. We restrict ourselves to the minimum search.

Before we engage off-the-shelf computer programs, we should obtain some rough idea of their function. The best way in most cases is a graphical presentation. It is not important for the user to know the details of the programs, but some knowledge of their underlying principles is helpful.

A.10.1 *Monte Carlo Search*

In order to obtain a rough impression of the function to be investigated, and of the approximate location of its minimum, we may randomly select parameter values. A starting region has to be selected. Usual programs will then further restrict the parameter space in dependence of the search results. An advantage of this method is that the probability to end up in a

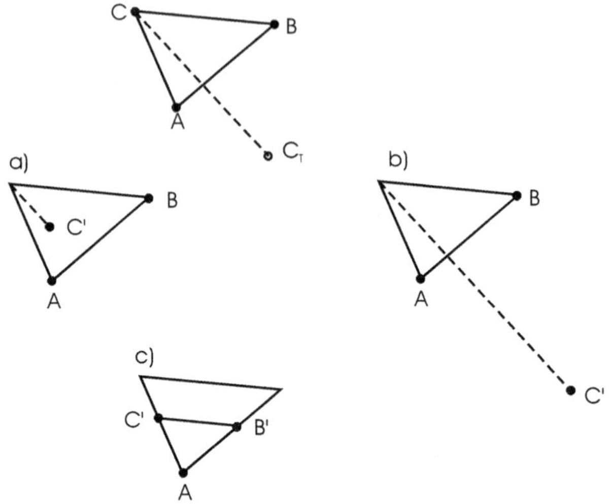

Fig. A.3 Siplex algorithm.

relative minimum is rather small. Since it is fairly inefficient, it should be used only for the first step of a minimization procedure.

A.10.2 *The Simplex Algorithm*

Simplex is a robust algorithm but it is quite slow but, as it ignores derivatives. For an n-dimensional parameter space $n+1$ starting points are selected, and for each point the function values calculated. The point which delivers the largest function value is rejected and replaced by a new point. How this point is found is demonstrated in two dimensions.

Figure A.3 shows in the upper picture three points. let us assume that A has the lowest function value and point C the largest $f(x_C, y_C)$. We want to eliminate C and to replace it by a superior point C'. We take its mirror image with respect to the center of gravity of points A, B and obtain the test point C_T. If $f(x_{C_T}, y_{C_T}) < f(x_C, y_C)$ we did find a better point, thus we replace C by C_T and continue with the new triplet. In the opposite case we double the step width (A.3b) with respect to the center of gravity and find C'. Again we accept C', if it is superior to C. If not, we compare it with the test point C_T and if $f(x_{C_T}, y_{C_T}) < f(x_{C'}, y_{C'})$ holds, the step width is halved and reversed in direction (A.3a). The point C' now moves to the inner region of the simplex triangle. If it is superior to C it replaces C as above. In all other cases the original simplex is shrunk by a factor two in the direction of the best point A (A.3c). In each case one of the four configurations is chosen and the iteration continued.

There exist many variants (see Refs. in [128] of the original version of Nelder and Mead [127]). Standard Simplex [127] has been used in most fits of this book. If the number of parameters is large, and especially if the parameters are correlated, Simplex fits have the tendency to stop without having reached the function minimum [128]. This situation occurs in fits of unfolded histograms. Simplex may choose shrinkage while a reflection of the worst parameter point could be the optimal choice. Finally, all points have almost equal parameter coordinates such that the convergence criterion is fulfilled. Further improvement steps are so small that reducing the convergence parameter does not change the result. The convergence problem is studied in great detail in Ref. [128] and a solution which introduces stochastic elements in the stepping process is proposed.

In this book a different approach is followed. After Simplex signals convergence, the fit is repeated where the best point so far obtained is kept and the remaining points are initialized in the same way are before. Alternatively these points are chosen randomly centered at the best value.

A.10.3 *Parabola Method*

Again we begin with starting points in parameter space. In the one-dimensional case we choose 3 points and put a parabola through them. The point with the largest function value is dropped and replaced by the minimum of the parabola and a new parabola is computed. In the general situation of an n-dimensional space, $2n + 1$ points are selected which determine a paraboloid. Again the worst point is replaced by the vertex of the paraboloid. The iteration converges for functions which are convex in the search region.

A.10.4 *Method of Steepest Descent*

A traveler, walking in a landscape unknown to him, who wants to find a lake, will chose a direction down-hill perpendicular to the curves of equal height (if there are no insurmountable obstacles). The same method is applied when searching for a minimum by the method of steepest descent. We consider this local method in more detail, as in some cases it has to be programmed by the user himself.

We start from a certain point $\boldsymbol{\lambda}_0$ in the parameter space, calculate the gradient $\nabla_\lambda f(\boldsymbol{\lambda})$ of the function $f(\boldsymbol{\lambda})$ which we want to minimize and move by $\Delta\boldsymbol{\lambda}$ downhill.

$$\Delta\boldsymbol{\lambda} = -\alpha\nabla_\lambda f(\boldsymbol{\lambda}).$$

The step length depends on the learning constant α which is chosen by the user. This process is iterated until the function remains essentially constant. The convexity of the function at the supposed minimum has to be checked.

The method of steepest descent has advantages as well as drawbacks:

- The decisive advantage is its simplicity which permits to handle a large number of parameters at the same time. If convenient, for the calculation of the gradient rough approximations can be used. Important is only that the function decreases with each step. As opposed to the simplex and parabola methods its complexity increases only linear with the number of parameters. Therefore problems with huge parameter sets can be handled.
- It is possible to evaluate a sample sequentially, element by element, which is especially useful for the back-propagation algorithm of neural networks.
- Unsatisfactory is that the learning constant is not dimensionless. In other words, the method is not independent of the parameter scales. For a space-time parameter set the gradient path will depend, for instance, on the choice whether to measure the parameters in meters or millimeters, respectively hours or seconds.
- In regions with flat parameter space the convergence is slow. In a narrow valley oscillations may appear. For too large values of α oscillations will make exact minimizing difficult.

The last mentioned problems can be reduced by various measures where the step length and direction partially depend on results of previous steps. When the function change is small and similar in successive steps α is increased. Oscillations in a valley can be avoided by adding to the gradient in step i a fraction of the gradient of step $i-1$:

$$\Delta\boldsymbol{\lambda}_i = \alpha\left(\nabla_\lambda f(\boldsymbol{\lambda}_i) + 0.5\nabla_\lambda f(\boldsymbol{\lambda}_{i-1})\right).$$

Oscillations near the minimum are easily recognized and removed by decreasing α.

The method of steepest descent is applied in ANN and useful in the updating alignment of tracking detectors [126].

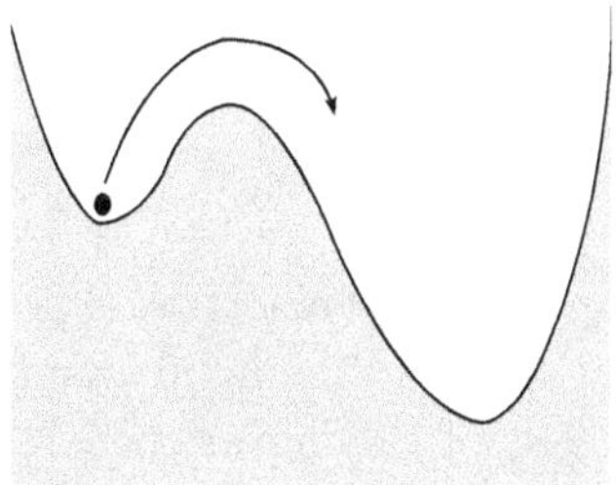

Fig. A.4 Stochastic annealing. A local minimum can be left with a certain probability.

A.10.5 *Stochastic Elements in Minimum Search*

A physical system which is cooled down to the absolute zero point will principally occupy an energetic minimum. When cooled down fast it may, though, be captured in a local (relative) minimum. An example is a particle in a potential wall. For somewhat higher temperature it may leave the local minimum, thanks to the statistical energy distribution (Fig. A.4). This is used for instance in the stimulated annealing of defects in solid matter.

This principle can be used for minimum search in general. A step in the wrong direction, where the function increases by Δf, can be accepted, when using the method of steepest descent, e.g. with a probability

$$P(\Delta f) = \frac{1}{1 + e^{\Delta f / T}} .$$

The scale factor T ("temperature") steers the strength of the effect. It has been shown that for successively decreasing T the absolute minimum will be reached.

A.11 Linear Regression with Constraints

We consider N measurements $\boldsymbol{y}$ at known locations $\boldsymbol{x}$, with a $N \times N$ covariance matrix C_N and a corresponding weight matrix $\mathsf{V}_N = \mathsf{C}_N^{-1}$. (We indicate the dimensions of quadratic matrices with an index).

In the linear model the measurements are described by $P < N$ parameters $\boldsymbol{\theta}$ in form of linear relations

$$\langle \boldsymbol{y} \rangle = \mathsf{A}(\boldsymbol{x})\boldsymbol{\theta}, \qquad (A.42)$$

with the rectangular $N \times P$ "design" matrix A.

In 6.8.1 we have found that the corresponding χ^2-expression is minimized by

$$\hat{\boldsymbol{\theta}} = (\mathsf{A}^T \mathsf{V}_N \mathsf{A})^{-1} \mathsf{A}^T \mathsf{V}_N \boldsymbol{y} .$$

We now include constraints between the parameters, expressed by $K < P$ linear relations:

$$\mathsf{H}\boldsymbol{\theta} = \boldsymbol{\rho},$$

with $\mathsf{H}(\boldsymbol{x})$ a given rectangular $K \times P$ matrix and $\boldsymbol{\rho}$ a K-dimensional vector.

This problem is solved by introducing K Lagrange multipliers $\boldsymbol{\alpha}$ and looking for a *stationary* point of the lagrangian

$$\Lambda = (\boldsymbol{y} - \mathsf{A}\boldsymbol{\theta})^T \mathsf{V}_N (\boldsymbol{y} - \mathsf{A}\boldsymbol{\theta}) + 2\boldsymbol{\alpha}^T (\mathsf{H}\boldsymbol{\theta} - \boldsymbol{\rho}).$$

Differentiating with respect to $\boldsymbol{\theta}$ and $\boldsymbol{\alpha}$ gives the normal equations

$$\mathsf{A}^T \mathsf{V}_N \mathsf{A}\boldsymbol{\theta} + \mathsf{H}^T \boldsymbol{\alpha} = \mathsf{A}^T \mathsf{V}_N \boldsymbol{y}, \tag{A.43}$$

$$\mathsf{H}\boldsymbol{\theta} = \boldsymbol{\rho} \tag{A.44}$$

to be solved for $\hat{\boldsymbol{\theta}}$ and $\hat{\boldsymbol{\alpha}}$. Note that Λ is minimized only with respect to $\boldsymbol{\theta}$, but maximized with respect to $\boldsymbol{\alpha}$: The stationary point is a saddle point, which complicates a direct extremum search. Solving (A.43) for $\boldsymbol{\theta}$ and inserting it into (A.44), we find

$$\hat{\boldsymbol{\alpha}} = \mathsf{C}_K^{-1} (\mathsf{H}\mathsf{C}_P \mathsf{A}^T \mathsf{V}_N \boldsymbol{y} - \boldsymbol{\rho})$$

and, re-inserting the estimates into (A.43), we obtain

$$\hat{\boldsymbol{\theta}} = \mathsf{C}_P [\mathsf{A}^T \mathsf{V}_N \boldsymbol{y} - \mathsf{H}^T \mathsf{C}_K^{-1} (\mathsf{H}\mathsf{C}_P \mathsf{A}^T \mathsf{V}_N \boldsymbol{y} - \boldsymbol{\rho})],$$

where the abbreviations $\mathsf{C}_P = (\mathsf{A}^T \mathsf{V}_N \mathsf{A})^{-1}$, $\mathsf{C}_K = \mathsf{H}\mathsf{C}_P \mathsf{H}^T$ have been used.

As in the case of no constraints 6.8.1 the estimate $\hat{\boldsymbol{\theta}}$ is linear in $\boldsymbol{y}$ and unbiased, which is easily seen by taking the expectation value in the above equation and using (A.42) and (A.44).

The covariance matrix is found from linear error propagation, after a somewhat lengthy calculation, as

$$\mathrm{cov}(\hat{\boldsymbol{\theta}}) = \mathsf{D}\mathsf{C}_N \mathsf{D}^T = (\mathsf{I}_P - \mathsf{C}_P \mathsf{H}^T \mathsf{C}_K^{-1} \mathsf{H}) \mathsf{C}_P,$$

where

$$\mathsf{D} = \mathsf{C}_P (\mathsf{I}_P - \mathsf{H}^T \mathsf{C}_K^{-1} \mathsf{H}\mathsf{C}_P) \mathsf{T}^T \mathsf{V}_N$$

has been used. The covariance matrix is symmetric positive definite. Without constraints, it equals C_P, the negative term is absent. Of course, the introduction of constraints reduces the errors and thus improves the parameter estimation.

A.12 Formulas Related to the Polynomial Approximation

A.12.0.1 *Errors of the Expansion Coefficients*

In Section 11.2.2 we have discussed the approximation of measurements by orthogonal polynomials and given the following formula for the error of the expansion coefficients a_k,

$$\mathrm{var}(a_k) = 1/\sum_{\nu=1}^{N} \frac{1}{\delta_\nu^2}$$

which is valid for all $k = 1, \ldots, K$. Thus all errors are equal to the error of the weighted mean of the measurements y_ν.

Proof: From linear error propagation we have, for independent measurements y_ν,

$$\begin{aligned}\mathrm{var}(a_k) &= \mathrm{var}\left(\sum_\nu w_\nu u_k(x_\nu) y_\nu\right) \\ &= \sum_\nu w_\nu^2 (u_k(x_\nu))^2 \delta_\nu^2 \\ &= \sum_\nu w_\nu u_k^2(x_\nu) / \sum_\nu \frac{1}{\delta_\nu^2} \\ &= 1/\sum_\nu \frac{1}{\delta_\nu^2},\end{aligned}$$

where in the third step we used the definition of the weights, and in the last step the normalization of the polynomials u_k.

A.12.0.2 *Polynomials for Data with Uniform Errors*

If the errors $\delta_1, \ldots, \delta_N$ are uniform, the weights become equal to $1/N$, and for certain patterns of the locations $x_1, \ldots, x_N$, for instance for an equidistant distribution, the orthogonal polynomials $u_k(x)$ can be calculated. They are given in mathematical handbooks, for instance in Ref. [129]. Although the general expression is quite involved, we reproduce it here for the convenience of the reader. For x defined in the domain $[-1, 1]$ (eventually after some linear transformation and shift), and $N = 2M + 1$ equidistant (with distance $\Delta x = 1/M$) measured points $x_\nu = \nu/M$, $\nu = 0, \pm 1, \ldots, \pm M$, they are given by

$$u_k(x) = \left(\frac{(2M+1)(2k+1)[(2M)!]^2}{(2M+k+1)!(2M-k)!}\right)^{1/2} \sum_{i=0}^{k} (-1)^{i+k} \frac{(i+k)^{[2i]}(M+t)^{[i]}}{(i!)^2 (2M)^{[i]}},$$

for $k = 0, 1, 2, \ldots 2M$, where we used the notation $t = x/\Delta x = xM$ and the definitions

$$z^{[i]} = z(z-1)(z-2)\cdots(z-i+1)$$
$$z^{[0]} = 1,\ z \geq 0,\ 0^{[i]} = 0,\ i = 1, 2, \ldots .$$

A.13 Formulas for B-Spline Functions

A.13.1 *Linear B-Splines*

Linear B-splines cover an interval $2b$ and overlap with both neighbors:

$$\begin{aligned} B(x; x_0) &= 2\frac{x - x_0 - b}{b} \quad \text{for} \quad x_0 - b \leq x \leq x_0, \\ &= 2\frac{-x - x_0 + b}{b} \quad \text{for} \quad x_0 \leq x \leq x_0 + b, \\ &= 0 \quad \text{else.} \end{aligned}$$

They are normalized to unit area. Since the central values are equidistant, we fix them by the lower limit $x_{\min}$ of the x-interval and count them as $x_0(k) = x_{\min} + kb$, with the index k running from $k_{\min} = 0$ to $k_{\max} = (x_{\max} - x_{\min})/b = K$.

At the borders only half of a spline is used.

Remark: The border splines are defined in the same way as the other splines. After the fit the part of the function outside of its original domain is ignored. In the literature the definition of the border splines is often different.

A.13.2 *Quadratic B-Splines*

The definition of quadratic splines is analogous:

$$\begin{aligned} B(x; x_0) &= \frac{1}{2b}\left(\frac{x - x_0 + 3/2b}{b}\right)^2 \quad \text{for} \quad x_0 - 3b/2 \leq x \leq x_0 - b/2, \\ &= \frac{1}{2b}\left[\frac{3}{2} - 2\left(\frac{x - x_0}{b}\right)^2\right] \quad \text{for } x_0 - b/2 \leq x \leq x_0 + b/2, \\ &= \frac{1}{2b}\left(\frac{x - x_0 - 3/2b}{b}\right)^2 \quad \text{for } x_0 + b/2 \leq x \leq x_0 + 3b/2, \\ &= 0 \quad \text{else.} \end{aligned}$$

The supporting points $x_0 = x_{\min} + (k - 1/2)b$ lie now partly outside of the x-domain. The index k runs from 0 to $k_{\max} = (x_{\max} - x_{\min})/b + 2$. Thus, the number K of splines is by two higher than the number of intervals. The relations (11.13) and (11.12) are valid as before.

A.13.3 *Cubic B-Splines*

Cubic B-splines are defined as follows:

$$
\begin{aligned}
B(x;x_0) &= \frac{1}{6b}\left(2+\frac{x-x_0}{b}\right)^3 \quad \text{for } x_0-2b \le x \le x_0-b,\\
&= \frac{1}{6b}\left[-3\left(\frac{x-x_0}{b}\right)^3 - 6\left(\frac{x-x_0}{b}\right)^2 + 4\right] \quad \text{for } x_0-b \le x \le x_0,\\
&= \frac{1}{6b}\left[3\left(\frac{x-x_0}{b}\right)^3 - 6\left(\frac{x-x_0}{b}\right)^2 + 4\right] \quad \text{for } x_0 \le x \le x_0+b,\\
&= \frac{1}{6b}\left(2-\frac{x-x_0}{b}\right)^3 \quad \text{for } x_0+b \le x \le x_0+2b,\\
&= 0 \quad \text{else}.
\end{aligned}
$$

The shift of the center of the spline is performed as before: $x_0 = x_{\min} + (k-1)b$. The index k runs from 0 to $k_{\max} = (x_{\max} - x_{\min})/b + 3$. The number $k_{\max} + 1$ of splines is equal to the number of intervals plus 3.

A.14 Support Vector Classifiers

Support vector machines are described some detail in Refs. [16, 106–108].

A.14.1 *Linear Classifiers*

Linear classifiers[11] separate the two training samples by a hyperplane. Let us initially assume that in this way a complete separation is possible. Then an optimal hyperplane is the plane which divides the two samples with the largest margin. This is shown in Fig. A.5. The hyperplane can be constructed in the following way: The shortest connection Δ between the convex hulls[12] of the two non-overlapping classes determines the direction $\boldsymbol{w}/|\boldsymbol{w}|$ of the normal $\boldsymbol{w}$ of this plane which cuts the distance at its center. We represent the hyperplane in the form

$$\boldsymbol{w}\cdot\boldsymbol{x} + b = 0, \tag{A.45}$$

[11] A linear classification scheme was already introduced in Section 11.4.1.

[12] The convex hull is the smallest polyhedron which contains all points and their connecting straight lines.

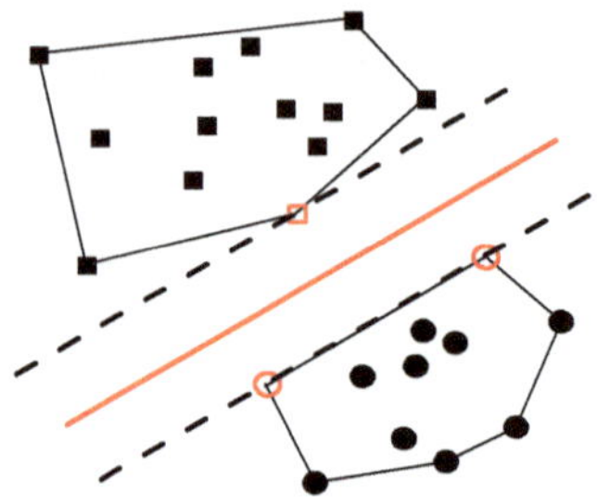

Fig. A.5 The red hyperplane separates squares from circles. Shown are the convex hulls and the support vectors in red.

where b fixes its distance from the origin. Note that $\boldsymbol{w}$ is not normalized, a common factor in $\boldsymbol{w}$ and b does not change condition (A.45). Once we have found the hyperplane $\{\boldsymbol{w}, b\}$ which separates the two classes $y_i = \pm 1$ of the training sample $\{(\boldsymbol{x}_1, y_1), \dots, (\boldsymbol{x}_N, y_N)\}$ we can use it to classify new input:

$$\hat{y} = f(\boldsymbol{x}) = \mathrm{sign}(\boldsymbol{w} \cdot \boldsymbol{x} + b). \tag{A.46}$$

To find the optimal hyperplane which divides Δ into equal parts, we define the two marginal planes which touch the hulls:

$$\boldsymbol{w} \cdot \boldsymbol{x} + b = \pm 1.$$

If $\boldsymbol{x}^+, \boldsymbol{x}^-$ are located at the two marginal hyperplanes, the following relations hold which also fix the norm of $\boldsymbol{w}$:

$$\boldsymbol{w} \cdot (\boldsymbol{x}^+ - \boldsymbol{x}^-) = 2 \Rightarrow \Delta = \frac{\boldsymbol{w}}{|\boldsymbol{w}|} \cdot (\boldsymbol{x}^+ - \boldsymbol{x}^-) = \frac{2}{|\boldsymbol{w}|}.$$

The optimal hyperplane is now found by solving a constrained quadratic optimization problem

$$|\boldsymbol{w}|^2 = \text{minimum, subject to } y_i(\boldsymbol{w} \cdot \boldsymbol{x_i} + b) \ge 1,\ i = 1, \dots, N.$$

For the solution, only the constraints with equals sign are relevant. The vectors corresponding to points on the marginal planes form the so-called active set and are called support vectors (see Fig. A.5). The optimal solution can be written as

$$\boldsymbol{w} = \sum_i \alpha_i y_i \boldsymbol{x}_i$$

with $\alpha_i > 0$ for the active set, else $\alpha_i = 0$, and furthermore $\sum \alpha_i y_i = 0$. The last condition ensures translation invariance: $\boldsymbol{w}(\boldsymbol{x}_i - \boldsymbol{a}) = \boldsymbol{w}(\boldsymbol{x}_i)$. Together with the active constraints, after substituting the above expression for $\boldsymbol{w}$, it provides just the required number of linear equations to fix α_i and b. Of course, the main problem is to find the active set. For realistic cases this requires the solution of a large quadratic optimization problem, subject to linear inequalities. For this purpose an extended literature as well as program libraries exist.

This picture can be generalized to the case of overlapping classes. Assuming that the optimal separation is still given by a hyperplane, the picture remains essentially the same, but the optimization process is substantially more complex. The standard way is to introduce so called *soft margin classifiers*. There some points on the wrong side of their marginal plane are tolerated, but with a certain penalty in the optimization process. It is chosen proportional to the sum of their distances or their square distance from their own territory. The proportionality constant is adjusted to the given problem.

A.14.2 *General Kernel Classifiers*

All quantities determining the linear classifier $\hat{y}$ (A.46) depend only on inner products of vectors of the input space. This concerns not only the dividing hyperplane, given by (A.45), but also the expressions for $\boldsymbol{w}$, b and the factors α_i associated to the support vectors. The inner product $\boldsymbol{x} \cdot \boldsymbol{x}'$ which is a bilinear symmetric scalar function of two vectors, is now replaced by another scalar function $K(\boldsymbol{x}, \boldsymbol{x}')$ of two vectors, the kernel, which need not be bilinear, but should also be symmetric, and is usually required to be positive definite. In this way a linear problem in an inner product space is mapped into a very non-linear problem in the original input space where the kernel is defined. We then are able to separate the classes by a hyperplane in the inner product space that may correspond to a very complicated hypersurface in the input space. This is the so-called *kernel trick*.

To illustrate how a non-linear surface can be mapped into a hyperplane, we consider a simple example. In order to work with a linear cut, i.e. with a dividing hyperplane, we transform our input variables $\boldsymbol{x}$ into new variables: $\boldsymbol{x} \to \boldsymbol{X}(\boldsymbol{x})$. For instance, if x_1, x_2, x_3 are momentum components and a cut in energy, $x_1^2 + x_2^2 + x_3^2 < r^2$, is to be applied, we could transform the momentum space into a space

$$\boldsymbol{X} = \{x_1^2, x_2^2, x_3^2, \ldots\}.$$

where the cut corresponds to the hyperplane $X_1 + X_2 + X_3 = r^2$. Such a mapping can be realized by substituting the inner product by a kernel:

$$\boldsymbol{x} \cdot \boldsymbol{x}' \to K(\boldsymbol{x}, \boldsymbol{x}') = \boldsymbol{X}(\boldsymbol{x}) \cdot \boldsymbol{X}(\boldsymbol{x}').$$

In our example a kernel of the so-called monomial form is appropriate:

$$\begin{aligned} &K(\boldsymbol{x}, \boldsymbol{x}') = (\boldsymbol{x} \cdot \boldsymbol{x}')^d \quad \text{with } d = 2, \\ &(\boldsymbol{x} \cdot \boldsymbol{x}')^2 = (x_1 x_1' + x_2 x_2' + x_3 x_3')^2 = \boldsymbol{X}(\boldsymbol{x}) \cdot \boldsymbol{X}(\boldsymbol{x}') \end{aligned} \tag{A.47}$$

with

$$\boldsymbol{X}(\boldsymbol{x}) = \{x_1^2, x_2^2, x_3^2, \sqrt{2}x_1x_2, \sqrt{2}x_1x_3, \sqrt{2}x_2x_3\}.$$

The sphere $x_1^2 + x_2^2 + x_3^2 = r^2$ in x-space is mapped into the 5-dimensional hyperplane $X_1 + X_2 + X_3 = r^2$ in 6-dimensional X-space. (A kernel inducing instead of monomials of order d (A.47), polynomials of all orders, *up to* order d is $K(\boldsymbol{x}, \boldsymbol{x}') = (1 + \boldsymbol{x} \cdot \boldsymbol{x}')^d$.)

The most common kernel used for classification is the Gaussian (see Section 11.2.1.3):

$$K(\boldsymbol{x}, \boldsymbol{x}') = \exp\left(-\frac{(\boldsymbol{x} - \boldsymbol{x}')^2}{2s^2}\right).$$

It can be shown that it induces a mapping into a space of infinite dimensions [108] and that nevertheless the training vectors can in most cases be replaced by a relatively small number of support vectors. The only free parameter is the penalty constant which regulates the degree of overlap of the two classes. A high value leads to a very irregular shape of the hypersurface separating the training samples of the two classes to a high degree in the original space whereas for a low value its shape is much smoother and more minority observations are tolerated.

In practice, this mapping into the inner product space is not performed explicitly, in fact it is even rarely known. All calculations are performed in x-space, especially the determination of the support vectors and their weights α. The kernel trick merely serves to prove that a classification with support vectors is feasible. The classification of new input then proceeds with the kernel K and the support vectors directly:

$$\hat{y} = \text{sign}\left(\sum_{y_i=+1} \alpha_i K(\boldsymbol{x}, \boldsymbol{x}_i) - \sum_{y_i=-1} \alpha_i K(\boldsymbol{x}, \boldsymbol{x}_i)\right).$$

The use of a relatively small number of support vectors (typically only about 5% of all α_i are different from zero) drastically reduces the storage requirement and the computing time for the classification. Remark that the result of the support vector classifier is not identical to that of the original kernel classifier but very similar.

A.15 The Bayes Factor

In Chapter 6 we have introduced the likelihood ratio to discriminate between simple hypotheses. For two composite hypotheses H_1 and H_2 with free parameters, in the Bayesian approach the simple ratio is to be replaced by the so-called Bayes factors.

Let us assume for a moment that H_1 applies. Then the actual parameters follow a p.d.f. proportional to $L_1(\boldsymbol{\theta}_1|\boldsymbol{x})\pi_1(\boldsymbol{\theta}_1)$ where $L_1(\boldsymbol{\theta}_1|\boldsymbol{x})$ is the likelihood function and $\pi_1(\boldsymbol{\theta}_1)$ the prior density of the parameters. The same reasoning is valid for H_2. The probability that H_1 (H_2) is true is proportional to the integral over the parameter space, $\int L_1(\boldsymbol{\theta}_1|\boldsymbol{x})\pi_1(\boldsymbol{\theta}_1)\mathrm{d}\boldsymbol{\theta}_1$ ($\int L_2(\boldsymbol{\theta}_2|\boldsymbol{x})\pi_2(\boldsymbol{\theta}_2)\mathrm{d}\boldsymbol{\theta}_2$). The relative betting odds thus are given by the Bayes factor B,

$$B = \frac{\int L_1(\boldsymbol{\theta}_1|\boldsymbol{x})\pi_1(\boldsymbol{\theta}_1)\mathrm{d}\boldsymbol{\theta}_1}{\int L_2(\boldsymbol{\theta}_2|\boldsymbol{x})\pi_2(\boldsymbol{\theta}_2)\mathrm{d}\boldsymbol{\theta}_2}. \tag{A.48}$$

In the case with no free parameters, B reduces to the simple likelihood ratio L_1/L_2.

The two terms forming the ratio are called *marginal likelihoods.* The integration automatically introduces a penalty for additional parameters and related overfitting: The higher the dimensionality of the parameter space is, the larger is in average the contribution of low likelihood regions to the integral. In this way the concept follows the philosophy of *Ockham's razor*[13] which in short states that *from different competing theories, the one with the fewest assumptions, i.e. the simplest, should be preferred.*

The Bayes factor is intended to replace the p-value of frequentist statistics.

H. Jeffreys [22] has suggested a classification of Bayes factors into different categories ranging from < 3 (barely worth mentioning) to > 100 (decisive).

In Example 10.11 we had computed the significance of a bump of normal shape above a uniform background (Fig. 10.18). The free parameters were the fraction t of events in the bump and its location μ. The interesting parameter ranges were $0 \leq t \leq 0.5$ and $0.2 \leq \mu \leq 0.8$ which also fix the integration limits in the numerator of (A.48. The width of the bump is fixed. For the computation of the Bayes factor the prior densities of the parameters were assumed to be constant. The Bayes factor in favor of the existence of a bump is $B = 54$ which is considered as very significant. This result is

[13] Postulated by William of Ockham, English logician in the 14th century.

inversely proportional to the parameter ranges as is expected because the probability to find a fake signal in a flat background is proportional to its range. In the cited example we had found a likelihood ratio of $1.1 \cdot 10^4$ taken at the MLE. The corresponding p-value was $p = 1.8 \cdot 10^{-4}$ for the hypothesis of a flat background, much smaller than the Bayesian betting odds of 1/54 for this hypothesis. While the Bayes factor indicates the significance for a resonance in the full parameter range of location and fraction of events, in Example 10.11 the p-value derived from the likelihood ratio had been taken simply at the MLE of the parameters. The *look-else-where effect* had been ignored. On the other hand, for the calculation of the Bayes factor at least partially subjective prior probabilities of the parameters have to be included. However, a Bayes factor conditioned on uniform priors

The Bayes factor is a very reasonable and conceptually attractive concept which requires little computational effort. It is to be preferred to the frequentist p-value approach in decision making. However, for the documentation of a measurement the Bayes factor has the typical Bayesian drawback that it depends on prior densities and unfortunately there is no objective way to fix those.

A.16 Robust Fitting Methods

A.16.1 *Introduction*

If one or a few observations in a sample are separated from the bulk of the data, we speak of outliers. The reasons for their existence range from trivial mistakes or detector failures to important physical effects. In any case, the assumed statistical model has to be questioned if one is not willing to admit that a large and very improbable fluctuation did occur.

Outliers are quite disturbing: They can change parameter estimates by large amounts and increase their errors drastically.

Frequently outliers can be detected simply by inspection of appropriate plots. It goes without saying, that simply dropping them is not a good advice. In any case at least a complete documentation of such an event is required. Clearly, objective methods for their detection and treatment are preferable.

In the following, we restrict our treatment to the simple one-dimensional case of Gaussian-like distributions, where outliers are located far from the average, and where we are interested in the mean value. If a possible outlier is contained in the allowed variate range of the distribution – which is always true for a Gaussian – a statistical fluctuation cannot be excluded as a logical possibility. Since the outliers are removed on the basis of a statistical procedure, the corresponding modification of results due to the possible removal of correct observations can be evaluated.

We distinguish three cases:

(1) The standard deviations of the measured points are known.
(2) The standard deviations of the measured points are unknown but known to be the same for all points.
(3) The standard deviations are unknown and different.

It is obvious that case 3 of unknown and unequal standard deviations cannot be treated.

The treatment of outliers, especially in situations like case 2, within the LS formalism is not really satisfying. If the data are of bad quality we may expect a sizeable fraction of outliers with large deviations. These may distort the LS fit to such an extend that outliers become difficult to define (*masking* of outliers). This kind of fragility of the LS method, and the fact that in higher dimensions the outlier detection becomes even more critical, has lead statisticians to look for estimators which are less disturbed by data not obeying the assumed statistical model (typical are deviations from the assumed normal distribution), even when the efficiency suffers. In a second – not robust – fit procedure with cleaned data it is always possible to optimize the efficiency.

In particle physics, a typical problem is the reconstruction of particle tracks from hits in wire or silicon detectors. Here outliers due to other tracks or noise are a common difficulty, and for a first rough estimate of the track parameters and the associated hit selection for the pattern recognition, robust methods are useful.

A.16.2 *Robust Methods*

A.16.2.1 *Truncated Least Square Fits*

The simplest method to remove outliers is to eliminate those measurements which contribute excessively to the χ^2 of a least square (LS) fit. In this truncated least square fit (LST) all observations that deviate by more than a certain number of standard deviations from the mean are excluded. Reasonable values lie between 1.5 and 2 standard deviations, corresponding to a χ^2 cut $\chi^2_{\max} = 2.25$ to 4. The optimal value of this cut depends on the expected amount of background or false measurements and the number of observations. In case 2 the variance has to be estimated from the data and the estimated variance $\hat{\delta}^2$ is, according to Chapter 3.2.3, given by

$$\hat{\delta}^2 = \sum (y_i - \hat{\mu})^2/(N-1).$$

This method can be improved by removing outliers sequentially (LSTS). In a first step we use all measurements $y_1, \dots, y_N$, with standard deviations $\delta_1, \dots, \delta_N$ to determine the mean value $\hat{\mu}$ which in our case is just the weighted mean. Then we compute the normalized residuals, also called pulls, $r_i = (y_i - \hat{\mu})/\delta_i$ and select the measurement with the largest value of r_i^2. The value of χ^2 is computed with respect to the mean and variance of the remaining observations and the measurement is excluded if it exceeds the parameter χ^2_{max}.[14] The fit is repeated until all measurements are within the margin. In case that all measurements are genuine Gaussian measurements, this procedure only marginally reduces the precision of the fit.

In both methods LST and LSTS a minimum fraction of measurements has to be retained. A reasonable value is 50 % but depending on the problem other values may be appropriate.

A.16.2.2 *The Sample Median*

A first step (already proposed by Laplace) in the direction to estimators more robust than the sample mean is the introduction of the sample median as estimator for location parameters. While the former follows to an extremely outlying observation up to $\pm\infty$, the latter stays nearly unchanged in this case. This change can be expressed as a change of the objective function, i.e. the function to be minimized with respect to μ, from $\sum_i (y_i - \mu)^2$ to $\sum_i |y_i - \mu|$ which is indeed minimized if $\hat{\mu}$ coincides with the sample median in case of N odd. For even N, $\hat{\mu}$ is the mean of the two innermost points. Besides the slightly more involved computation (sorting instead of summing), the median is not an optimal estimator for a pure Gaussian distribution:

$$\text{var}(median) = \frac{\pi}{2}\,\text{var}(mean) = 1.571\,\text{var}(mean),$$

but it weights large residuals less and therefore performs better than the arithmetic mean for distributions which have longer tails than the Gaussian. Indeed for large N we find for the Cauchy distribution $\text{var}(median) =$

[14] If the variance has to be estimated from the data its value is biased towards smaller values because for a genuine Gaussian distribution eliminating the measurement with the largest pull reduces the expected variance.

$\pi^2/(4N)$, while var$(mean) = \infty$ (see 3.6.9), and for the two-sided exponential (Laplace) distribution var$(median) =$ var$(mean)/2$.

A.16.2.3 *M-Estimators*

The objective function of the LS approach can be generalized to

$$\sum_i \rho\left(\frac{y_i - t(x_i, \boldsymbol{\theta})}{\delta_i}\right) \tag{A.49}$$

with $\rho(z) = z^2$ for the LS method which is optimal for Gaussian errors. For the Laplace distribution mentioned above the optimal objective function is based on $\rho(z) = |z|$, derived from the likelihood analog which suggests $\rho \propto \ln f$. To obtain a more robust estimation the function ρ can be modified in various ways but we have to retain the symmetry, $\rho(z) = \rho(-z)$ and to require a single minimum at $z = 0$. This kind of estimators with objective functions ρ different from z^2 are called M-estimators, "M" reminding maximum likelihood. The best known example is due to Huber, [130]. His proposal is a kind of mixture of the appropriate functions of the Gauss and the Laplace cases:

$$\rho(z) = \begin{cases} z^2/2 & \text{if } |z| \leq c \\ c(|z| - c/2) & \text{if } |z| > c. \end{cases}$$

The constant c has to be adapted to the given problem. For a normal population the estimate is of course not efficient. For example with $c = 1.345$ the inverse of the variance is reduced to 95% of the standard value. Obviously, the fitted objective function (A.49) no longer follows a χ^2-distribution with appropriate degrees of freedom.

A.16.2.4 *Estimators with High Breakdown Point*

In order to compare different estimators with respect to their robustness, the concept of the *breakdown point* has been introduced. It is the smallest fraction ε of corrupted data points which can lead the fitted values to differ by an arbitrary large amount from the correct ones. For LS, ε approaches zero, but for M-estimators or truncated fits, changing a single point would be not sufficient to shift the fitted parameter by a large amount. The maximal value of ε is smaller than 50% if the outliers are the minority. It is not difficult to construct estimators which approach this limit, see [131]. This is achieved, for instance, by ordering the residuals according to their absolute value (or ordering the squared residuals, resulting in the same ranking) and

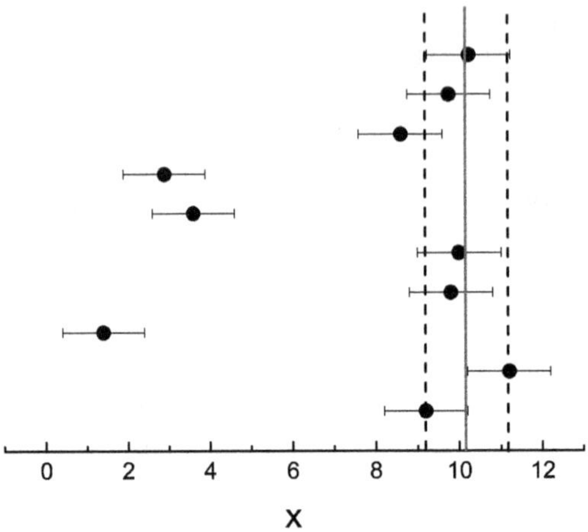

Fig. A.6 Estimates of the location parameter for a sample with three outliers.

retaining only a certain fraction, at least 50%, for the minimization. This so-called *least trimmed squares* (LTS) fit is to be distinguished from truncated least square fit (LST, LSTS) with a fixed cut against large residuals.

An other method relying on rank order statistics is the so-called *least median of squares* (LMS) method. It is defined as follows: Instead of minimizing with respect to the parameters μ the *sum* of squared residuals, $\sum_i r_i^2$, one searches the minimum of the *sample median* of the squared residuals:

$$minimize_\mu \left\{\text{median}(r_i^2(\mu))\right\} .$$

This definition implies that for N data points, $N/2+1$ points enter for N even and $(N+1)/2$ for N odd. Assuming equal errors, this definition can be illustrated geometrically in the one-dimensional case considered here: $\hat{\mu}$ is the center of the smallest interval (vertical strip in Fig. A.6) which covers half of all x values. The width 2Δ of this strip can be used as an estimate of the error. Many variations are of course possible: Instead of requiring 50% of the observations to be covered, a larger fraction can be chosen. Usually, in a second step, a LS fit is performed with the retained observations, thus using the LMS only for outlier detection. This procedure is chosen, since it can be shown that, at least in the case of normal distributions, ranking methods are statistically inferior as compared to LS fits.

Example A.5. Fitting a mean value in the presence of outliers

In Fig. A.6 a simple example is presented. Three data points, representing the outliers, are taken from $\mathcal{N}(3, 1)$ and seven from $\mathcal{N}(10, 1)$. The LS fit (7.7 ± 1.1) is quite disturbed by the outliers. The sample median is here initially 9.5, and becomes 10.2 after excluding the outliers. It is less disturbed by the outliers. The LMS fit corresponds to the thick line, and the minimal strip of width 2Δ to the dashed lines. It prefers the region with largest point density and is therefore a kind of mode estimator. While the median is a location estimate which is robust against large symmetric tails, the mode is also robust against asymmetric tails, i.e. skew distributions of outliers. A more quantitative comparison of different fitting methods is presented in the following table.

Method	**Background**		
	asymm.	uniform	none
simple LS	2.12	0.57	0.32
median	0.72	0.49	0.37
LS trimmed	0.60	0.52	0.37
LS sequentially truncated	0.56	0.62	0.53
least median of squares	0.55	0.66	0.59

We have generated 100000 samples with a 7 point signal given by $\mathcal{N}(10, 1)$ and 3 points of background, a) asymmetric: $\mathcal{N}(3, 1)$ (the same parameters as used in the example before), b) uniform in the interval $[5, 15]$, and in addition c) pure normally distributed points following $\mathcal{N}(10, 1)$ without background. The table contains the root mean squared deviation of the mean values from the nominal value of 10. To make the comparison fair, as in the LMS method also in the trimmed LS fit 6 points have been retained and in the sequentially truncated LS fit a minimum of 6 points was used. With the asymmetric background, the first three methods lead to biased mean values (7.90 for the simple LS, 9.44 for the median and 9.57 for the trimmed LS) and thus the corresponding r.m.s. values are relatively large. As expected the median suffers much less from the background than a standard LS fit. The results of the other two methods, LMS and LS sequentially truncated perform reasonable in this situation, they succeed to eliminate the background completely without biasing the result but are rather weak when little or no background is present. The result of LMS is not improved in our example when a least square fit is performed with the retained data.

The methods can be generalized to the multi-parameter case. Essentially, the r.m.s deviation is replaced by χ^2. In the least square fits, truncated or trimmed, the measurements with the largest χ^2-values are excluded. The LMS method searches for the parameter set where the median of the χ^2-values is minimal.

More information than presented in this short and simplified introduction into the field of robust methods can be found in the literature cited above and the newer book of R. Maronna, D. Martin and V. Yohai [132].

Table of Symbols

Symbol	*Explanation*
A , B	Events
$\overline{A}$	Negation of A
Ω / $\emptyset$	Certain / impossible event
$A \cup B, A \cap B, A \subset B$	A OR B, A AND B, A implies B etc.
$P\{A\}$	Probability of A
$P\{A\|B\}$	Conditional probability of A (for given B)
x, y, z ; k, l, m	(Continuous; discrete) random variable (variate)
θ, μ, σ	Parameter of distributions
$f(x)$, $f(x\|\theta)$	Probability density function
$F(x)$, $F(x\|\theta)$	Integral (probability-) distribution function (for parameter value θ, respectively)(p. 17)
$f(\boldsymbol{x})$, $f(\boldsymbol{x}\|\boldsymbol{\theta})$	Respective multidimensional generalizations (p. 49)
A, A_{ji}	Matrix, matrix element in column i, row j
A^T, $A^T_{ji} = A_{ij}$	Transposed matrix
$\boldsymbol{a}$, $\boldsymbol{a} \cdot \boldsymbol{b} \equiv \boldsymbol{a}^T \boldsymbol{b}$	Column vector, inner (dot-) product
$L(\theta)$, $L(\theta\|x_1, \ldots, x_N)$	Likelihood function (p. 160)
$L(\boldsymbol{\theta}\|x_1, \ldots, x_N)$	Generalization to more dimensional parameter space
$\hat{\theta}$	Statistical estimate of the parameter θ(p. 168)
$E(u(x))$, $\langle u(x) \rangle$	Expected value of $u(x)$
$\overline{u(x)}$	Arithmetic sample mean, average (p. 23)
δx	Measurement error of x (p. 95)

Symbol	*Explanation*
σ_x	Standard deviation of x
σ_x^2, var(x)	Variance (dispersion) of x (p. 24)
cov(x, y), σ_{xy}	Covariance (p. 52)
ρ_{xy}	Correlation coefficient
μ_i	Moment of order i with respect to origin 0, initial moment
μ'_i	Central moment (p. 36)
μ_{ij}, μ'_{ij} etc.	Two-dimensional generalizations (p. 51)
γ_1, β_2, γ_2	Skewness , kurtosis , excess (p. 28)
κ_i	Semiinvariants (cumulants) of order i, (p. 40)

Bibliography

[1] M. G. Kendall and W. R. Buckland, *A Dictionary of Statistical Terms*, Longman, London (1982).

[2] L. Lyons, *Bayes and frequentism: A particle physicist's perspective*, arXiv:1301.1273v1 (2013).

[3] M. G. Kendall and A. Stuart, *The Advanced Theory of Statistics*, Griffin, London (1979).

[4] S. Brandt, *Data Analysis*, Springer, Berlin (1999).

[5] A. G. Frodesen, O. Skjeggestad, M. Tofte, *Probability and Statistics in Particle Physics*, Universitetsforlaget, Bergen (1979).

[6] R. Barlow, *Statistics*, Wiley, Chichester (1989).

[7] L. Lyons, *Statistics for Nuclear and Particle Physicists*, Cambridge University Press (1992).

[8] W. T. Eadie *et al.*, *Statistical Methods in Experimental Physics*, North-Holland, Amsterdam, (1982).

[9] F. James, *Statistical Methods in Experimental Physics*, World Scientific Publishing, Singapore (2007).

[10] *Data Analysis in High Energy Physics: A Practical Guide to Statistical Methods*, ed. O. Behnke *et al.*, J. Wiley & Sons (2013).

[11] *Statistical Analysis Techniques in Particle Physics: Fits, Density Estimation and Supervised Learning*, J. Wiley & Sons (2013).

[12] V. Blobel, E. Lohrmann, *Statistische und numerische Methoden der Datenanalyse*, Teubner, Stuttgart (1998).

[13] B. P. Roe, *Probability and Statistics in Experimental Physics*, Springer, Berlin (2001).

[14] G. Cowan, *Statistical Data Analysis*, Clarendon Press, Oxford (1998).

[15] G. D'Agostini, *Bayesian Reasoning in Data Analysis: A Critical Introduction*, World Scientific Pub., Singapore (2003).

[16] T. Hastie, R. Tibshirani, J. H. Friedman, *The Elements of Statistical Learning*, Springer, Berlin (2001).

[17] G. E. P. Box and G. C. Tiao, *Bayesian Inference in Statistical Analysis*, Addison-Weseley, Reading (1973).

[18] R. A. Fisher, *Statistical Methods, Experimental Design and Scientific Inference*, Oxford University Press (1990). (First publication 1925).

[19] A. W. F. Edwards, *Likelihood*, The John Hopkins University Press, Baltimore (1992).
[20] I. J. Good, *Good Thinking, The Foundations of Probability and its Applications*, Univ. of Minnesota Press, Minneapolis (1983).
[21] L. J. Savage, *The Writings of Leonard Jimmie Savage - A Memorial Selection*, ed. American Statistical Association, Washington (1981).
[22] H. Jeffreys, *Theory of Probability*, Clarendon Press, Oxford (1983).
[23] L. J. Savage, *The Foundation of Statistical Inference*, Dover Publ., New York (1972).
[24] Proceedings of PHYSTAT03, *Statistical Problems in Particle Physics, Astrophysics and Cosmology* ed. L. Lyons *et al.*, SLAC, Stanford (2003) Proceedings of PHYSTAT05, *Statistical Problems in Particle Physics, Astrophysics and Cosmology* ed. L. Lyons *et al.*, Imperial College Press, Oxford (2005).
[25] M. Abramowitz, I.A. Stegun, *Handbook of Mathematical Functions*, Dover Publications,inc., New York (1970).
[26] G. Bohm, G. Zech, *Statistics of weighted Poisson events and its applications*, Nucl. Instr. and Meth. A 748 (2014) 1.
[27] P. Sinervo, *Definition and Treatment of Systematic Uncertainties in High Energy Physics and Astrophysics*, Proceedings of PHYSTAT2003, P123, ed. L. Lyons, R. Mount, R. Reitmeyer, Stanford, Ca (2003).
[28] R. J. Barlow, *Systematic Errors, Fact and Fiction*, hep-ex/0207026 (2002).
[29] R. Wanke, *How to Deal with Sytematic Uncertainties* in Data Analysis in High Energy Physics: A Practical Guide to Statistical Methods, ed. O. Behnke *et al.*, J. Wiley Sons (2013).
[30] https://en.wikipedia.org/wiki/*Prior probability.*
[31] A. B. Balantekin *et al.*, *Review of Particle Physics*, J. of Phys. G 33 (2006),1.
[32] R. M. Neal, *Probabilistic Inference Using Markov Chain Monte Carlo Methods*, University of Toronto, Department of Computer Science, Tech, Rep. CRG-TR-9 3-1 (1993).
[33] N. Metropolis, A. W. Rosenbluth, M. N. Rosenbluth, A. H. Teller, and E. Teller *Equation of state calculations by fast computing machines*, J. Chem. Phys. 21 (1953) 1087.
[34] P. E. Condon and P. L. Cowell, *Channel likelihood: An extension of maximum likelihood to multibody final states*, Phys. Rev. D 9 (1974)2558.
[35] R. J. Barlow, *Extended maximum likelihood*, Nucl. Instr. and Meth. A 297 (1990) 496.
[36] M. Casarsa *et al.*, *A statistical prescription to estimate properly normalized distributions of different particle species*, Nucl. Instr. and Meth. A640 (2010) 219.
[37] G. Bohm and G. Zech, *Comparing statistical data to Monte Carlo simulation with weighted events*, Nucl. Instr. and Meth. A691 (2012) 171.
[38] V. S. Kurbatov and A. A. Tyapkin, in Russian edition of W. T. Eadie *et al.*, *Statistical Methods in Experimental Physics*, Atomisdat, Moscow (1976).

[39] B. List, *Constrained Fits* in *Data Analysis in High Energy Physics: A Practical Guide to Statistical Methods*, ed. O. Behnke *et al.*, J. Wiley & Sons (2013).

[40] G. Zech, *Reduction of Variables in Parameter Inference*, Proceedings of PHYSTAT2005, ed. L. Lyons, M. K. Unel, Oxford (2005).

[41] G. Zech, *A Monte Carlo Method for the Analysis of Low Statistic Experiments*, Nucl. Instr. and Meth. 137 (1978) 551.

[42] M. Diehl and O. Nachtmann, *Optimal observables for the measurement of the three gauge boson couplings in* $e^+e^- \rightarrow W^+W^-$, Z. f. Phys. C 62, (1994, 397.

[43] O. E. Barndorff-Nielsen, *On a formula for the distribution of a maximum likelihood estimator*, Biometrika 70 (1983), 343.

[44] D. R. Cox and N. Reid, *Parameter Orthogonality and Approximate Conditional Inference*, J. R. Statist. Soc. B 49, No 1, (1987) 1, D. Fraser and N. Reid, *Likelihood inference with nuisance parameters*, Proc. of PHYSTAT2003, ed. L. Lyons, R. Mount, R. Reitmeyer, SLAC, Stanford (2003) 265.

[45] G. A. Barnard, G. M. Jenkins and C. B. Winstein, *Likelihood inference and time series*, J. Roy. Statist. Soc. A 125 (1962).

[46] A. Birnbaum, *More on the concepts of statistical evidence*, J. Amer. Statist. Assoc. 67 (1972), 858.

[47] D. Basu, *Statistical Information and Likelihood*, Lecture Notes in Statistics 45, ed. J. K. Ghosh, Springer, Berlin (1988).

[48] J. O. Berger and R. L. Wolpert, *The Likelihood Principle*, Lecture Notes of Inst. of Math. Stat., Hayward, Ca, ed. S. S. Gupta (1984).

[49] CERN computer library, root.cern.ch.

[50] R. Barlow, *Asymmtric Errors*, arXiv:physics/0401042v1 (2004), Proceedings of PHYSTAT2005, ed. L. Lyons, M. K. Unel, Oxford (2005),56.

[51] S. Chiba and D. L. Smith, *Impacts of data transformations on least-squares solutions and their significance in data analysis and evaluation*, J. Nucl. Sci. Technol. 31 (1994) 770.

[52] H. J. Behrendt *et al.*, *Determination of* α_s, *and* $sin^2\theta$ *from measurements of the total hadronic cross section in* e^+e^- *annihilation at PETRA*, Phys. Lett. 183B (1987) 400.

[53] R. W. Peelle, *Peelle's Pertinent Puzzle*, Informal memorandum dated October 13, 1987, ORNL, USA (1987)

[54] G. Zech, *Analysis of distorted measurements - parameter estimation and unfolding*, arXiv (2016)

[55] G. Bohm, G. Zech, *Comparison of experimental data to Monte Carlo simulation - Parameter estimation and goodness-of-fit testing with weighted events*, Nucl. Instr. and Meth. A691 (2012), 171.

[56] V. B. Anykeyev, A. A. Spiridonov and V. P. Zhigunov, *Comparative investigation of unfolding methods*, Nucl. Instr. and Meth. A303 (1991) 350.

[57] V. Blobel, *Unfolding methods in high-energy physics experiments*, CERN Yellow Report 85-09 (1985) 88.

[58] G. Zech, *Comparing statistical data to Monte Carlo simulation - parameter fitting and unfolding*, Desy 95-113 (1995).

[59] Proceedings of the PHYSTAT 2011 Workshop on Statistical Issues Related to Discovery Claims in Search Experiments and Unfolding, CERN, Geneva, Switzerland, ed. H. B. Prosper and L. Lyons (2011).

[60] V. Blobel, *Unfolding methods in particle physics*, Proceedings of the PHYSTAT 2011 Workshop on Statistical Issues Related to Discovery Claims in Search Experiments and Unfolding, CERN, Geneva, Switzerland, ed. H. B. Prosper and L. Lyons (2011).

[61] V. Blobel, *Unfolding* in Data Analysis in High Energy Physics: A Practical Guide to Statistical Methods, ed. O. Behnke *et al.*, J. Wiley and Sons (2013).

[62] H. N. Mülthei and B. Schorr, *On an iterative method for the unfolding of spectra*, Nucl. Instr. and Meth. A257 (1987) 371.

[63] M. Schmelling, *The method of reduced cross-entropy - a general approach to unfold probability distributions*, Nucl. Instr. and Meth. A340 (1994) 400.

[64] L. Lindemann and G. Zech, *Unfolding by weighting Monte Carlo events*, Nucl. Instr. and Meth. A354 (1994) 516.

[65] G. D'Agostini, *A multidimensional unfolding method based on Bayes' theorem*, Nucl. Instr. and Meth. A 362 (1995) 487.

[66] A. Hoecker and V. Kartvelishvili, *SVD approach to data unfolding*, Nucl. Instr. and Meth. A 372 (1996), 469.

[67] N. Milke *et al. Solving inverse problems with the unfolding program TRUEE: Examples in astroparticle physics*, Nucl. Instr. and Meth. A697 (2013) 133.

[68] A. P. Dempster, N. M. Laird, D. B. Rubin, *Maximum Likelihood from Incomplete Data via the EM Algorithm*, J. R. Statist.Soc. B 39 (1977) 1.

[69] W. H. Richardson, *Bayesian based Iterative Method of Image restoration* Journal. of the Optical Society of America 62 (1972) 55.

[70] L. B. Lucy, *An iterative technique for the rectification of observed distributions*, Astron. Journ. 79 (1974) 745

[71] L. A. Shepp and Y. Vardi, *Maximum likelihood reconstruction for emission tomography*, IEEE trans. Med. Imaging MI-1 (1982) 113

[72] A. Kondor, *Method of converging weights - an iterative procedure for solving Fredholm's integral equations of the first kind*, Nucl. Instr. and Meth. 216 (1983) 177.

[73] Y. Vardi, L. A. Shepp and L. Kaufmann, *A statistical model for positron emission tomography*, J. Am. Stat. Assoc.80 (1985) 8, Y. Vardi and D. Lee, *From image deblurring to optimal investments: Maximum likelihood solution for positive linear inverse problems (with discussion)*, J. R. Statist. Soc. B55, 569 (1993).

[74] H. N. Mülthei, B. Schorr, *On properties of the iterative maximum likelihood reconstruction method*, Math. Meth. Appl. Sci. 11 (1989) 331.

[75] D. M. Titterington, *Some aspects of statistical image modeling and restoration*, Proceedings of PHYSTAT 05, ed. L. Lyons and M. K. ÃIJnel, Oxford (2005).

[76] P. C. Hansen, *Discrete Inverse Problems: Insight and Algorithms*, SIAM, (2009)

[77] B. Efron and R. T. Tibshirani, *An Introduction to the Bootstrap*, Chapman & Hall, London (1993).

[78] M. Kuusela and V. M. Panaretos, *Statistical unfolding of elementary particle spectra: Empirical Bayes estimation and bias-corrected uncertainty quantification*, Annals of Applied Statistics 9 (2015) 1671. M. Kuusela, *Statistical Issues in Unfolding Methods for High Energy Physics*, Master's thesis, Aalto University, Finnland (2912).

[79] G. D'Agostini, *Improved iterative Bayesian unfolding*, arXiv:1010.632v1 (2010).

[80] I. Volobouev, *On the Expectation-Maximization Unfolding with smoothing*, arXiv:1408.6500v2 (2015).

[81] A. N. Tichonoff, *Solution of incorrectly formulated problems and the regularization method*, translation of the original article (1963) in Soviet Mathematics 4, 1035.

[82] D. W. Scott, and S. R. Sain, *Multi-Dimensional Density Estimation*, in Handbook of Statistics, Vol 24: Data Mining and Computational Statistics, ed. C.R. Rao and E.J. Wegman, Elsevier, Amsterdam (2004).

[83] R. Narayan, R. Nityananda, *Maximum entropy image restoration in astronomy*, Ann. Rev. Astrophys. 24 (1986) 127.

[84] P. Magan, F. Courbin and S. Sohy, *Deconvolution with correct sampling*, Astrophys. J. 494 (1998) 472.

[85] H. P. Dembinski, M. Roth, *An algorithm for automatic unfolding of one-dimensional distributions*, Nucl. Instr. and Meth. A 729 (2013) 725.

[86] G. Zech, *Iterative unfolding with the Richardson-Lucy algorithm*, Nucl. Instr. and Meth. A716 (2013) 1.

[87] B. Aslan and G.Zech, *Statistical energy as a tool for binning-free, multivariate goodness-of-fit tests, two-sample comparison and unfolding*, Nucl. Instr. and Meth. A 537 (2005) 626.

[88] G. Schott, *Hypothesis Testing*, in Ref. 10.

[89] R. B. D'Agostino and M. A. Stephens (editors), *Goodness of Fit Techniques*, M. Dekker, New York (1986).

[90] L. Demortier, *P Values: What They Are and How to Use Them*, CDF/MEMO/STATISTICS/PUBLIC (2006).

[91] D. S. Moore, *Tests of Chi-Squared Type* in Goodness of Fit Techniques, ed. R. B. D'Agostino and M. A. Stephens, M. Dekker, New York (1986).

[92] T. W. Anderson, D. A. Darling,*Asyptotic Theory of Certain Goodness of Fit Criteria Based on Stochastic Processes*, Ann. of Math. Stat. 23(2) (1952) 193.

[93] N. H. Kuiper, *Tests concerning random points in a circle*, Proceedings of the Koninklijke Netherlandse Akademie van Wetenschappen, A63 (1960) 38.

[94] J. Neyman, *"Smooth test" for goodness of fit*, Scandinavisk Aktuaristidskrift 11 (1037),149.

[95] E. S. Pearson, *The Probability Integral Transformation for Testing Goodness of Fit and Combining Independent Tests of Significance*, Biometrika 30 (1938), 134.

[96] A. W. Bowman, *Density based tests for goodness-of-fit*, J. Statist. Comput. Simul. 40 (1992) 1.

[97] B. Aslan and G. Zech, *New Test for the Multivariate Two-Sample Problem based on the Concept of Minimum Energy*, J. Statist. Comput. Simul. 75, 2 (2004), 109.

[98] G. J. Szekely and M. L. Rizzo, *E-statistics: Statistics based on distance*, Journal of statistical Planning and Inference 143,8(1913) 1249 and Refs. therein.

[99] G. Zech, *Scaling Property of the Two-Sample Statistical Energy Test*, arXiv:1804.10599v1 (2018).

[100] S. G. Mallat, *A Wavelet Tour of Signal Processing*, Academic Press, New York (1999), A. Graps, *An introduction to wavelets*, IEEE, Computational Science and Engineering, 2 (1995) 50 und www.amara.com /IEEEwave/IEEEwavelet.html.

[101] A. W. Bowman and A. Azzalini, *Applied Smoothing Techniques for Data Analysis: The Kernel Approach with S-Plus Illustration*, Oxford University Press (1997).

[102] I. T. Jolliffe, *Principal Component Analysis*, Springer, Berlin (2002).

[103] P. M. Schulze, *Beschreibende Statistik*, Oldenburg Verlag (2003).

[104] R. Rojas, *Theorie der neuronalen Netze*, Springer, Berlin (1991).

[105] M. Feindt and U. Kerzel, *The NeuroBayes neural network package*, Nucl. Instr. and Meth A 559 (2006) 190

[106] V. Vapnik, *The Nature of Statistical Learning Theory*, Springer, Berlin (1995), V. Vapnik, *Statistical Lerning Theory*, Wiley, New York (1998), B. Schölkopf and A. Smola, *Lerning with Kernels*, MIT press, Cambridge, MA (2002).

[107] B. Schölkopf, *Statistical Learning and Kernel Methods*, http://research.microsoft.com.

[108] J. H. Friedman, *Recent Advances in Predictive (Machine) Learning* Proceedings of PHYSTAT03, *Statistical Problems in Particle Physics, Astrophysics and Cosmology* ed. L. Lyons *et al.*, SLAC, Stanford (2003).

[109] Y. Freund and R. E. Schapire, *Experiments with a new boosting algorithm*, Proc. COLT, Academic Press, New York (1996) 209.

[110] B. P. Roe et. al., *Boosted decision trees as an alternative to artificial neural networks for particle identification*, Nucl. Instr. and Meth. A543 (2005) 577.

[111] L. Breiman, *Bagging predictors*, Technical Report No. 421, Department of Statistics, University of California, Berkeley, Ca, (1994).

[112] L. Breiman, *Random Forests*, Technical Report, Department of Statistics, University of California, Berkeley, Ca (2001).

[113] J. S. Simonoff, *Smoothing Methods in Statistics*, Springer, Berlin (1996).

[114] D. W. Scott, *Multivariate Density Estimation: Theory, Practice, and Visualization*, John Wiley, New York (1992).

[115] E. L. Lehmann, *Elements of Large-Sample Theory*,

[116] W. HÃďrdle, M. MÃijller, S. Sperlich, A. Werwatz, *Nonparametric and Semiparametric Models*, Springer, Berlin (2004).

[117] T. A. Bancroft and Chien-Pai Han, *Statistical Theory and Inference in Research*, ed. D. B. Owen, Dekker, New York (1981).

[118] M. Fisz, *Probability Theorie and Mathematical Statistics*, R.E. Krieger Publishing Company, Malabre, Florida (1980) 464.

[119] G. J. Feldman, R. D. Cousins, *Unified approach to the classical statistical analysis of small signals.* Phys. Rev. D 57 (1998) 3873.

[120] B. Efron, *Why isn't Everyone a Bayesian?* Am. Stat. 40 (1986) 1,
R. D. Cousins, *Why Isn't Every Physicist a Bayesian?* Am. J. Phys. 63 (1995) 398,
D. V. Lindley, Wald Lectures: *Bayesian Statistics*, Statistical Science, 5 (1990) 44.

[121] G. Zech, *Frequentist and Bayesian confidence intervals*,EPJdirect C12 (2002) 1, ArXiv:hep-ex/0106023 (2002).

[122] D. Inosov, UniversitÃd't Dresden, private information (2021).

[123] M. A. Stephens, *Tests based on EDF Statistics*, in Goodness of Fit Techniques, ed. R. B. d'Agostino and M. A. Stephens, Dekker, New York (1986).

[124] G. J. Babu *et al.*, *Second-order correctness of the Poisson bootstrap*,The Annals of Statistics Vol 27, No. 5 (1999) 1666-1683).

[125] https://en.wikipedia.org/wiki/Expectation-maximization-algorithm.

[126] G. Zech, *A Simple Iterative Alignment Method using Gradient Descending Minimum Search*, Proceedings of PHYSTAT03, *Statistical Problems in Particle Physics, Astrophysics and Cosmology*, ed. L. Lyons *et al.*, SLAC, Stanford (2003), 226.

[127] J. A. Nelder and R. Mead, *A simplex method for function minimization*, The Computer Journal, 7 (1965) 308.

[128] J. J. Tomik, *On Convergence of the Nelder-Mead Simplex algorithm for unconstrained stochastic optimization*, PhD Thesis, Pensylvania State university, Department of Statistics (1995).

[129] G. A. Korn and Th. M. Korn, *Mathematical Handbook for Scientists and Engineers*, McGraw-Hill, New York (1961)

[130] P. J. Huber, *Robust Statistics*, John Wiley, New York (1981).

[131] P. J. Rousseeuw, *Robust Regression and Outlier Detection*, John Wiley, New York (1987).

[132] R. Maronna, D. Martin and V. Yohai, *Robust Statistics – Theory and Methods*, John Wiley, New York (2006).

[133] Some useful internet links:
- http://www.stats.gla.ac.uk/steps/glossary/basic-definitions.html, *Statistics Glossary* (V. J. Easton and J. H. McColl)
- http://www.nu.to.infn.it/Statistics/, *Useful Statistics Links for Particle Physicists*
- http://www.statsoft.com/textbook/stathome.html, *Electonic Textbook Statsoft*
- http://wiki.stat.ucla.edu/socr/index.php/EBook, *Electronic Statistics Book*
- http://www.york.ac.uk/depts /maths/histstat/lifework.htm,*Life and Work of Statisticians* (University of York, Dept. of Mathematics)

- http://www.montefiore.ulg.ac.be/services/stochastic/pubs/2000/Geu00/geurts-pkdd2000-bagging.pdf, *Some enhancements of Decision Tree Bagging* (P. Geurts):
- http://www.math.ethz.ch/ blatter/Waveletsvortrag.pdf, *Wavelet script* (in German)
- http://www.xmarks.com/site/www.umiacs.umd.edu/ joseph/support-vector-machines4.pdf, *A Tutorial on Support Vector Machines for Pattern Recognition* (Ch. J. C. Burges)
- http://www-stat.stanford.edu /~jhf/ftp/machine.pdf, *Recent Advances in Predictive (Machine) Learning* (J. B. Friedman)

Index